Modern Algebra
with Applications

Modern Algebra with Applications

WILLIAM J. GILBERT

University of Waterloo

A WILEY-INTERSCIENCE PUBLICATION

JOHN WILEY & SONS

New York/Chichester/Brisbane/Toronto

Library of Congress Cataloging in Publication Data:

Gilbert, William J   1941–
  Modern algebra with applications.

  "A Wiley-Interscience publication."
  Bibliography: p.
  Includes index.
  1. Algebra, Abstract. I. Title.

QA162.G53   512'.02   76-22756
ISBN 0-471-29891-3

Printed in the United States of America

10 9 8 7 6

Preface

Until recently the applications of modern algebra were mainly confined to other branches of mathematics. However, the importance of modern algebra and discrete structures to many areas of science and technology is now growing rapidly. It is being used extensively in computing science, physics, chemistry, and data communication as well as in new areas of mathematics such as combinatorics. We believe that the fundamentals of these applications can now be taught at the junior level. This book therefore constitutes a one-year course in modern algebra for those students who have been exposed to some linear algebra. It contains the essentials of a first course in modern algebra together with a wide variety of applications.

Modern algebra is usually taught from the point of view of its intrinsic interest, and students are told that applications will appear in later courses. Many students lose interest when they do not see the relevance of the subject and often become skeptical of the perennial explanation that the material will be used later. However, we believe that, by providing interesting and nontrivial applications as we proceed, the student will better appreciate and understand the subject.

We cover all the group, ring, and field theory that is usually contained in a standard modern algebra course; the exact sections containing this material are indicated in the Table of Contents. We stop short of the Sylow theorems and Galois theory. These topics could only be touched on in a first course, and we feel that more time should be spent on them if they are to be appreciated.

In Chapter 2 we discuss Boolean algebras and their application to switching circuits. These provide a good example of algebraic structures whose elements are nonnumerical. However, many instructors may prefer to postpone or omit this chapter and start with the group theory in Chapters 3 and 4. Groups are viewed as describing symmetries in nature and in mathematics. In keeping with this view, the rotation groups of the regular solids are investigated in Chapter 5. This material provides a good starting point for students interested in applying group theory to physics and chemistry. Chapter 6 introduces the Pólya-Burnside method of enumerating equivalence classes of sets of symmetries and provides a very practical application of group theory to combinatorics. Monoids are becoming more important algebraic structures today; these are discussed in Chapter 7 and are applied to finite-state machines.

The ring and field theory is covered in Chapters 8–11. This theory is motivated by the desire to extend the familiar number systems to obtain the Galois fields and to discover the structure of various subfields of the real and complex numbers. Groups are used in Chapter 12 to construct Latin squares, whereas Galois fields are used to construct orthogonal Latin squares. These can be used to design statistical experiments. We also indicate the close relationship between orthogonal Latin squares and finite geometries. In Chapter 13 field extensions are used to show that some famous geometrical constructions, such as the trisection of an angle and the squaring of the circle, are impossible to perform using only a straightedge and compass. Finally, Chapter 14 gives an introduction to coding theory using polynomial and matrix techniques.

We do not give exhaustive treatments of any of the applications. We only go so far as to give the flavor without becoming too involved in technical complications. The interested reader may delve further into any topic by consulting the books in the bibliography.

It is important to realize that the study of these applications is not the only reason for learning modern algebra. These examples illustrate the varied uses to which algebra has been put in the past, and it is extremely likely that many more different applications will be found in the future.

One cannot understand mathematics without doing numerous examples. There are a total of over 600 exercises of varying difficulty, at the end of the chapters. Answers to the odd-numbered exercises are given at the back of the book.

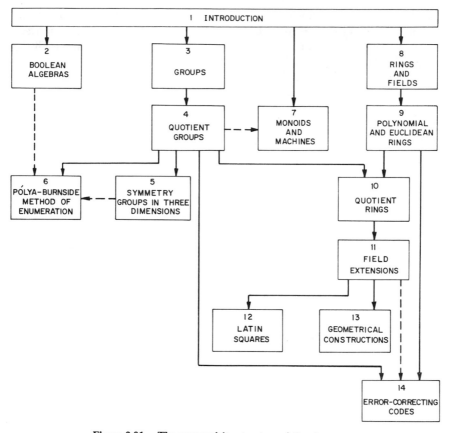

Figure 0.01.   The prerequisite structure of the chapters.

Figure 0.01 illustrates the interdependence of the various chapters. A solid line indicates a necessary prerequisite for the whole chapter, and a dotted line indicates a prerequisite for one section of the chapter. Since the book contains more than sufficient material for a two-term course, various sections or chapters may be omitted. The choice of topics will depend on the interests of the students and the instructor. However, to preserve the essence of the book, the instructor should be careful not to devote most of the course to the theory, but should leave sufficient time for the applications to be appreciated.

I would like to thank all my students and colleagues at the University of Waterloo, especially Harry Davis, D. Ž. Djoković, Denis Higgs, and Keith Rowe, who offered helpful suggestions during the various stages of the manuscript. I am very grateful to Michael Boyle, Ian McGee, Juris Stepŕans, and Jack Weiner for their help in preparing and proofreading the preliminary versions and the final draft. Finally, I would like to thank Sue Cooper, Annemarie DeBrusk, Lois Graham, and Denise Stack for their excellent typing of the different drafts, and Nadia Bahar for tracing all the figures.

WILLIAM J. GILBERT

Waterloo, Ontario, Canada
April 1976

Contents

Those sections that constitute the core of a modern algebra course are indicated by the symbol†

Modern Algebra
with Applications

Introduction | 1

Algebra can be defined as the manipulation of symbols. Its history falls into two distinct parts, with the dividing date being approximately 1800. The algebra done before the nineteenth century is called "classical algebra," whereas most of that done later is called "modern algebra" or "abstract algebra."

CLASSICAL ALGEBRA

The technique of introducing a symbol, such as x, to represent an unknown number in solving problems was known to the ancient Greeks. This symbol could be manipulated just like the arithmetic symbols until a solution was obtained. *Classical algebra* can be characterized by the fact that each symbol *always* stood for a number. This number could be integral, real, or complex. However, in the seventeenth and eighteenth centuries, mathematicians were not quite sure whether the square root of minus one was a number. It was not until the nineteenth century and the beginning of modern algebra that a satisfactory explanation of the complex numbers was given.

The main goal of classical algebra was to use algebraic manipulation to solve polynomial equations. Classical algebra succeeded in producing algorithms for solving all polynomial equations in one variable of degree less than or equal to four. It was shown by Niels Henrik Abel (1802–1829), by modern algebraic methods, that it was not always possible to solve a polynomial equation of degree five or higher in terms of nth roots. Classical algebra also developed methods for dealing with linear equations containing several variables, but little was known about the solution of nonlinear equations.

Classical algebra provided a powerful tool for tackling many scientific problems, and it is still extremely important for working out today's problems. Perhaps the mathematical tool most useful in science, engineering, and the social sciences is the method of solution of a system of linear equations together with all its allied linear algebra.

MODERN ALGEBRA

In the nineteenth century, it was gradually realized that mathematical symbols did not necessarily have to stand for numbers; in fact, it was not necessary that they stand for anything at all! From this realization emerged what is now known as *modern algebra* or *abstract algebra*.

For example, the symbols could be interpreted as symmetries of an object, as the position of a switch, as an instruction to a machine, or as a way to design a statistical experiment. The symbols could be manipulated using some of the usual rules for numbers. For example, the polynomial $3x^2 + 2x - 1$ could be added to and multiplied by other polynomials without ever having to interpret the symbol x as a number.

Modern algebra has two basic uses. Its first is to describe patterns or symmetries that occur in nature and in mathematics. For example, it can describe the different crystal formations in which certain chemical substances are found and can be used to show the similarity between the logic of switching circuits and the algebra of subsets of a set. The second basic use of modern algebra is to naturally extend the common number systems to other useful systems.

BINARY OPERATIONS

The symbols that are to be manipulated are elements of some set, and the manipulation is done by performing certain operations on elements of that set. Examples of such operations are addition and multiplication on the set of real numbers.

As shown in Figure 1.01, we can visualize an operation as a "black box" with various inputs coming from a set S and one output, which combines the inputs in some specified way.

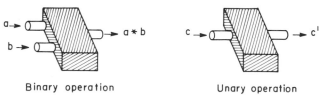

Binary operation   Unary operation

Figure 1.01

If the black box has two inputs, the operation combines *two* elements of the set to form a third. Such an operation is called a binary operation. If there is only one input, the operation is called unary. An example of a unary operation is finding the reciprocal of a nonzero real number.

A *binary operation*, $\star$, on a set S is really just a particular function from $S \times S$ to S. We denote the image of the pair (a,b) under this function by $a \star b$. In other words, the binary operation $\star$ assigns to any two elements a and b of S the element $a \star b$ of S. We often refer to an operation $\star$ as being *closed* to emphasize that each element $a \star b$ belongs to the set S and not to a possibly larger set. Many symbols are used for binary operations; the most common are $+$, $\cdot$, $-$, $\circ$, $\div$, $\cup$, $\cap$, $\wedge$ and $\vee$.

A *unary operation* on S is just a function from S to S. The image of c under a unary operation is usually denoted by a symbol such as c', $\bar{c}$, c^{-1} or $(-c)$.

Let $\mathbf{P} = \{1, 2, 3, \ldots\}$ be the set of positive integers. Addition and multiplication are both binary operations on $\mathbf{P}$, because, if $x, y \in \mathbf{P}$, then $x + y$ and $x \cdot y \in \mathbf{P}$. However, subtraction is *not* a binary operation on $\mathbf{P}$ because, for instance, $1 - 2 \notin \mathbf{P}$. Other natural binary operations on $\mathbf{P}$ are exponentiation and the greatest common divisor, since, for any two positive integers x and y, x^y and $\mathrm{GCD}(x, y)$ are well-defined elements of $\mathbf{P}$.

Let $\mathbf{R}$ be the set of all real numbers. Addition, multiplication, and subtraction are all binary operations on $\mathbf{R}$ because $x + y$, $x \cdot y$, and $x - y$ are real numbers for every pair of real numbers x and y. The symbol $-$ stands for a binary operation when used in an expression such as $x - y$, but it stands for the unary operation of taking the negative when used in the expression $-x$. Division is not a binary operation on $\mathbf{R}$ because division by zero is undefined. However, division is a binary operation on $\mathbf{R} - \{0\}$, the set of nonzero real numbers.

A binary operation on a finite set can often be conveniently presented by means of a *table*. For example, consider the set $T = \{a, b, c\}$, containing three elements. A binary operation $\star$ on T is defined by Table 1.1. In this table, $x \star y$ is the element in row x and column y. For example, $b \star c = b$ and $c \star b = a$.

Table 1.1.   A binary operation on $\{a, b, c\}$

$\star$	a	b	c
a	b	a	a
b	c	a	b
c	c	a	b

One important binary operation is the composition of symmetries of a given figure or object. Consider a square lying in a plane. The set S of symmetries of this square is the set of mappings of the square to itself that preserve distances. Figure 1.02 illustrates the composition of two such symmetries to form a third symmetry.

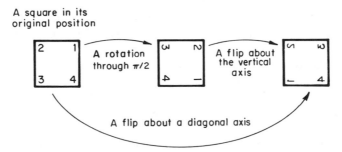

Figure 1.02.   Composition of symmetries of a square.

Most of the binary operations we use have one or more of the following special properties. Let $\star$ be a binary operation on a set S. This operation is called *associative* if $a \star (b \star c) = (a \star b) \star c$ for all $a, b, c \in S$. The operation $\star$ is called *commutative* if $a \star b = b \star a$ for all $a, b \in S$. The element $e \in S$ is said to be an *identity* for $\star$ if $a \star e = e \star a = a$ for all $a \in S$.

If $\star$ is a binary operation on S that has an identity e, then b is called the *inverse* of a with respect to $\star$ if $a \star b = b \star a = e$. We usually denote the inverse of a by a^{-1}; however, if the operation is addition, the inverse is denoted by $-a$.

If $\star$ and $\circ$ are two binary operations on S, then $\circ$ is said to be *distributive* over $\star$ if $a \circ (b \star c) = (a \circ b) \star (a \circ c)$ and $(b \star c) \circ a = (b \circ a) \star (c \circ a)$ for all $a, b, c \in S$.

Addition and multiplication are both associative and commutative operations on the set of real numbers, **R**. The identity for addition is 0, whereas the multiplicative identity is 1. Every real number, a, has an inverse under addition, namely, its negative, $-a$. Every nonzero real number a has a multiplicative inverse, a^{-1}. Furthermore, multiplication is distributive over addition because $a \cdot (b + c) = (a \cdot b) + (a \cdot c)$ and $(b + c) \cdot a = (b \cdot a) + (c \cdot a)$; however, addition is not distributive over multiplication because $a + (b \cdot c) \neq (a + b) \cdot (a + c)$ in general.

Denote the set of $n \times n$ real matrices by $\mathfrak{M}(n \times n; \textbf{R})$. Matrix multiplication is an associative operation on $\mathfrak{M}(n \times n; \textbf{R})$ but it is not commutative (unless $n = 1$). The matrix I, whose (i,j)th entry is 1 if $i = j$ and 0 otherwise, is the multiplicative identity. Matrices that have inverses under multiplication are called *nonsingular*.

ALGEBRAIC STRUCTURES

A set, together with one or more operations on the set, is called an *algebraic structure*. The set is called the *underlying set* of the structure. Modern algebra is the study of these structures; in later chapters, we examine various types of algebraic structures. For example, a field is an algebraic structure consisting of a set **F** together with two binary operations, usually denoted by $+$ and $\cdot$, that satisfy certain conditions. We denote such a structure by $(\textbf{F}, +, \cdot)$.

In order to understand a particular structure, we usually begin by examining its *substructures*. The underlying set of a substructure is a subset of the underlying set of the structure, and the operations in both structures are the same. For example, the set of complex numbers, **C**, contains the set of real numbers, **R**, as a subset . The operations of addition and multiplication on **C** restrict to the same operations on **R**, and therefore $(\textbf{R}, +, \cdot)$ is a substructure of $(\textbf{C}, +, \cdot)$.

Two algebraic structures of a particular type may be compared by means of structure-preserving functions called morphisms. This concept of morphism is one of the fundamental notions of modern algebra. We encounter it among every algebraic structure we consider.

More precisely, let $(S, \star)$ and $(T, \circ)$ be two algebraic structures consisting of the sets S and T, together with the binary operations $\star$ on S and $\circ$ on T. Then a function $f: S \to T$ is said to be a *morphism* from $(S, \star)$ to $(T, \circ)$ if, for every $x, y \in S$,

$$f(x \star y) = f(x) \circ f(y).$$

If the structures contain more than one operation, the morphism must preserve all these operations. Furthermore, if the structures have identities, these must be preserved too.

As an example of a morphism, consider the set of all integers, $\mathbf{Z}$, under the operation of addition and the set of positive real numbers, $\mathbf{R}_{>0}$, under multiplication. The function $f: \mathbf{Z} \to \mathbf{R}_{>0}$ defined by $f(x) = e^x$ is a morphism from $(\mathbf{Z}, +)$ to $(\mathbf{R}_{>0}, \cdot)$. Multiplication of the exponentials e^x and e^y corresponds to addition of their exponents x and y.

A vector space is an algebraic structure whose underlying set is a set of vectors. Its operations consist of the binary operation of addition and, for each scalar λ, a unary operation of multiplication by λ. A function $f: S \to T$, between vector spaces, is a morphism if $f(\mathbf{x} + \mathbf{y}) = f(\mathbf{x}) + f(\mathbf{y})$ and $f(\lambda \mathbf{x}) = \lambda f(\mathbf{x})$ for all vectors $\mathbf{x}$ and $\mathbf{y}$ in the domain S and all scalars λ. Such a vector space morphism is usually called a *linear transformation*.

A morphism preserves some, but not necessarily all, of the properties of the domain structure. However, if a morphism between two structures is a bijective function (that is, one-to-one and onto), it is called an *isomorphism*, and the structures are called *isomorphic*. Isomorphic structures have identical properties, and they are indistinguishable from an algebraic point of view. For example, two vector spaces of the same finite dimension over a field $\mathbf{F}$ are isomorphic.

One important method of constructing new algebraic structures from old ones is by means of equivalence relations. If $(S, \star)$ is a structure consisting of the set S with the binary operation $\star$ on it, the equivalence relation $\sim$ on S is said to be compatible with $\star$ if, whenever $a \sim b$ and $c \sim d$, it follows that $a \star c \sim b \star d$. Such a compatible equivalence relation allows us to construct a new structure called the *quotient structure*, whose underlying set is the set of equivalence classes. For example, the quotient structure of the integers, $(\mathbf{Z}, +, \cdot)$, under the congruence relation modulo n, is the set of integers modulo n, $(\mathbf{Z}_n, +, \cdot)$.

Extending Number Systems

In the words of Leopold Kronecker (1823–1891), "God created the natural numbers; everything else was man's handiwork." Starting with the set of natural numbers under addition and multiplication, we show how this can be extended to other algebraic systems that satisfy properties not held by the natural numbers. The integers $(\mathbf{Z}, +, \cdot)$ is the smallest system containing the natural numbers, in which addition has an identity (the zero) and every element has an inverse under addition (its negative). The integers have an identity under multiplication (the element 1), but 1 and -1 are the only elements with multiplicative inverses. A standard construction will produce the field of fractions of the integers, which is the rational number system $(\mathbf{Q}, +, \cdot)$, and we show that this is the smallest field containing $(\mathbf{Z}, +, \cdot)$. We can now divide by nonzero elements in $\mathbf{Q}$ and solve every linear equation of the form $ax = b$ $(a \neq 0)$. However, not all quadratic equations have solutions in $\mathbf{Q}$; for example, $x^2 - 2 = 0$ has no rational solution.

The next step is to extend the rationals to the real number system $(\mathbf{R}, +, \cdot)$. The construction of the real numbers requires the use of non-algebraic concepts such as Dedekind cuts or Cauchy sequences, and we will not pursue this, being content to assume that they have been constructed. Even though many polynomial equations have real solutions, there are some, such as $x^2 + 1 = 0$, that do not. We show how to extend the real number system by adjoining a root of $x^2 + 1$ to obtain the complex number system $(\mathbf{C}, +, \cdot)$. The complex number system is really the end of the line, because Carl Friedrich Gauss (1777–1855), in his doctoral thesis, proved that any nonconstant polynomial with real or complex coefficients has a root in the complex numbers. This result is now known as the Fundamental Theorem of Algebra.

However, the classical number system can be generalized in a different way. We can look for fields that are not subfields of $(\mathbf{C}, +, \cdot)$. An example of such a field is the system of integers modulo a prime p, $(\mathbf{Z}_p, +, \cdot)$. All the usual operations of addition, subtraction, multiplication, and division by nonzero elements can be performed in $\mathbf{Z}_p$. We show that these fields can be extended and that, for each prime p and positive integer n, there is a field $(\mathbf{GF}(p^n), +, \cdot)$ with p^n elements. These finite fields are called *Galois fields* after the French mathematician Évariste Galois. We use these Galois fields in the construction of orthogonal latin squares and in coding theory.

Boolean Algebras | 2

A Boolean algebra is a good example of a type of algebraic structure in which the symbols usually represent nonnumerical objects. This algebra is modeled after the algebra of subsets of a set under the binary operations of union and intersection and the unary operation of complementation. However, Boolean algebra has important applications to switching circuits, where each symbol represents a particular electrical circuit or switch. The origin of Boolean algebra dates back to 1847, when the English mathematician George Boole published a slim volume entitled "The Mathematical Analysis of Logic," which showed how algebraic symbols could be applied to logic. The manipulation of logical propositions by means of Boolean algebra is now called the propositional calculus.

At the end of this chapter, we show that any finite Boolean algebra is equivalent to the algebra of subsets of a set; in other words, there is a Boolean algebra isomorphism between the two algebras.

ALGEBRA OF SETS

In this section, we develop some properties of the basic operations on sets, assuming that the reader is familiar with elementary set theory.

2.01 DEFINITION. Let X be any set. The set of all subsets of X is called the *power set* of X and is denoted by $\mathscr{P}(X)$. Hence $\mathscr{P}(X)=\{A\,|\,A\subseteq X\}$.

If $X=\{a,b\}$, then $\mathscr{P}(X)=\{\varnothing,\{a\},\{b\},X\}$. If $X=\{1,2,3\}$, then $\mathscr{P}(X)=\{\varnothing,\{1\},\{2\},\{3\},\{1,2\},\{1,3\},\{2,3\},X\}$. The empty set is denoted by $\varnothing$.

If A and B are subsets of X, their *intersection* is defined by $A\cap B=\{x\,|\,x\in A$ and $x\in B\}$; this is the set of elements common to A and B. The *union* of A and B is $A\cup B=\{x\,|\,x\in A$ or $x\in B\}$; this is the set of elements in A or B (or both). The *complement* of A in X is $\bar{A}=\{x\,|\,x\in X$ and $x\notin A\}$ and is the set of elements in X that are not in A. The shaded areas of the *Venn diagrams* in Figure 2.01 illustrate these operations.

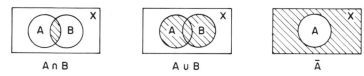

Figure 2.01. Venn diagrams.

Union and intersection are both *binary* operations on the power set $\mathscr{P}(X)$, whereas complementation is a *unary* operation on $\mathscr{P}(X)$. For example, with $X=\{a,b\}$, the tables for the structures $(\mathscr{P}(X),\cap)$, $(\mathscr{P}(X),\cup)$ and $(\mathscr{P}(X),^{-})$ are given in Table 2.1, where we write A for $\{a\}$ and B for $\{b\}$.

Table 2.1. Intersection, union and complements in $\mathscr{P}(\{a,b\})$

$\cap$	$\varnothing$	A	B	X	$\cup$	$\varnothing$	A	B	X	Subset	Complement
$\varnothing$	$\varnothing$	$\varnothing$	$\varnothing$	$\varnothing$	$\varnothing$	$\varnothing$	A	B	X	$\varnothing$	X
A	$\varnothing$	A	$\varnothing$	A	A	A	A	X	X	A	B
B	$\varnothing$	$\varnothing$	B	B	B	B	X	B	X	B	A
X	$\varnothing$	A	B	X	X	X	X	X	X	X	$\varnothing$

2.02 PROPOSITION. The following are some of the more important relations involving the operations $\cap$, $\cup$ and $^{-}$, holding for all $A,B,C\in\mathscr{P}(X)$.

(i) $A\cap(B\cap C)=(A\cap B)\cap C.$

(ii) $A\cup(B\cup C)=(A\cup B)\cup C.$

(iii) $A\cap B=B\cap A.$

(iv) $A\cup B=B\cup A.$

(v) $A\cap(B\cup C)=(A\cap B)\cup(A\cap C).$

(vi) $A\cup(B\cap C)=(A\cup B)\cap(A\cup C).$

(vii) $A\cap X=A.$

(viii) $A\cup\varnothing=A.$

(ix) $A\cap\bar{A}=\varnothing.$

(x) $A\cup\bar{A}=X.$

(xi) $A \cap \varnothing = \varnothing.$ (xii) $A \cup X = X.$
(xiii) $A \cap (A \cup B) = A.$ (xiv) $A \cup (A \cap B) = A.$
(xv) $A \cap A = A.$ (xvi) $A \cup A = A.$
(xvii) $\overline{(A \cap B)} = \overline{A} \cup \overline{B}.$ (xviii) $\overline{(A \cup B)} = \overline{A} \cap \overline{B}.$
(xix) $\overline{X} = \varnothing.$ (xx) $\overline{\varnothing} = X.$
(xxi) $\overline{\overline{A}} = A.$

PROOF.   We shall prove the relations (v) and (x) and leave the proofs of the others to the reader.

(v) $A \cap (B \cup C) = \{x \mid x \in A \text{ and } x \in B \cup C\}$
$\qquad = \{x \mid x \in A \text{ and } (x \in B \text{ or } x \in C)\}$
$\qquad = \{x \mid (x \in A \text{ and } x \in B) \text{ or } (x \in A \text{ and } x \in C)\}$
$\qquad = \{x \mid x \in A \cap B \text{ or } x \in A \cap C\}$
$\qquad = (A \cap B) \cup (A \cap C).$

The Venn diagrams in Figure 2.02 illustrate this result.

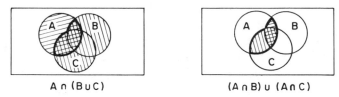

A ∩ (B∪C)   (A∩B) ∪ (A∩C)

Figure 2.02.   Venn diagrams illustrating a distributive law.

(x) $A \cup \overline{A} = \{x \mid x \in A \text{ or } x \in \overline{A}\}$
$\qquad = \{x \mid x \in A \text{ or } (x \in X \text{ and } x \notin A)\}$
$\qquad = \{x \mid (x \in X \text{ and } x \in A) \text{ or } (x \in X \text{ and } x \notin A)\}, \text{ since } A \subseteq X$
$\qquad = \{x \mid x \in X \text{ and } (x \in A \text{ or } x \notin A)\}$
$\qquad = \{x \mid x \in X\}, \text{ since it is always true that } x \in A \text{ or } x \notin A$
$\qquad = X. \quad \square$

   Relations (i)–(iv), (vii), and (viii) show that $\cap$ and $\cup$ are associative and commutative operations on $\mathcal{P}(X)$ with identities X and $\varnothing$, respectively. The only element with an inverse under $\cap$ is its identity X, and the only element with an inverse under $\cup$ is its identity $\varnothing$.
   Note the duality between $\cap$ and $\cup$. If these operations are interchanged in any relation, the resulting relation is also true.
   Another operation on $\mathcal{P}(X)$ is the *difference* of two subsets. It is defined by

$$A - B = \{x \mid x \in A \text{ and } x \notin B\} = A \cap \overline{B}.$$

Since this operation is neither associative nor commutative, we introduce another operation called the *symmetric difference* illustrated in Figure 2.03; this is defined by

$$A \triangle B = (A \cap \overline{B}) \cup (\overline{A} \cap B) = (A \cup B) - (A \cap B) = (A - B) \cup (B - A).$$

The symmetric difference of A and B is the set of elements in A or B, but not in both. This is often referred to as the *exclusive OR* function of A and B.

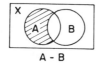

A - B   A △ B

Figure 2.03. The difference and symmetric difference of sets.

2.03 EXAMPLE. Write down the table for the structure $(\mathcal{P}(X), \triangle)$ when $X = \{a, b\}$.

SOLUTION. The table is given in Table 2.2, where we write A for $\{a\}$ and B for $\{b\}$. □

Table 2.2. The symmetric difference in $\mathcal{P}(\{a, b\})$

$\triangle$	$\varnothing$	A	B	X
$\varnothing$	$\varnothing$	A	B	X
A	A	$\varnothing$	X	B
B	B	X	$\varnothing$	A
X	X	B	A	$\varnothing$

2.04 PROPOSITION. The operation $\triangle$ is associative and commutative on $\mathcal{P}(X)$; it has an identity $\varnothing$, and each element is its own inverse. That is, the following relations hold for all $A, B, C \in \mathcal{P}(X)$.

(i) $A \triangle (B \triangle C) = (A \triangle B) \triangle C.$ (ii) $A \triangle B = B \triangle A.$
(iii) $A \triangle \varnothing = A.$ (iv) $A \triangle A = \varnothing.$

Three further properties of the symmetric difference are:

(v) $A \triangle X = \overline{A}.$ (vi) $A \triangle \overline{A} = X.$
(vii) $A \cap (B \triangle C) = (A \cap B) \triangle (A \cap C).$

PROOF.   (ii) follows because the definition of $A \Delta B$ is symmetric in A and B.

(i)   Using Proposition 2.02 and the definition of the symmetric difference we have

$$A \Delta (B \Delta C) = \left(A \cap \overline{(B \Delta C)} \right) \cup \left(\overline{A} \cap (B \Delta C) \right)$$

$$= \left(A \cap \overline{\left((B \cap \overline{C}) \cup (\overline{B} \cap C) \right)} \right) \cup \left(\overline{A} \cap \left((B \cap \overline{C}) \cup (\overline{B} \cap C) \right) \right)$$

$$= \left(A \cap \left((\overline{B} \cup C) \cap (B \cup \overline{C}) \right) \right) \cup \left((\overline{A} \cap B \cap \overline{C}) \cup (\overline{A} \cap \overline{B} \cap C) \right)$$

$$= (A \cap \overline{B} \cap B) \cup (A \cap \overline{B} \cap \overline{C}) \cup (A \cap C \cap B) \cup (A \cap C \cap \overline{C})$$

$$\cup (\overline{A} \cap B \cap \overline{C}) \cup (\overline{A} \cap \overline{B} \cap C)$$

$$= (A \cap B \cap C) \cup (A \cap \overline{B} \cap \overline{C}) \cup (\overline{A} \cap B \cap \overline{C}) \cup (\overline{A} \cap \overline{B} \cap C).$$

This expression is symmetric in A, B, and C, so

$$A \Delta (B \Delta C) = C \Delta (A \Delta B) = (A \Delta B) \Delta C, \qquad \text{using (ii)}.$$

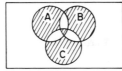

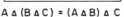

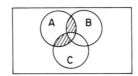

A Δ (B Δ C) = (A Δ B) Δ C   A ∩ (B Δ C) = (A ∩ B) Δ (A ∩ C)

Figure 2.04.   Venn diagrams.

We leave the proof of the other parts to the reader. Parts (i) and (vii) are illustrated in Figure 2.04. ☐

Relation (vii) of the previous proposition is a distributive law and states that $\cap$ is distributive over Δ. It is natural to ask whether $\cup$ is distributive over Δ.

2.05 EXAMPLE.   Is it true that $A \cup (B \Delta C) = (A \cup B) \Delta (A \cup C)$ for all $A, B, C \in \mathcal{P}(X)$?

SOLUTION.   The Venn diagrams for each side of the equation are given in Figure 2.05. If the shaded areas are not the same, we will be able to find a counterexample.

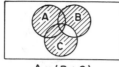
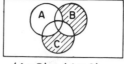

| $A \cup (B \Delta C)$ | $(A \cup B) \Delta (A \cup C)$ |

Figure 2.05.   Venn diagrams of unequal expressions.

We see from the diagrams that the result will be false if A is nonempty. If $A = X$ and $B = C = \varnothing$, then $A \cup (B \Delta C) = A$, whereas $(A \cup B) \Delta (A \cup C)$ $= \varnothing$; thus union is not distributive over symmetric difference. $\square$

NUMBER OF ELEMENTS IN A SET

If a set X contains 2 or 3 elements, we have seen that $\mathscr{P}(X)$ contains 2^2 or 2^3 elements, respectively. This suggests the following general result on the number of subsets of a finite set.

2.06 THEOREM.   If X is a finite set with n elements, then $\mathscr{P}(X)$ contains 2^n elements.

PROOF.   Each of the n elements of X is either in a given subset A or not in A. Hence, in choosing a subset of X, we have two choices for each element, and these choices are independent. Therefore, the number of choices is 2^n, and this is the number of subsets of X.

If $n = 0$, then $X = \varnothing$ and $\mathscr{P}(X) = \{\varnothing\}$ which contains one element. $\square$

Denote the number of elements of a set X by $\#X$. If A and B are finite *disjoint* sets, then

$$\#(A \cup B) = \#A + \#B.$$

2.07 PROPOSITION.   For any two finite sets A and B,

$$\#(A \cup B) = \#A + \#B - \#(A \cap B).$$

PROOF.   We can express $A \cup B$ as the disjoint union of A and $B - A$; also B can be expressed as the disjoint union of $B - A$ and $A \cap B$, as shown in Figure 2.06. Hence $\#(A \cup B) = \#A + \#(B - A)$ and $\#B = \#(B - A) + \#(A \cap B)$. It follows that $\#(A \cup B) = \#A + \#B - \#(A \cap B)$. $\square$

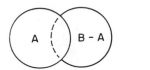

Figure 2.06

2.08 PROPOSITION.   For any three finite sets A, B, and C,

$$\#(A \cup B \cup C) = \#A + \#B + \#C - \#(A \cap B) - \#(A \cap C)$$

$$- \#(B \cap C) + \#(A \cap B \cap C).$$

PROOF.   Write $A \cup B \cup C$ as $(A \cup B) \cup C$. Then, by Proposition 2.07,

$$\#(A \cup B \cup C) = \#(A \cup B) + \#C - \#[(A \cup B) \cap C]$$

$$= \#A + \#B - \#(A \cap B) + \#C - \#[(A \cap C) \cup (B \cap C)]$$

$$= \#A + \#B + \#C - \#(A \cap B) - \#(A \cap C) - \#(B \cap C)$$

$$+ \#((A \cap C) \cap (B \cap C)).$$

The result follows because $(A \cap C) \cap (B \cap C) = A \cap B \cap C$. $\square$

2.09 EXAMPLE.   A survey of 1000 smokers reported that 850 smoked cigarettes, 200 smoked pipes, and 350 smoked marijuana, whereas 130 smoked cigarettes and a pipe, 220 smoked cigarettes and marijuana, 30 smoked a pipe and marijuana, and 20 smoked all three. Are these figures consistent?

SOLUTION.   Let C, P, and M be the sets of cigarette, pipe, and marijuana smokers, respectively. Then

$$\#(C \cup P \cup M) = \#C + \#P + \#M - \#(C \cap P) - \#(C \cap M)$$

$$- \#(P \cap M) + \#(C \cap P \cap M)$$

$$= 850 + 200 + 350 - 130 - 220 - 30 + 20$$

$$= 1040.$$

Since this number is greater than 1000, the figures must be inconsistent. The breakdown of the reported figures into their various classes is illustrated in Figure 2.07. The sum of all these numbers is 1040. $\square$

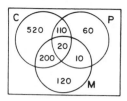

Figure 2.07.   The different classes of smokers.

2.10 EXAMPLE.   If 47% of the people in a community voted in a local election and 75% voted in a federal election, what is the least percentage that voted in both?

SOLUTION.   Let L and F be the sets of people who voted in the local and federal elections, respectively. Let n be the total number of voters in the community.
   Then

$$\#(L \cup F) \leqslant n \quad \text{and} \quad \#(L \cup F) = \#L + \#F - \#(L \cap F) \quad \text{so}$$

$$n \geqslant \#L + \#F - \#(L \cap F) = \frac{47n}{100} + \frac{75n}{100} - \#(L \cap F).$$

Hence $\dfrac{100}{n} \#(L \cap F) \geqslant 47 + 75 - 100 = 22$, and at least 22% voted in both elections.   □

Boolean Algebras

We now give the definition of an abstract Boolean algebra in terms of a set with two binary operations and one unary operation on it. We show that various algebraic structures, such as the algebra of sets, the logic of propositions, and the algebra of switching circuits are all Boolean algebras. It then follows that any general result derived from the axioms will hold in all our examples of Boolean algebras.

   It should be noted that this axiom system is only one of many equivalent ways of defining a Boolean algebra. Another common way is to define a Boolean algebra as a lattice satisfying certain properties. This is mentioned on page 28.

2.11 DEFINITION.   A *Boolean algebra* $(K, \wedge, \vee, ')$ is a set K together with two binary operations $\wedge$ and $\vee$, and a unary operation $'$ on K satisfying the following axioms for all $A, B, C \in K$.

(i) $A \wedge (B \wedge C) = (A \wedge B) \wedge C.$ (ii) $A \vee (B \vee C) = (A \vee B) \vee C.$

                                                       (Associative Laws)

(iii) $A \wedge B = B \wedge A.$ (iv) $A \vee B = B \vee A.$

                                                      (Commutative Laws)

(v) $A \wedge (B \vee C) = (A \wedge B) \vee$ (vi) $A \vee (B \wedge C) = (A \vee B) \wedge$

     $(A \wedge C).$ $(A \vee C).$

                                                        (Distributive Laws)

(vii)   There is a *zero element* 0 in K such that $A \vee 0 = A.$

(viii) There is a *unit element* 1 in K such that $A \wedge 1 = A.$

(ix) $A \wedge A' = 0.$ (x) $A \vee A' = 1.$

We call the operations $\wedge$ and $\vee$, *meet* and *join*, respectively. The element A' is called the *complement* of A.

The associative axioms (i) and (ii) are redundant in the above system because, with a little effort, they can be deduced from the other axioms. However, since associativity is such an important property, we keep these properties as axioms.

It follows from Proposition 2.02 that $(\mathscr{P}(X), \cap, \cup, ^{-})$ is a Boolean algebra with $\varnothing$ as zero and X as unit. When $X = \varnothing$, this Boolean algebra of subsets contains one element, and this is both the zero and unit. It can be proved (see Exercise 17) that, if the zero element and the unit element are the same, the Boolean algebra must have only one element.

We can define a two-element Boolean algebra $(\{0, 1\}, \wedge, \vee, ')$ by means of Table 2.3.

Table 2.3.   A two-element Boolean algebra

A	B	$A \wedge B$	$A \vee B$	A	A'
0	0	0	0	0	1
0	1	0	1	1	0
1	0	0	1		
1	1	1	1		

2.12 PROPOSITION.   If the binary operation $\star$ on the set K has an identity e such that $a \star e = e \star a = a$ for all $a \in K$, then this identity is unique.

PROOF.   Suppose that e and e' are both identities. Then

$$e = e \star e', \quad \text{since } e' \text{ is an identity}$$
$$= e', \quad \text{since } e \text{ is an identity}.$$

Therefore, the identity must be unique. $\square$

2.13 COROLLARY. The zero and unit elements in a Boolean algebra are unique.

PROOF. This follows directly from the above proposition, because the zero and unit elements are the identities for the join and meet operations, respectively. $\square$

2.14 PROPOSITION. The complementary operation in a Boolean algebra is unique; that is, for each $A \in K$ there is only one element $A' \in K$ satisfying axioms (ix) and (x).

PROOF. Suppose B and C are both complements of A so that $A \wedge B = 0$, $A \vee B = 1$, $A \wedge C = 0$, and $A \vee C = 1$. Then

$$B = B \vee 0 = B \vee (A \wedge C) = (B \vee A) \wedge (B \vee C)$$

$$= (A \vee B) \wedge (B \vee C) = 1 \wedge (B \vee C) = B \vee C.$$

Similarly $C = C \vee B$ and so $B = B \vee C = C \vee B = C$. $\square$

If we interchange $\wedge$ and $\vee$ and interchange 0 and 1 in the system of axioms for a Boolean algebra, we obtain the same system. Therefore, if any proposition is derivable from the axioms, so is the proposition obtained by interchanging $\wedge$ and $\vee$ and interchanging 0 and 1. This is called the *duality principle*. For example, in the following proposition, there are four pairs of dual statements. If one member of each pair can be proved, the other will follow directly from the duality principle.

If $(K, \wedge, \vee, ')$ is a Boolean algebra with 0 as zero and 1 as unit, then $(K, \vee, \wedge, ')$ is also a Boolean algebra with 1 as zero and 0 as unit.

2.15 PROPOSITION. If A, B, and C are elements of a Boolean algebra $(K, \wedge, \vee, ')$, the following relations hold.

(i) $A \wedge 0 = 0$. (ii) $A \vee 1 = 1$.

(iii) $A \wedge (A \vee B) = A$. (iv) $A \vee (A \wedge B) = A$.
(Absorption Laws)

(v) $A \wedge A = A$. (vi) $A \vee A = A$.
(Idempotent Laws)

(vii) $(A \wedge B)' = A' \vee B'$. (viii) $(A \vee B)' = A' \wedge B'$.
(De Morgan's Laws)

(ix) $(A')' = A$.

PROOF.   (i) $A \wedge 0 = (A \wedge 0) \vee 0 = (A \wedge 0) \vee (A \wedge A') = A \wedge (0 \vee A') = A \wedge A'$
$= 0$.

(ii) will follow by applying the duality principle to (i). That is,

$$A \vee 1 = (A \vee 1) \wedge 1 = (A \vee 1) \wedge (A \vee A') = A \vee (1 \wedge A') = A \vee A' = 1.$$

(iii) $A \wedge (A \vee B) = (A \vee 0) \wedge (A \vee B) = A \vee (0 \wedge B)$
$$\qquad\qquad\qquad\qquad = A \vee 0 \qquad\qquad\qquad\quad \text{by part (i) of this proposition}$$
$$\qquad\qquad\qquad\qquad = A.$$

(v) $A = A \wedge 1 = A \wedge (A \vee A') = (A \wedge A) \vee (A \wedge A') = (A \wedge A) \vee 0 = A \wedge A.$

(vii)   We prove this result by showing $A' \vee B'$ is *a* complement of $A \wedge B$; that is, we show that

$$(A \wedge B) \wedge (A' \vee B') = 0 \qquad \text{and} \qquad (A \wedge B) \vee (A' \vee B') = 1.$$

The uniqueness of the complement (Proposition 2.14) will then imply that $A' \vee B'$ is *the* complement of $A \wedge B$, namely, $(A \wedge B)'$. Now

$$(A \wedge B) \wedge (A' \vee B') = [(A \wedge B) \wedge A'] \vee [(A \wedge B) \wedge B']$$
$$= [(A \wedge A') \wedge B] \vee [A \wedge (B \wedge B')]$$
$$= [0 \wedge B] \vee [A \wedge 0]$$
$$= 0 \vee 0 \qquad\qquad \text{by part (i) of this proposition}$$
$$= 0 \qquad\qquad\qquad \text{by Definition 2.11 (vii).}$$

Also

$$(A \wedge B) \vee (A' \vee B') = [A \vee (A' \vee B')] \wedge [B \vee (A' \vee B')]$$
$$= [(A \vee A') \vee B'] \wedge [(B \vee B') \vee A']$$
$$= [1 \vee B'] \wedge [1 \vee A']$$
$$= 1 \wedge 1 \qquad\qquad \text{by part (ii) of this proposition}$$
$$= 1 \qquad\qquad\qquad \text{by Definition 2.11 (viii).}$$

(iv), (vi), and (viii) follow from the duality principle.

(ix)   It follows from Definition 2.11 that

$$A' \wedge A = 0 \qquad \text{and} \qquad A' \vee A = 1.$$

Therefore, A is a complement of A', and, since the complement is unique, $A = (A')'$. $\square$

2.16 EXAMPLE. Let $K=\{1,2,5,10,11,22,55,110\}$ be the set of positive divisors of 110. Show that $(K,\mathrm{GCD},\mathrm{LCM},')$ is a Boolean algebra where $x'=110/x$, the zero is 1, and the unit is 110.

SOLUTION. The binary operations greatest common divisor, GCD, and least common multiple, LCM, are both associative and commutative. Also $\mathrm{LCM}(1,x)=x$ and $\mathrm{GCD}(110,x)=x$ for all $x\in K$; thus 1 and 110 are the zero and unit elements, respectively.

The first distributive law states that

$$\mathrm{GCD}(x,\mathrm{LCM}(y,z))=\mathrm{LCM}(\mathrm{GCD}(x,y),\mathrm{GCD}(x,z)).$$

To prove this, we factor x, y, and z into prime factors to obtain

$$x=2^{\alpha_2}5^{\alpha_5}11^{\alpha_{11}},\quad y=2^{\beta_2}5^{\beta_5}11^{\beta_{11}},\quad \text{and}\quad z=2^{\gamma_2}5^{\gamma_5}11^{\gamma_{11}}$$

where the exponents are zero or one. The exponent of the prime p in $\mathrm{GCD}(x,\mathrm{LCM}(y,z))$ is

$$\min(\alpha_p,\max(\beta_p,\gamma_p))=\max(\min(\alpha_p,\beta_p),\min(\alpha_p,\gamma_p)).$$

This is the same as the exponent of p in $\mathrm{LCM}(\mathrm{GCD}(x,y),\mathrm{GCD}(x,z))$. Hence the first distributive law holds. The other distributive law can be proved in a similar way.

In the factorization of 110, no prime occurs as a square or higher power; thus each of the primes 2, 5, and 11 occur exactly once in one of the numbers x and $110/x$. Hence $\mathrm{GCD}(x,110/x)=1$, and $\mathrm{LCM}(x,110/x)=110$. This proves that the integer $110/x$ is the complement of x, and $(K,\mathrm{GCD},\mathrm{LCM},')$ is a Boolean algebra. $\square$

2.17 EXAMPLE. Show that $(K,\mathrm{GCD},\mathrm{LCM},')$ cannot be a Boolean algebra for any operation $'$ if $K=\{1,2,3,6,9,18\}$, the set of positive divisors of 18.

SOLUTION. The zero element must be 1 and the unit element 18. Let the complement of 6 be $6'$ so that $\mathrm{GCD}(6,6')=1$. By checking all the elements, we see that $6'=1$. However, $\mathrm{LCM}(6,6')=\mathrm{LCM}(6,1)=6$. Since this is not the unit, axiom (x) fails to hold, and K cannot be a Boolean algebra. $\square$

The difference between these last two examples is that $110=2\cdot5\cdot11$ contains no repeated factors, whereas $18=2\cdot3\cdot3$ does contain the repeated prime factor 3.

We now show briefly how algebra can be applied to the logic of propositions. Consider two sentences "*A*" and "*B*", which may either be true or false. For example, "*A*" could be "This apple is red", and "*B*" could be "This pear is green". We can combine these to form other sentences like "*A* and *B*", which would be "This apple is red, and this pear is green". We could also form the sentence "not *A*", which would be "This apple is not red". Let us now compare the truth or falsity of the derived sentences with the truth or falsity of the original ones. We illustrate the relationship by means of a diagram called a *truth table*. Table 2.4 shows the truth tables for the expressions "*A* and *B*", "*A* or *B*", and "not *A*". In these tables, T stands for "true," and F stands for "false".

Table 2.4. Truth tables

A	*B*	*A* and *B*	*A* or *B*		*A*	not *A*
F	F	F	F		F	T
F	T	F	T		T	F
T	F	F	T			
T	T	T	T			

For example, if the statement "*A*" is true, while "*B*" is false, the statement "*A* and *B*" will be false, and the statement "*A* or *B*" will be true.

We can have two seemingly different sentences with the same meaning; for example, "This apple is not red or this pear is not green" has the same meaning as "It is not true that this apple is red and that this pear is green". If two sentences, *P* and *Q*, have the same meaning, we say that *P* and *Q* are *logically equivalent*, and we write *P* = *Q*. The above example concerning the apples and pears implies that

$$(\text{not } A) \text{ or } (\text{not } B) = \text{not } (A \text{ and } B).$$

This equation corresponds to De Morgan's Law in a Boolean algebra.

It appears that a set of sentences behaves like a Boolean algebra. To be more precise, let us consider a set of sentences that are closed under the operations of "and", "or", and "not". Let *K* be the set, each element of which consists of all the sentences that are logically equivalent to a particular sentence. Then it can be verified that (*K*, and, or, not) is indeed a Boolean algebra. The zero element is called a *contradiction*; that is, a statement that is always false, such as "This apple is red and this apple is not red". The unit element is called a *tautology*; that is, a statement that is always true, such as "This apple is red or this apple is not red". This allows us to manipulate logical propositions using formulae derived from the axioms of a Boolean algebra.

An important method of combining two statements, A and B, in a sentence is by a *conditional*, such as "If A then B", or equivalently, "A implies B", which we shall write as "$A \Rightarrow B$". How does the truth or falsity of such a conditional depend on that of A and B? Consider the following sentences.

(i)   If $x > 4$ then $x^2 > 16$.
(ii) If $x > 4$ then $x^2 = 2$.
(iii) If $2 = 3$ then $0.2 = 0.3$.
(iv) If $2 = 3$ then the moon is made of green cheese.

Clearly, if A is true, then B must also be true for the sentence "$A \Rightarrow B$" to be true. However, if A is not true, then the sentence "If A, then B" has no standard meaning in everyday language. Let us take "$A \Rightarrow B$" to mean that we cannot have A true and B not true. This implies that the truth value of the statement "$A \Rightarrow B$" is the same as that of "not (A and not B)". Let us write $\wedge$, $\vee$ and $'$ for "and", "or", and "not", respectively. Then "$A \Rightarrow B$" is equivalent to $(A \wedge B')' = A' \vee B$. Thus "$A \Rightarrow B$" is true if A is false or if B is true. Using this definition, statements (i), (iii), and (iv) above are all true, while (ii) is false.

We can combine two conditional statements to form a *biconditional* statement of the form "A if and only if B" or "$A \Leftrightarrow B$." This has the same truth value as "$(A \Rightarrow B)$ and $(B \Rightarrow A)$" or equivalently $(A \wedge B) \vee (A' \wedge B')$. Another way of expressing this biconditional is to say that "A is a necessary and sufficient condition for B."

Table 2.5.   **Truth tables for conditional and biconditional statements**

A	B	$A \Rightarrow B$	$A \Leftarrow B$	$A \Leftrightarrow B$
F	F	T	T	T
F	T	T	F	F
T	F	F	T	F
T	T	T	T	T

It is seen from Table 2.5 that the statement "$A \Leftrightarrow B$" is true if either A and B are both true or A and B are both false.

2.18 EXAMPLE. Apply this propositional calculus to determine whether a certain politician's arguments are consistent. In one speech he states that, if taxes are raised, the rate of inflation will drop if and only if the value of the dollar does not fall. On television, he says that, if the rate of inflation decreases or the value of the dollar does not fall, taxes will not be raised. In a speech abroad, he states that either taxes must be raised or the value of the dollar will fall and the rate of inflation will decrease. His conclusion is that taxes will be raised, but the rate of inflation will decrease, and the value of the dollar will not fall.

SOLUTION.   We write

A to mean "Taxes will be raised",

B to mean "The rate of inflation will decrease",

C to mean "The value of the dollar will not fall".

The politician's three statements can be written symbolically as

(i) $A \Rightarrow (B \Leftrightarrow C)$.

(ii) $(B \lor C) \Rightarrow A'$.

(iii) $A \lor (C' \land B)$.

His conclusion is (iv) $A \land B \land C$.

The truth values of the first two statements are equivalent to those of the following.

(i) $A' \lor ((B \land C) \lor (B' \land C'))$.

(ii) $(B \lor C)' \lor A'$.

Table 2.6.   Truth tables for the politician's arguments

A	B	C	(i)	(ii)	(iii)	(i)$\land$(ii)$\land$(iii)	(iv)	(i)$\land$(ii)$\land$(iii)$\Rightarrow$(iv)
F	F	F	T	T	F	F	F	T
F	F	T	T	T	F	F	F	T
F	T	F	T	T	T	T	F	F
F	T	T	T	T	F	F	F	T
T	F	F	T	T	T	T	F	F
T	F	T	F	F	T	F	F	T
T	T	F	F	F	T	F	F	T
T	T	T	T	F	T	F	T	T

It follows from Table 2.6 that (i)$\land$(ii)$\land$(iii)$\Rightarrow$(iv) is not a tautology, that is, it is not always true. Therefore, the politician's arguments are incorrect. They break down when A and C are false and B is true, and when B and C are false and A is true.   □

SWITCHING CIRCUITS

In this section we use Boolean algebra to analyze some simple switching circuits. A *switch* is a device with two states; state 1 is the "on" state, and state 0 the "off" state. An ordinary household light switch is such a device, but the theory holds equally well for more sophisticated electronic or magnetic two-state devices. We analyze circuits with two terminals, and the circuit is said to be *closed* if current can pass between the terminals, and *open* if current cannot pass.

We denote a switch A by the symbol in Figure 2.08. We assign the value 1 to A if the switch A is closed and the value 0 if it is open. We denote two switches by the same letter if they open and close simultaneously. If B is a switch that is always in the opposite position to A (that is, if B is open when A is closed and B is closed when A is open), denote switch B by A'.

o————/A————o   **Figure 2.08.   The switch A.**

The two switches A and B in Figure 2.09 are said to be connected in *series*. If we connect this circuit to a power source and a light as in Figure 2.10, we see that the light will be on if and only if A *and* B are both switched on; we denote this series circuit by $A \wedge B$. Its effect is shown in Table 2.7.

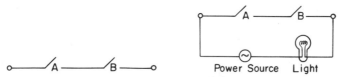

o————/A————/B————o

Figure 2.09.   Switches in series.

Power Source   Light

Figure 2.10.   A series circuit.

Table 2.7.   The effect of the series circuit

Switch A	Switch B	Circuit $A \wedge B$	Light
0 (off)	0 (off)	0 (open)	off
0 (off)	1 (on)	0 (open)	off
1 (on)	0 (off)	0 (open)	off
1 (on)	1 (on)	1 (closed)	on

The switches A and B in Figure 2.11 are said to be in *parallel*, and this circuit is denoted by $A \vee B$ because the circuit is closed if either A *or* B is switched on.

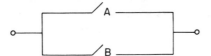

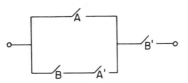 Figure 2.11.   Switches in parallel.

The reader should be aware that many books on switching theory use the notation $+$ and $\cdot$ instead of $\vee$ and $\wedge$, respectively.

Series and parallel circuits can be combined together to form circuits like the one in Figure 2.12. This circuit would be denoted by $(A \vee (B \wedge A')) \wedge B'$. Such circuits are called *series-parallel switching circuits*.

In actual practice, the wiring diagram may not look at all like Figure 2.12, because we would want switches A and A' together and B and B' together. Figure 2.13 illustrates one particular form that the wiring diagram could take.

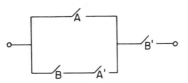

Figure 2.12.   A series-parallel circuit.   Figure 2.13.   A wiring diagram of the circuit.

Two circuits $\mathcal{C}_1$ and $\mathcal{C}_2$ involving the switches $A, B, \ldots$ are said to be *equivalent* if the positions of the switches $A, B, \ldots$, which allow current to pass, are the same for both circuits. We write $\mathcal{C}_1 = \mathcal{C}_2$ to mean that the circuits are equivalent. It can be verified that all the axioms for a Boolean algebra are valid when interpreted as series-parallel switching circuits. For example, Figure 2.14 illustrates a distributive law. The zero corresponds to a circuit that is always open, and the unit corresponds to a circuit that is always closed. The complement $\mathcal{C}'$ of a circuit $\mathcal{C}$ is open whenever $\mathcal{C}$ is closed and closed when $\mathcal{C}$ is open. The interpretations of the various Boolean algebra terms are given in Table 2.8.

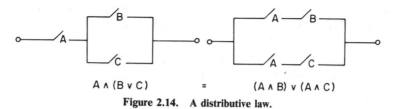

$$A \wedge (B \vee C) \qquad = \qquad (A \wedge B) \vee (A \wedge C)$$

Figure 2.14. A distributive law.

Table 2.8. A dictionary of Boolean algebra terms

Boolean algebra	Algebra of subsets of X	Series-parallel switching circuits	Propositional logic
$\wedge$	$\cap$	series	and
$\vee$	$\cup$	parallel	or
$'$	$-$	opposite	not
0	$\varnothing$	open	contradiction
1	X	closed	tautology
$=$	$=$	equivalent circuit	logically equivalent

POSETS AND LATTICES

Boolean algebras were derived from the algebra of sets, and there is one important relation between sets that we have neglected to generalize to Boolean algebras, namely, the inclusion relation. This relation can be defined in terms of the union operation by

$$A \subseteq B \text{ if and only if } A \cap B = A.$$

We can define a corresponding relation on any Boolean algebra using the meet operation.

2.19 DEFINITION. The relation $\leqslant$ is defined on a Boolean algebra $(K, \wedge, \vee, ')$ by

$$A \leqslant B \text{ if and only if } A \wedge B = A.$$

If the Boolean algebra is the algebra of subsets of X, this relation is the usual inclusion relation.

2.20 PROPOSITION. $A \wedge B = A$ if and only if $A \vee B = B$. Hence either of the these conditions will define the relation $\leqslant$.

PROOF.   If $A \wedge B = A$, it follows from the absorption law that $A \vee B$ $= (A \wedge B) \vee B = B$. Similarly, if $A \vee B = B$, it follows that $A \wedge B = A$.   □

2.21 PROPOSITION.   If A, B, and C are elements of a Boolean algebra, K, the following properties of the relation $\leqslant$ hold.

(i) $A \leqslant A$.   (Reflexivity)
(ii)   If $A \leqslant B$ and $B \leqslant A$, then $A = B$.   (Antisymmetry)
(iii)   If $A \leqslant B$ and $B \leqslant C$, then $A \leqslant C$.   (Transitivity)

A relation satisfying the above three properties is called a *partial order relation*, and a set with a partial order on it is called a *partially ordered set* or *poset* for short. The interpretation of the partial order in various Boolean algebras is given in Table 2.9.

Table 2.9.   The partial order relation in various Boolean algebras

Boolean algebra	Algebra of subsets	Series-parallel switching circuits	Propositional logic	Divisors of a square-free integer
$A \wedge B = A$	$A \cap B = A$	$A \wedge B = A$	A and $B = A$	$\text{GCD}(a,b) = a$
$A \leqslant B$	$A \subseteq B$	$A \Rightarrow B$	$A \Rightarrow B$	$a \mid b$
A is less than or equal to B	A is a subset of B	If A is closed then B is closed	A implies B	a divides b

PROOF.

(i) $A \wedge A = A$ is an idempotent law.
(ii)   If $A \wedge B = A$ and $B \wedge A = B$, then $A = A \wedge B = B \wedge A = B$.
(iii)   If $A \wedge B = A$ and $B \wedge C = B$, then $A \wedge C = (A \wedge B) \wedge C = A \wedge (B \wedge C)$ $= A \wedge B = A$.   □

A partial order on a finite set, K, can be conveniently displayed in a *poset diagram* in which the elements of K are represented by small circles. Lines are drawn connecting these elements so that there is a path from A to B that is always directed upwards if and only if $A \leqslant B$. Figure 2.15 illustrates the poset diagram of the Boolean algebra of subsets $(\mathcal{P}(\{a,b\},) \cap, \cup, ^-)$. Figure 2.16 illustrates the Boolean algebra of divisors of 110 considered in Example 2.16. The partial order relation is divisibility, so that there is an upwards path from a to b if and only if a divides b.

The following proposition shows that $\leqslant$ has properties similar to those of the inclusion relation in sets.

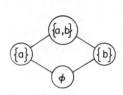

Figure 2.15. Poset diagram of $\mathcal{P}(\{a,b\})$.

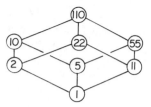

Figure 2.16. Poset diagram of the divisors of 110.

2.22 PROPOSITION. If A, B, C are elements of a Boolean algebra $(K, \wedge, \vee, ')$ then the following relations hold.

(i) $A \wedge B \leqslant A$.
(ii) $A \leqslant A \vee B$.
(iii) $A \leqslant C$ and $B \leqslant C$ implies $A \vee B \leqslant C$.
(iv) $A \leqslant B$ if and only if $A \wedge B' = 0$.
(v) $0 \leqslant A$ and $A \leqslant 1$ for all A.

PROOF.

(i) $(A \wedge B) \wedge A = (A \wedge A) \wedge B = A \wedge B$ so $A \wedge B \leqslant A$.
(ii) $A \wedge (A \vee B) = A$ so $A \leqslant A \vee B$.
(iii) $(A \vee B) \wedge C = (A \wedge C) \vee (B \wedge C) = A \vee B$.
(iv) If $A \leqslant B$, then $A \wedge B = A$ and $A \wedge B' = A \wedge B \wedge B' = A \wedge 0 = 0$.

If $A \wedge B' = 0$, then $A \leqslant B$ because

$$A = A \wedge 1 = A \wedge (B \vee B') = (A \wedge B) \vee (A \wedge B') = (A \wedge B) \vee 0 = A \wedge B.$$

(v) $0 \wedge A = 0$ and $A \wedge 1 = A$. $\square$

Not all posets are derived from Boolean algebras. A Boolean algebra is an extremely special kind of poset. We now determine conditions that insure that a poset is indeed a Boolean algebra. Given a partial order $\leqslant$ on a set K, we have to find two binary operations that correspond to the meet and join.

2.23 DEFINITION.   An element d is said to be the *greatest lower bound* of the elements a and b in a partially ordered set if $d \leqslant a$, $d \leqslant b$, and, if x is another element, for which $x \leqslant a$, $x \leqslant b$, then $x \leqslant d$. We denote the greatest lower bound of a and b by $a \wedge b$. Similarly, we can define the *least upper bound* and denote it by $\vee$. It follows from the antisymmetry of the partial order relation that each pair of elements a and b can have at most one greatest lower bound and at most one least upper bound.

2.24 DEFINITION. A *lattice* is a partially ordered set in which every two elements have a greatest lower bound and a least upper bound.

We can now give an alternative definition of a Boolean algebra in terms of a lattice. It can be verified that this definition is equivalent to our original one.

2.25 DEFINITION. A *Boolean algebra* is a lattice that has universal bounds (that is, elements 0 and 1 such that $0 \leqslant a$ and $a \leqslant 1$ for all elements a) and is distributive and complemented (that is, the distributive laws for $\wedge$ and $\vee$ hold, and complements exist).

In Figure 2.17, the elements c and d have a least upper bound b but no greatest lower bound. Figure 2.18 illustrates the lattice of divisors of 18. We saw in Example 2.17 that this did not have complements and therefore cannot be a Boolean algebra.

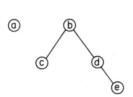
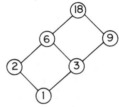

Figure 2.17. A poset that is not a lattice.   **Figure 2.18. A lattice that is not a Boolean algebra.**

For further reading on lattices in applied algebra, consult Birkhoff and Bartee [9] or Stone [12].

NORMAL FORMS AND SIMPLIFICATION OF CIRCUITS

If we have a complicated switching circuit represented by a Boolean expression, such as

$$(A \wedge (B \vee C')') \vee ((B \wedge C') \vee A'),$$

we would like to know if we can build another simpler circuit that would perform the same function. In other words, we would like to reduce this Boolean expression to a simpler form. In actual practice, it is usually desirable reduce the circuit to the one that is cheapest to build, and the form this takes will depend on the state of the technology at the time;

however, for our purposes we take the simplest form to mean the one with the fewest switches. It is a hard problem to find the simplest form for circuits with many switches, and there is no one method that will lead to that form. However, we do have methods for determining whether two Boolean expressions are equivalent. We can reduce the expressions to a certain normal form, and the expressions will be the same if and only if their normal forms are the same. We shall look at one such form, called the *disjunctive normal form*.

In the Boolean algebra of subsets of a set, every subset can be expressed as a union of singleton sets, and this union is unique to within the ordering of the terms. We shall obtain a corresponding result for arbitrary finite Boolean algebras. The elements that play the role of singleton sets are called atoms.

2.26 DEFINITION. An *atom* in a Boolean algebra $(K, \wedge, \vee, ')$ is a nonzero element B for which

$$B \wedge Y = B \quad \text{or} \quad B \wedge Y = 0 \quad \text{for each} \quad Y \in K.$$

B is an atom if $Y \leqslant B$ implies that $Y = 0$ or $Y = B$. This implies that the atoms are the elements immediately above the zero element in the poset diagram. In the case of the algebra of divisors of square-free integers, the atoms are the primes, because $y|b$, implying $y = 1$ or $y = b$ is the definition of b being prime.

We now give a more precise description of the algebra of switching circuits. The atoms of the algebra and the disjunctive normal form of an expression will become clear from this description.

An n-variable switching circuit can be viewed as a black box containing n independent switches $A_1, A_2, \ldots, A_n$, as shown in Figure 2.19, where each switch can either be on or off. The effect of such a circuit can be tested by trying all the 2^n different combinations of the n switches and observing when the box allows current to pass. In this way, each circuit defines a function of n variables $A_1, A_2, \ldots, A_n$

$$f: \{0, 1\}^n \rightarrow \{0, 1\},$$

which we call the *switching function of the circuit*. Two circuits give rise to the same switching function if and only if they are equivalent.

Figure 2.19. An n-variable switching circuit.

For example, the circuit in Figure 2.20, corresponding to the expression $(A \vee B') \wedge (C \vee A')$, gives rise to the switching function $f: \{0,1\}^3 \rightarrow \{0,1\}$ given in Table 2.10.

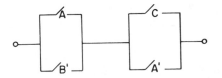

Figure 2.20. The circuit $(A \vee B') \wedge (C \vee A')$.

Table 2.10. Switching function

A	B	C	$f = (A \vee B') \wedge (C \vee A')$
0	0	0	1
0	0	1	1
0	1	0	0
0	1	1	0
1	0	0	0
1	0	1	1
1	1	0	0
1	1	1	1

Denote the set of all n-variable switching functions from $\{0,1\}^n$ to $\{0,1\}$ by $\mathcal{F}_n$. Each of the 2^n elements in the domain of such a function can be mapped to either of the two elements in the codomain. Therefore, the number of different n-variable switching functions, and hence the number of different circuits with n switches, is 2^{2^n}.

Let f and g be the switching functions of two circuits of the n-variables $A_1, A_2, \ldots, A_n$. When these circuits are connected in series or in parallel, they give rise to the switching functions $f \wedge g$ or $f \vee g$, respectively, where

$$(f \wedge g)(A_1, \ldots, A_n) = f(A_1, \ldots, A_n) \wedge g(A_1, \ldots, A_n)$$

and

$$(f \vee g)(A_1, \ldots, A_n) = f(A_1, \ldots, A_n) \vee g(A_1, \ldots, A_n).$$

The switching function of the opposite circuit to that defining f is f' where

$$f'(A_1, \ldots, A_n) = (f(A_1, \ldots, A_n))'.$$

2.27 THEOREM. The set of n-variable switching functions forms a Boolean algebra $(\mathcal{F}_n, \wedge, \vee, ')$ that contains 2^{2^n} elements.

PROOF.   It can be verified that $(\mathcal{F}_n, \wedge, \vee, ')$ satisfies all the axioms of a Boolean algebra. The zero element is the function whose image is always 0, and the unit element is the function whose image is always 1.   □

The Boolean algebra of switching functions of two variables contains 16 elements, which are displayed in Table 2.11.

Table 2.11.   The two-variable switching functions

A	0	0	1	1	Expressions in A and B
B	0	1	0	1	representing the function
f_0	0	0	0	0	0
f_1	0	0	0	1	$A \wedge B$
f_2	0	0	1	0	$A \wedge B'$ or $A \not\Rightarrow B$
f_3	0	0	1	1	A
f_4	0	1	0	0	$A' \wedge B$ or $A \not\Leftarrow B$
f_5	0	1	0	1	B
f_6	0	1	1	0	$A \Delta B$ or Exclusive OR(A, B)
f_7	0	1	1	1	$A \vee B$
f_8	1	0	0	0	$A' \wedge B'$ or NOR(A, B)
f_9	1	0	0	1	$A \Delta B'$ or $A \Leftrightarrow B$
f_{10}	1	0	1	0	B'
f_{11}	1	0	1	1	$A \vee B'$ or $A \Leftarrow B$
f_{12}	1	1	0	0	A'
f_{13}	1	1	0	1	$A' \vee B$ or $A \Rightarrow B$
f_{14}	1	1	1	0	$A' \vee B'$ or NAND(A, B)
f_{15}	1	1	1	1	1

For example, $f_6(A, B) = 0$, if $A = B$, and 1, if $A \neq B$. This function is the Exclusive OR function or a modulo 2 adder. It is also the symmetric difference function, where the symmetric difference of A and B in a Boolean algebra is defined by

$$A \Delta B = (A \wedge B') \vee (A' \wedge B).$$

The operations NAND and NOR stand for "not and" and "not or," respectively; these are discussed further in the next section on transistor gates.

As an example of the operations in the Boolean algebra $\mathcal{F}_2$, we calculate the meet and join of f_{10} and f_7 and the complement of f_{10} in Table 2.12. We see that $f_{10} \wedge f_7 = f_2$, $f_{10} \vee f_7 = f_{15}$ and $f'_{10} = f_5$. These correspond to the relations $B' \wedge (A \vee B) = A \wedge B'$, $B' \vee (A \vee B) = 1$ and $(B')' = B$.

Table 2.12.   Some operations in $\mathcal{F}_2$

A	B	f_{10}	f_7	$f_{10} \wedge f_7$	$f_{10} \vee f_7$	f'_{10}
0	0	1	0	0	1	0
0	1	0	1	0	1	1
1	0	1	1	1	1	0
1	1	0	1	0	1	1

In the Boolean algebra $\mathcal{F}_n$, $f \leqslant g$ if and only if $f \wedge g = f$, which happens if $g(A_1,\ldots,A_n) = 1$ whenever $f(A_1,\ldots,A_n) = 1$. The atoms of $\mathcal{F}_n$ are therefore the functions whose image contains precisely one nonzero element. $\mathcal{F}_n$ contains 2^n atoms, and the expressions that realize these atoms are of the form $A_1^{\alpha_1} \wedge A_2^{\alpha_2} \wedge \ldots \wedge A_n^{\alpha_n}$ where each $A_i^{\alpha_i} = A_i$ or A'_i.

The 16 elements of $\mathcal{F}_2$ are illustrated in Figure 2.21, and the four atoms are f_1, f_2, f_4, and f_8, which are defined in Table 2.11.

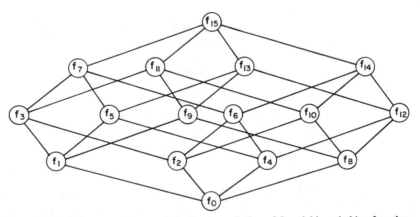

Figure 2.21.   The poset diagram of the Boolean algebra of 2-variable switching functions.

In order to show that every element of a finite Boolean algebra can be written as a join of atoms, we need three preliminary lemmas.

2.28 LEMMA.   If $A, B_1, \ldots, B_r$ are atoms in a Boolean algebra, then $A \leqslant (B_1 \vee \ldots \vee B_r)$ if and only if $A = B_i$, for some i with $1 \leqslant i \leqslant r$.

PROOF.   If $A \leqslant (B_1 \vee \ldots \vee B_r)$, then $A \wedge (B_1 \vee \ldots \vee B_r) = A$; thus $(A \wedge B_1) \vee \ldots \vee (A \wedge B_r) = A$. Since each B_i is an atom, $A \wedge B_i = B_i$ or 0. Not all the elements $A \wedge B_i$ can be 0, for this would imply $A = 0$. Hence there is some i, with $1 \leqslant i \leqslant r$, for which $A \wedge B_i = B_i$. But A is also an atom and so $A = A \wedge B_i = B_i$.

The implication the other way is straightforward. □

2.29 LEMMA. If Z is a nonzero element of a finite Boolean algebra, there exists an atom B with $B \leqslant Z$.

PROOF. If Z is an atom, take $B = Z$. If not, then it follows from the definition of atoms that there exists a nonzero element Z_1, different from Z, with $Z_1 \leqslant Z$. If Z_1 is not an atom, we continue in this way to obtain a sequence of distinct nonzero elements $\ldots \leqslant Z_3 \leqslant Z_2 \leqslant Z_1 \leqslant Z$, which, because the algebra is finite, must terminate in an atom B. □

2.30 LEMMA. If $B_1, \ldots, B_n$ are all the atoms of a finite Boolean algebra, then $Y = 0$ if and only if $Y \wedge B_i = 0$ for all i such that $1 \leqslant i \leqslant n$.

PROOF. Suppose $Y \wedge B_i = 0$ for each i. If Y is nonzero, it follows from the previous lemma that there is an atom B_j with $B_j \leqslant Y$. Hence $B_j = Y \wedge B_j = 0$, which is a contradiction, and so $Y = 0$. The converse implication is trivial. □

2.31 DISJUNCTIVE NORMAL FORM. Each element X of a finite Boolean algebra can be written as a join of atoms

$$X = B_\alpha \vee B_\beta \vee \ldots \vee B_\omega.$$

Moreover, this expression is unique up to the order of the atoms.

PROOF. Let $B_\alpha, B_\beta, \ldots, B_\omega$ be all the atoms less than or equal to X in the partial order. It follows from Proposition 2.22 (iii) that the join $Y = B_\alpha \vee B_\beta \vee \ldots \vee B_\omega \leqslant X$.

We will show that $X \wedge Y' = 0$, which, by Proposition 2.22 (iv), is equivalent to $X \leqslant Y$. We have

$$X \wedge Y' = X \wedge B_\alpha' \wedge \ldots \wedge B_\omega'.$$

If B is an atom in the join Y, say $B = B_\alpha$, it follows that $X \wedge Y' \wedge B = 0$, since $B_\alpha' \wedge B_\alpha = 0$. If B is an atom that is not in Y, then $X \wedge Y' \wedge B = 0$ also, because $X \wedge B = 0$. Therefore, by Lemma 2.30, $X \wedge Y' = 0$, which is equivalent to $X \leqslant Y$. The antisymmetry of the partial order relation implies that $X = Y$.

To show uniqueness, suppose that X can be written as the join of two sets of atoms

$$X = B_\alpha \vee \ldots \vee B_\omega = B_a \vee \ldots \vee B_z.$$

Now $\dot{B}_\alpha \leqslant X$; thus, by Lemma 2.28, B_α is equal to one of the atoms on the right-hand side, $B_a, \ldots, B_z$. Repeating this argument, we see that the two sets of atoms are the same, except possibly for their order. $\square$

In the Boolean algebra of n-variable switching functions, the atoms are realized by expressions of the form $A_1^{\alpha_1} \wedge A_2^{\alpha_2} \wedge \ldots \wedge A_n^{\alpha_n}$, where the α_is are 0 or 1 and $A_i^{\alpha_i} = A_i$, if $\alpha_i = 1$, whereas $A_i^{\alpha_i} = A_i'$, if $\alpha_i = 0$. The expression $A_1^{\alpha_1} \wedge A_2^{\alpha_2} \wedge \ldots \wedge A_n^{\alpha_n}$ is included in the disjunctive normal form of the function f if and only if $f(\alpha_1, \alpha_2, \ldots, \alpha_n) = 1$. Hence there is one atom in the disjunctive normal form for each time the element 1 occurs in the image of the switching function.

2.32 EXAMPLE. Find the disjunctive normal form for the expression $(B \vee (A \wedge C)) \wedge ((A \vee C) \wedge B)'$, and check the result by using the axioms to reduce the expression to that form.

SOLUTION.

Table 2.13. Switching function

A	B	C	$(B \vee (A \wedge C)) \wedge ((A \vee C) \wedge B)'$
0	0	0	0
0	0	1	0
0	1	0	1
0	1	1	0
1	0	0	0
1	0	1	1
1	1	0	0
1	1	1	0

We see from the values of the switching function in Table 2.13 that the disjunctive normal form is $(A' \wedge B \wedge C') \vee (A \wedge B' \wedge C)$.

From the axioms we have

$$
\begin{aligned}
(B \vee (A \wedge C)) \wedge ((A \vee C) \wedge B)' &= (B \vee (A \wedge C)) \wedge ((A' \wedge C') \vee B') \\
&= ((B \vee (A \wedge C)) \wedge (A' \wedge C')) \vee \\
&\quad ((B \vee (A \wedge C)) \wedge B') \\
&= (B \wedge A' \wedge C') \vee (A \wedge C \wedge A' \wedge C') \vee \\
&\quad (B \wedge B') \vee (A \wedge C \wedge B') \\
&= (A' \wedge B \wedge C') \vee 0 \vee 0 \vee (A \wedge B' \wedge C) \\
&= (A' \wedge B \wedge C') \vee (A \wedge B' \wedge C). \quad \square
\end{aligned}
$$

2.33 EXAMPLE. Are any of the three expressions $(A \vee B) \wedge B'$, $(A \vee B) \wedge (A \wedge B)'$, and $(A \wedge B)' \wedge (A \wedge B')$ equivalent?

SOLUTION.

Table 2.14. Switching function

A	B	$(A \vee B) \wedge B'$	$(A \vee B) \wedge (A \wedge B)'$	$(A \wedge B)' \wedge (A \wedge B')$
0	0	0	0	0
0	1	0	1	0
1	0	1	1	1
1	1	0	0	0

We see from Table 2.14 that $(A \vee B) \wedge B' = (A \wedge B)' \wedge (A \wedge B')$ and that these are both equal to $A \wedge B'$. □

The atoms in the Boolean algebra $\mathcal{F}_2$ are realized by the expressions $A' \wedge B'$, $A' \wedge B$, $A \wedge B'$, and $A \wedge B$. These atoms partition the Venn diagram in Figure 2.22 into four disjoint regions. The disjunctive normal form for any Boolean expression involving the variables A and B can be calculated by shading the region of the Venn diagram corresponding to the expression and then taking the join of the atoms in the shaded region. Figure 2.23 illustrates the eight regions of the corresponding Venn diagram for three variables.

By looking at the shaded region of a Venn diagram corresponding to a Boolean expression, it is often possible to see how to simplify the expression. Furthermore, the disjunctive normal form provides a method of proving hypotheses derived from these Venn diagrams.

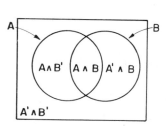

Figure 2.22.   Venn diagram for $\mathcal{F}_2$.

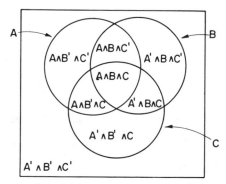

Figure 2.23.   Venn diagram for $\mathcal{F}_3$.

However, Venn diagrams become too complicated and impracticable for functions of more than four variables. For other general methods of simplifying circuits, consult a book on Boolean algebras such as Mendelson [17].

2.34 EXAMPLE. Find the disjunctive normal form and simplify the circuit in Figure 2.24.

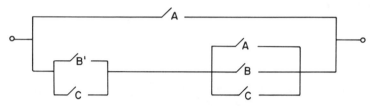

Figure 2.24.   Series-parallel circuit.

SOLUTION.   This circuit is represented by the Boolean expression

$$f = A \vee ((B' \vee C) \wedge (A \vee B \vee C)).$$

The Boolean function $f: \{0, 1\}^3 \rightarrow \{0, 1\}$ that this expression defines is given in Table 2.15.

Table 2.15.   Switching function

A	B	C	f
0	0	0	0
0	0	1	1
0	1	0	0
0	1	1	1
1	0	0	1
1	0	1	1
1	1	0	1
1	1	1	1

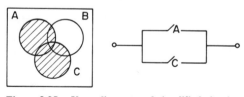

Figure 2.25.   Venn diagram and simplified circuit.

It follows from Table 2.15 that the disjunctive normal form is

$$(A' \wedge B' \wedge C) \vee (A' \wedge B \wedge C) \vee (A \wedge B' \wedge C') \vee (A \wedge B' \wedge C) \vee (A \wedge B \wedge C')$$
$$\vee (A \wedge B \wedge C),$$

which is certainly not simpler than the original. However, by looking at the

Venn diagram in Figure 2.25, we see that this expression is equivalent to just $A \vee C$; thus a simpler equivalent circuit is given in Figure 2.25. $\square$

In building a computer, one of the most important pieces of equipment needed is a circuit that will add two numbers in binary form. Consider the problem of adding the numbers 15 and 5. Their binary forms are 1111 and 101, respectively. The binary and decimal additions are shown below. In general, if we add the number $\ldots a_2 a_1 a_0$ to $\ldots b_2 b_1 b_0$, we have to carry the digits $\ldots c_2 c_1$ to obtain the sum $\ldots s_2 s_1 s_0$.

1111	15	$\ldots a_2 a_1 a_0$
101	5	$\ldots b_2 b_1 b_0$
1111 $\longleftarrow$ carry digits $\longrightarrow$ 1		$\ldots c_2 c_1$
10100	20	$\ldots s_2 s_1 s_0$
Binary Addition	*Decimal Addition*	

Let us first design a circuit to add a_0 and b_0 to obtain s_0 and the carry digit c_1. This is called a *half adder*. The digits s_0 and c_1 are functions of a_0 and b_0 which are given by Table 2.16. For example, in binary arithmetic, $1 + 1 = 10$, which means that, if $a_0 = 1$ and $b_0 = 1$, $s_0 = 0$, and we have to carry $c_1 = 1$.

Table 2.16. Switching functions for the half adder

a_0	b_0	c_1	s_0
0	0	0	0
0	1	0	1
1	0	0	1
1	1	1	0

We see from Table 2.16 that $c_1 = a_0 \wedge b_0$ and $s_0 = (a_0' \wedge b_0) \vee (a_0 \wedge b_0')$. These circuits are shown in Figure 2.26.

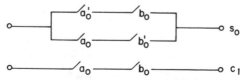

Figure 2.26. Circuits for the half adder.

A circuit that adds a_i, b_i, and the carry digit, c_i, to obtain s_i, with c_{i+1} to carry, is called a *full adder*. The functions c_{i+1} and s_i are defined by Table 2.17, and their Venn diagrams are given in Figure 2.27. Notice that $s_i = a_i \Delta b_i \Delta c_i$.

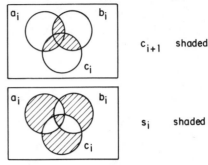

Table 2.17. Switching functions for a full adder

a_i	b_i	c_i	c_{i+1}	s_i
0	0	0	0	0
0	0	1	0	1
0	1	0	0	1
0	1	1	1	0
1	0	0	0	1
1	0	1	1	0
1	1	0	1	0
1	1	1	1	1

Figure 2.27. Venn diagrams for a full adder.

Suitable expressions for a full adder are as follows. The corresponding circuits are shown in Figure 2.28.

$$s_i = (a_i' \wedge b_i' \wedge c_i) \vee (a_i' \wedge b_i \wedge c_i') \vee (a_i \wedge b_i' \wedge c_i') \vee (a_i \wedge b_i \wedge c_i)$$

$$= (a_i' \wedge ((b_i' \wedge c_i) \vee (b_i \wedge c_i'))) \vee (a_i \wedge ((b_i' \wedge c_i') \vee (b_i \wedge c_i))).$$

$$c_{i+1} = (a_i \wedge b_i) \vee (a_i \wedge c_i) \vee (b_i \wedge c_i)$$

$$= (a_i \wedge (b_i \vee c_i)) \vee (b_i \wedge c_i).$$

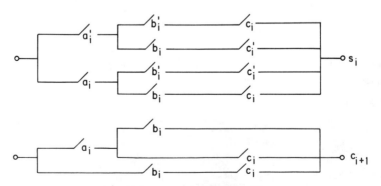

Figure 2.28. Circuits for a full adder.

Using one half adder and $(n-1)$ full adders, we can design a circuit that will add two numbers that, in a binary form, have n of fewer digits (that is, numbers less than 2^n).

Transistor Gates

The switches we have been dealing with so far have been simple two-state devices. Transistor technology, however, allows us to construct basic switches with multiple inputs. These are called *transistor gates*. Transistor gates can be used to implement the logical operations AND, OR, NOT, and modulo 2 addition (that is, exclusive OR). Gates for the composite operations NOT-AND and NOT-OR are also easily built from transistors; these are called NAND and NOR gates, respectively. Figure 2.29 illustrates the symbols and outputs for these gates when there are two inputs. However, any number of inputs is possible. Note that the inversion operation is indicated by a small circle.

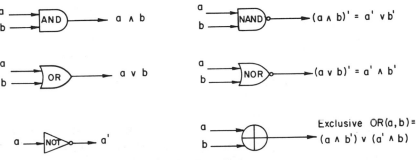

Figure 2.29.   Transistor gates.

Transistor gates can be combined in series and in parallel to form more complex circuits. Any circuit with n inputs and one output defines an n-variable switching function. The set of all such n-variable functions again forms the Boolean algebra $(\mathcal{F}_n, \wedge, \vee, ')$.

It follows from the disjunctive normal form that any Boolean function can be constructed from AND, OR, and NOT gates. What is not so obvious is that any Boolean function can be constructed solely from NOR gates (or solely from NAND gates). This is of interest because, with certain types of transistors, it is easier to build NOR gates (and NAND gates) than it is to build the basic operations. Figure 2.30 illustrates how 2-input

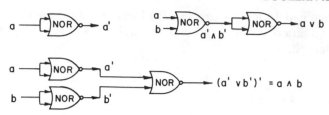

Figure 2.30. The basic operations constructed from NOR gates.

NOR gates can be used to construct 2-input AND and OR gates as well as a NOT gate.

2.35 EXAMPLE. Verify that the circuit in Figure 2.31 is indeed a full adder.

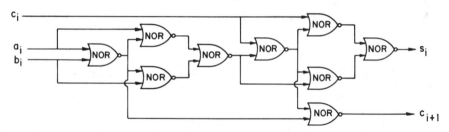

Figure 2.31. A full adder using NOR gates.

SOLUTION. We analyze this circuit by breaking it up into component parts as illustrated in Figure 2.32. Consider the subcircuit consisting of four NOR gates in Figure 2.32 with inputs a and b and outputs l and m. If u and v are the intermediate functions as shown in the figure, then

$$m = \text{NOR}(a,b) = a' \wedge b',$$

$$u = \text{NOR}(a, a' \wedge b') = a' \wedge (a' \wedge b')' = a' \wedge (a \vee b)$$

$$= (a' \wedge a) \vee (a' \wedge b) = 0 \vee (a' \wedge b) = a' \wedge b,$$

and $v = a \wedge b'$, similarly. Therefore

$$l = \text{NOR}(u,v) = (a' \wedge b)' \wedge (a \wedge b')' = (a \vee b') \wedge (a' \vee b)$$

$$= (a \wedge a') \vee (a \wedge b) \vee (b' \wedge a') \vee (b' \wedge b) = 0 \vee (a \wedge b) \vee (b' \wedge a') \vee 0$$

$$= (a \wedge b) \vee (a' \wedge b') = a \Delta b'.$$

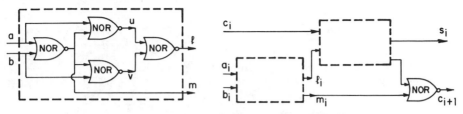

Figure 2.32. The component parts of the full adder.

The whole circuit can now be constructed from two of these identical subcircuits together with one NOR gate, as shown in Figure 2.32. The switching functions for the subcircuit and the full adder are calculated in Table 2.18.

Table 2.18. Switching functions for the NOR circuit

a	b	l	m	a_i	b_i	c_i	l_i	m_i	$NOR(c_i, l_i)$	c_{i+1}	s_i
0	0	1	1	0	0	0	1	1	0	0	0
0	1	0	0	0	0	1	1	1	0	0	1
1	0	0	0	0	1	0	0	0	1	0	1
1	1	1	0	0	1	1	0	0	0	1	0
				1	0	0	0	0	1	0	1
				1	0	1	0	0	0	1	0
				1	1	0	1	0	0	1	0
				1	1	1	1	0	0	1	1

We have $c_{i+1} = NOR(m_i, NOR(c_i, l_i))$, while $s_i = c_i \Delta l_i'$. We see from Table 2.18 that the circuits do perform the addition of a_i, b_i, and c_i correctly. □

Instead of using many individual transistors, complex circuits can now be made on a single semiconductor "chip," such as the one in Figure 2.33. This chip may contain thousands of gates. (See the Scientific American articles [15] and [16].) Simplification of a circuit may not mean the reduction of the circuit to the smallest number of gates. It could mean simplification to standard modules or the reduction of the number of cross-over points in the circuit. The design of Large Scale Integrated (LSI) chips is mainly limited by the number of entry and exit pins that a chip can hold. In the design of high-speed computers, it is also important to reduce the time that a circuit will take to perform a given set of operations.

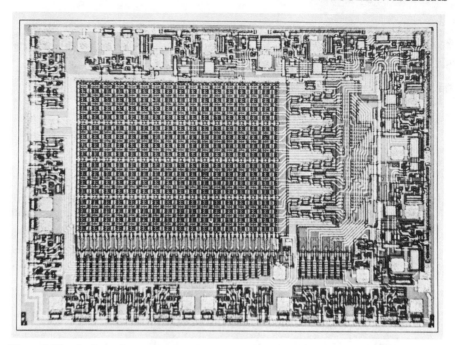

Figure 2.33. A photomicrograph of a 40-pin integrated circuit. It contains thousands of gates and is mounted in a package about 15 mm long. (Photograph courtesy of Texas Instruments Incorporated.)

REPRESENTATION THEOREM

A Boolean algebra is a generalization of the notion of the algebra of sets. However, we now show that every *finite* Boolean algebra is in fact essentially the same as the algebra of subsets of some finite set. To be more precise in what we mean by algebras being essentially the same, we introduce the notion of morphism and isomorphism of Boolean algebras. A morphism between two Boolean algebras is a function between their elements that preserves the two binary operations and the unary operation.

2.36 DEFINITION. If $(K, \wedge, \vee, ')$ and $(L, \cap, \cup, ^-)$ are two Boolean algebras, the function $f: K \rightarrow L$ is called a *Boolean algebra morphism* if the following conditions hold for all $A, B \in K$.

(i) $f(A \wedge B) = f(A) \cap f(B)$.
(ii) $f(A \vee B) = f(A) \cup f(B)$.
(iii) $f(A') = \overline{f(A)}$.

A *Boolean algebra isomorphism* is a bijective Boolean algebra morphism.

Isomorphic Boolean algebras have identical properties. For example, their poset diagrams are the same, except for the labeling of the elements. Furthermore, the atoms of one algebra must correspond to the atoms in the isomorphic algebra.

If we wish to find an isomorphism between any Boolean algebra, K, and an algebra of sets, the atoms of K must correspond to the singleton elements of the algebra of sets. This suggests that we try to define an isomorphism from K to the algebra of subsets, $\mathscr{P}(\mathcal{Q})$, where $\mathcal{Q}$ is the set of atoms of K. The following theorem shows that, if K is finite, we can set up such an isomorphism.

2.37 REPRESENTATION THEOREM FOR FINITE BOOLEAN ALGEBRAS.

Let $\mathcal{Q}$ be the set of atoms of the finite Boolean algebra $(K, \wedge, \vee, ')$. Then there is a Boolean algebra isomorphism between $(K, \wedge, \vee, ')$ and the algebra of subsets $(\mathscr{P}(\mathcal{Q}), \cap, \cup, ^-)$.

PROOF.   We already have a natural correspondence between the atoms of K and the atoms of $\mathscr{P}(\mathcal{Q})$. We use the disjunctive normal form to extend this correspondence to all the elements of K.

By the disjunctive normal form, any element of K can be written as a join of atoms of K, say $B_\alpha \vee \ldots \vee B_\omega$. Define the function $f: K \to \mathscr{P}(\mathcal{Q})$ by

$$f(B_\alpha \vee \ldots \vee B_\omega) = \{B_\alpha\} \cup \ldots \cup \{B_\omega\}.$$

The uniqueness of the normal form implies that each element of K has a unique image in $\mathscr{P}(\mathcal{Q})$ and that f is a bijection.

We still have to show that f is a morphism of Boolean algebras. If X and Y are two elements of K, the atoms in the normal forms of $X \vee Y$ and $X \wedge Y$ are, respectively, the atoms in the forms of X or Y and the atoms common to the forms of X and Y. Therefore, $f(X \vee Y) = f(X) \cup f(Y)$, and $f(X \wedge Y) = f(X) \cap f(Y)$. An atom B is in the normal form for X' if and only if $B \leqslant X'$, which, by Proposition 2.22 (iv), happens if and only if $B \wedge X = 0$. Therefore, the atoms in X' are all the ·atoms that are not in X and $f(X') = \overline{f(X)}$. This proves that f is a Boolean algebra isomorphism. $\square$

2.38 COROLLARY.
If $(K, \wedge, \vee, ')$ is a finite Boolean algebra, then K has 2^n elements, where n is the number of atoms in K.

PROOF.   This follows from Theorem 2.06. $\square$

Consider the Representation Theorem applied to the Boolean algebra in Example 2.16, which consists of the divisors of 110. The atoms of this algebra are the prime divisors, 2, 5, and 11. The Representation Theorem defines a Boolean algebra isomorphism to the algebra of subsets of {2, 5, 11}. This isomorphism, f, maps a number onto the subset consisting of its prime divisors; for example, $f(11) = \{11\}$ and $f(10) = \{2, 5\}$.

2.39 EXAMPLE. Do the divisors of 12 form a Boolean algebra under GCD and LCM?

SOLUTION. The set of divisors of 12 is $\{1, 2, 3, 4, 6, 12\}$. Since the number of elements is not a power of 2, it cannot form a Boolean algebra. $\square$

2.40 EXAMPLE. Do the divisors of 24 form a Boolean algebra under GCD and LCM?

SOLUTION. There are $8 = 2^3$ divisors of 24, namely 1, 2, 3, 4, 6, 8, 12, and 24. However, the poset diagram in Figure 2.34 shows that 2 and 3 are the only atoms. Hence, by Corollary 2.39, it cannot be a Boolean algebra because it does not have $2^2 = 4$ elements. $\square$

Figure 2.34.   Poset diagram of the divisors of 24.

An *infinite* Boolean algebra is not necessarily isomorphic to the algebra of *all* subsets of a set, but is isomorphic to the algebra of *some* subsets of a set. This result is known as Stone's Representation Theorem, and a proof can be found in Mendelson [17; Section 5.7].

Exercises

1–6. If A, B, and C are subsets of a set X under what conditions do the following equalities hold?

1. $A \cup B = A \,\triangle\, B \,\triangle\, (A \cap B)$.
2. $A \cap (B \cup C) = (A \cap B) \cup C$.
3. $A - (B \cup C) = (A - B) \cup (A - C)$.

4. $A \Delta (B \cap C) = (A \Delta B) \cap (A \Delta C)$.

5. $A \Delta (B \cup C) = (A \Delta B) \cup (A \Delta C)$.

6. $A \cup (B \cap A) = A$.

7. Prove the remaining parts of Proposition 2.02.

8. Prove the remaining parts of Proposition 2.04.

9. Prove Theorem 2.06 by induction on n. That is, if X is a finite set with n elements, prove that $\mathscr{P}(X)$ contains 2^n elements.

10. Prove or give a counterexample to the following statements.

(i) $\mathscr{P}(X) \cap \mathscr{P}(Y) = \mathscr{P}(X \cap Y)$.

(ii) $\mathscr{P}(X) \cup \mathscr{P}(Y) = \mathscr{P}(X \cup Y)$.

11. *(Cantor's Theorem)* Prove that there is no surjective (onto) function from X to $\mathscr{P}(X)$ for any finite or infinite set X. This shows that $\mathscr{P}(X)$ always contains more elements than X.

12. Write down the table for $(\mathscr{P}(X), -)$, under the difference operation, when $X = \{a, b\}$.

13. If A, B, C, and D are finite sets, find an expression for $\#(A \cup B \cup C \cup D)$ in terms of the number of elements in their intersections.

14. Of the Chelsea Pensioners who returned from a war, *at least* 70% had lost an eye, 75% an ear, 80% an arm, and 85% a leg. What percentage, *at least* must have lost all four? (From Lewis Carroll, "A Tangled Tale").

15. One hundred students were questioned about their study habits. Seventy said they sometimes studied during the day, 55 said they sometimes studied during the night, and 45 said they sometimes studied during the weekend. Also, 36 studied during the day and night, 24 during the day and at weekends, 17 during the night and at weekends, and 3 during the day, night, and weekends. How many did not study at all?

16. Prove that the associative laws in Definition 2.11 follow from the other axioms of a Boolean algebra.

17. If the zero element is the same as the unit element in a Boolean algebra, prove that the algebra has only one element. Is this algebra isomorphic to the algebra of subsets of some set?

18. Draw the poset diagram for $\mathscr{F}_1$, the Boolean algebra of switching functions of one variable.

19–24. If A, B, and C are elements of a Boolean algebra $(K, \wedge, \vee, ')$ and $\leqslant$ is the related partial order, prove the following from the axioms and Propositions 2.15, 2.21 and 2.22.

19. $0' = 1$.

20. $A \wedge (A' \vee B) = A \wedge B$.

21. $(A \wedge B) \vee (B \wedge C) \vee (C \wedge A) = (A \vee B) \wedge (B \vee C) \wedge (C \vee A)$.

22. $A \leqslant B \wedge C$ implies that $A \leqslant B$ and $A \leqslant C$.

23. $(A \wedge B') \vee C = (B \wedge C) \vee (B' \wedge (A \vee C))$.

24. $A \leqslant B$ if and only if $B' \leqslant A'$.

25. Write down the truth tables for the following propositions. Which of these propositions are equivalent?

(i) $A \Rightarrow B$.   (ii) $B' \Rightarrow A'$.

(iii) $(A \wedge B) \Leftrightarrow B$.   (iv) $(A \vee B) \Leftrightarrow A$.

26. Is the proposition $[(A' \Rightarrow B) \wedge (B \Leftrightarrow C)']$ equivalent to $[B \vee (A \wedge C)]$ or $[(C \Rightarrow B) \vee (B \Rightarrow (A \wedge C))]$?

27. Which of the following are tautologies and which are contradictions?

(i) $(A \wedge B) \Leftrightarrow (A \Rightarrow B')$.   (ii) $A \Rightarrow (B \Rightarrow A)$.

(iii) $(A \wedge B') \Leftrightarrow (A \Rightarrow B)'$.   (iv) $(A \Rightarrow B) \Rightarrow ((B \Rightarrow C) \Rightarrow (A \Rightarrow C))$.

28. Harry broke the window if and only if he ran away and John was lying. John said that either Harry broke the window or Harry did not run away. If Harry ran away, then he did not break the window. What conclusions can you come to?

29–30. Draw circuits to realize the following expressions.

29. $(A \wedge (B \vee C' \vee D)) \vee B'$.

30. $(A \wedge B' \wedge C') \vee (A' \wedge B \wedge C) \vee (A \wedge B \wedge C)$.

31. Simplify the following expression and then draw a circuit for it.

$$((A \wedge B) \vee C') \wedge (B' \vee (C \wedge A')) \vee (A' \wedge B' \wedge C').$$

32–36. Give a Boolean expression for each of the circuits in Figure 2.35, find their disjunctive normal forms, and then try to simplify the circuits.

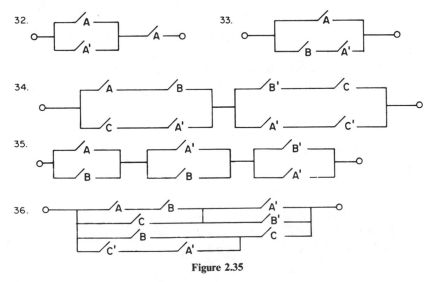

Figure 2.35

37. By looking at all the possible paths through the bridge circuit in Figure 2.36, show that it corresponds to the Boolean expression

$$(A \wedge D) \vee (B \wedge E) \vee (A \wedge C \wedge E) \vee (B \wedge C \wedge D).$$

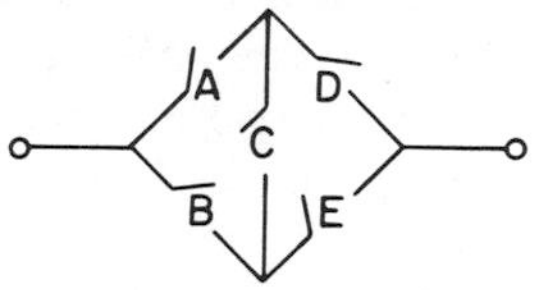

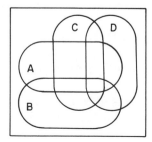

Figure 2.36

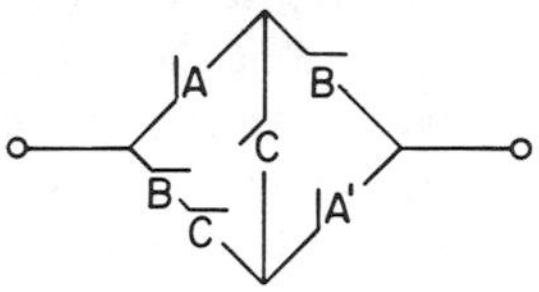

Figure 2.37

38. Find a series-parallel circuit that is equivalent to the bridge circuit in Figure 2.37 and simplify your circuit.

39. A hall light is controlled by two switches, one upstairs and one downstairs. Design a circuit so that the light can be switched on or off from the upstairs or the downstairs.

40. A large room has three separate entrances, and there is a light switch by each entrance. Design a circuit that will allow the lights to be turned on or off by throwing any one switch.

41. A voting machine for three people contains three YES-NO switches and allows current to pass if and only if there is a majority of YES votes. Design and simplify such a machine.

42. Design and simplify a voting machine for five people.

43. Design a circuit for a light that is controlled by two independent switches *A* and *B* and a master switch *C*. *C* must always be able to turn the light on. When *C* is off, the light should be able to be turned on and off using *A* or *B*.

44. A committee consists of a chairman *A*, and three other members *B*, *C*, and *D*. If the three members, *B*, *C*, and *D*, are not unanimous in their voting, the chairman decides the vote. Design a voting machine for this committee and simplify it as much as possible.

45. Verify that the Venn diagram in Figure 2.38 illustrates the 16 atoms

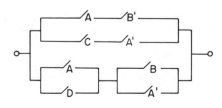

Figure 2.38

for a Boolean expression in four variables. Then use the diagram to simplify the circuit in Figure 2.38.

46. Design four series-parallel circuits to multiply two numbers in binary form that have at most two digits each.

47. Design a circuit that will turn an orange light on if exactly one of the four switches A, B, C, and D is on and a green light when all four are on.

48. Five switches are set to correspond to a number in binary form that has at most five digits. Design and simplify a circuit that will switch a light on if and only if the binary number is a perfect square.

49. In Chapter 11 we will construct a finite field $F = \{0, 1, \alpha, \beta\}$ whose multiplication table is given in Table 2.19.

Table 2.19. Multiplication in a four-element field

·	0	1	α	β
0	0	0	0	0
1	0	1	α	β
α	0	α	β	1
β	0	β	1	α

Writing 00 for 0, 01 for 1, 10 for α, and 11 for β, design and simplify circuits to perform this multiplication.

50. A swimming pool has four relay switches that open when the water temperature is above the maximum allowable, when the water temperature is below the minimum, when the water level is too high, and when the level is too low. These relays are used to control the valves that add cold water, that let water out, and that heat the water in the pool. Design and simplify a circuit that will perform the following tasks. If the temperature is correct but the level too high, it is to let water out. If the temperature is correct but the level too low, it is to let in cold water and heat the water. If the pool is too warm, add cold water and, if the level is also too high, let water out at the same time. If the pool is too cold but the level correct, heat the water; if the level is too low, heat the water and add cold water, and, if the level is too high, just let out the water.

51. In a dual fashion to the disjunctive normal form, every Boolean expression in n-variables can be written in its *conjunctive normal form*. What are the conjunctive normal forms for $A \triangle B$ and $A \wedge B'$?

52–57. Draw poset diagrams for the following sets with divisibility as the partial order and determine whether the systems are lattices or Boolean algebras.

52. $\{1,2,3,4,5,6\}$.

53. $\{2,4,6,12\}$.

54. Positive divisors of 54.

55. $\{1,2,4,8\}$.

56. Positive divisors of 42.

57. $\{1,2,3,5,6,10,30,60\}$.

58. Let $K=\{x\in\mathbf{R}|0\leqslant x\leqslant1\}$ and let $x\wedge y$ and $x\vee y$ be the smaller and larger of x and y, respectively. Show that it is not possible to define a complement $'$ on K so that $(K,\wedge,\vee,')$ is a Boolean algebra. However, if we define $x'=1-x$, which of the properties in Definition 2.11 and Proposition 2.15 still remain true? This is the kind of algebraic model that would be required to deal with transistor switching gates under transient conditions. The voltage or current varies continuously between the levels 0 and 1, while an AND gate performs the operation $x\wedge y$, an OR gate performs $x\vee y$, and a NOT gate produces x'.

59. If f is a Boolean algebra morphism from K to L, prove that $f(0_K)=0_L$ and $f(1_K)=1_L$, where 0_K, 0_L, 1_K, 1_L are the respective zero and unit elements.

60. Write down the tables for the NOR and NAND operations on the set $\mathscr{P}(\{a,b,c\})$.

61. Can every switching circuit be built out of AND and NOT gates?

62. (i) Design a half adder using 5 NOR gates.

(ii) Design a half adder using 5 NAND gates.

63. Analyze the effect of the circuit in Figure 2.39.

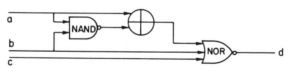

Figure 2.39

64. Design a NOR circuit that will produce a parity check symbol for four binary input digits; that is, the circuit must produce a 0 if the inputs contain an even number of 1s, and it must produce 1 otherwise.

65. One of the basic types of components of a digital computer is a *flip-flop*. This is a device that can be in one of two states (corresponding to outputs 0 and 1) and it will remain in a particular state Q until an input changes the state to the next state Q^*. One important use of a flip-flop is to store a binary digit. An RS flip-flop is a circuit with two inputs, R and S, and one output, Q, corresponding to the state of the flip-flop. An input $R=1$ resets the next state Q^* to 0, and an input $S=1$ sets the next state to 1. If both R and S are 0, the next state is the same as the previous state Q. It is assumed that R and S cannot both be 1 simultaneously. Verify that the NOR circuit in Figure 2.40 is indeed an RS flip-flop. In order to eliminate spurious effects due to the time it takes a transistor to operate, this circuit should be controlled by a "clock." The output Q should be read only when the clock "ticks," whereas the inputs are free to change between "ticks."

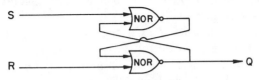

Figure 2.40.   An *RS* flip-flop.

66.   A *JK* flip-flop is similar to an *RS* flip-flop except that both inputs are allowed to be 1 simultaneously, and in this case the state Q changes to its opposite state. Design a *JK* flip-flop using NOR and NAND gates.

Groups 3

Symmetries and permutations in nature and in mathematics can be conveniently described by an algebraic object called a group. In Chapter 5, we use group theory to determine all the symmetries that can occur in two- or three-dimensional space. This can be used, for example, to classify all the forms that chemical crystals can take. If we have a large class of objects, some of which are equivalent under permutations or symmetries, we show, in Chapter 6, how groups can be used to count the nonequivalent objects. For example, we count the number of different switching functions of n variables, if we allow permutations of the inputs.

Historically, the basic ideas of group theory arose with the investigation of permutations of finite sets in the theory of equations. One of the aims of mathematicians at the beginning of the nineteenth century was to find methods for solving polynomial equations of degree five and higher. Algorithms, involving the elementary arithmetical operations and the extraction of roots, were already known for solving all polynomial equations of degree less than five; the formulae for solving quadratic equations had been known since Babylonian times, and cubic and quartic equations had been solved by various Italian mathematicians in the sixteenth century. However, in 1829, using the rudiments of group theory, the Norwegian Niels Abel showed that some equations of the fifth degree could not be solved by any such algorithm. Just before he was mortally wounded in a duel, at the age of 20, the brilliant mathematician Évariste Galois (1811–1832) developed a whole theory that connected the solvability of an equation with the permutation group of its roots. This theory, now called Galois Theory, is beyond the scope of this text, but interested students should look at Stewart [45] after reading Chapter 11 of this book.

It was not until the 1880s that the abstract definition of a group that we use today began to emerge. However, Cayley's Theorem, proved at the end of this chapter, shows that every abstract group can be considered as a group of permutations. It was soon discovered that this concept of a group was so universal that it cropped up in many different branches of mathematics and science.

GROUPS AND SYMMETRIES

3.01 DEFINITION. A *group* $(G, \cdot)$ is a set G together with a binary operation $\cdot$ satisfying the following axioms.

   (i) G is *closed* under the operation $\cdot$; that is, $a \cdot b \in G$ for all $a, b \in G$.
   (ii)   The operation $\cdot$ is *associative*; that is $(a \cdot b) \cdot c = a \cdot (b \cdot c)$ for all $a, b, c \in G$.
   (iii)   There is an *identity element* $e \in G$ such that $e \cdot a = a \cdot e = a$ for all $a \in G$.
   (iv)   Each element $a \in G$ has an *inverse element* $a^{-1} \in G$ such that $a^{-1} \cdot a = a \cdot a^{-1} = e$.

The closure axiom is already implied by the definition of a binary operation; however, it is included because it is often overlooked otherwise.

If the operation is commutative, that is, if $a \cdot b = b \cdot a$ for all $a, b \in G$, the group is called *commutative* or *Abelian*, in honour of the mathematician Niels Abel.

Let G be the set of complex numbers $\{1, -1, i, -i\}$ and let $\cdot$ be the standard multiplication of complex numbers. Then $(G, \cdot)$ is an Abelian group. The product of any two of these elements is an element of G; thus G is closed under the operation. Multiplication is associative and commutative in G because multiplication of complex numbers is always associative and commutative. The identity element is 1, and the inverse of each element a is the element $1/a$. Hence $1^{-1} = 1$, $(-1)^{-1} = -1$, $i^{-1} = -i$ and $(-i)^{-1} = i$. The multiplication of any two elements of G can be represented by Table 3.1.

Table 3.1 The group $\{1, -1, i, -i\}$

·	1	-1	i	$-i$
1	1	-1	i	$-i$
-1	-1	1	$-i$	i
i	i	$-i$	-1	1
$-i$	$-i$	i	1	-1

The set of all rational numbers, **Q**, forms an Abelian group (**Q**, $+$) under addition. The identity is 0, and the inverse of each element is its negative. Similarly, (**Z**, $+$), (**R**, $+$), and (**C**, $+$) are all Abelian groups under addition.

If **Q***, **R***, and **C*** denote the set of nonzero rational, real, and complex numbers, respectively, (**Q***, $\cdot$), (**R***, $\cdot$), and (**C***, $\cdot$) are all Abelian groups under multiplication.

For any set $X, (\mathscr{P}(X), \Delta)$ is an Abelian group. The group axioms follow from Proposition 2.04; the empty set, $\varnothing$, is the identity, and each element is its own inverse.

Every group must have at least one element, namely, its identity, e. A group with only this one element is called *trivial*. A trivial group takes the form $(\{e\}, \cdot)$, where $e \cdot e = e$.

A translation of the plane $\mathbf{R}^2$ in the direction of the vector (a, b) is a function $f: \mathbf{R}^2 \rightarrow \mathbf{R}^2$ defined by $f(x, y) = (x + a, y + b)$. The composition of this translation with a translation g in the direction of (c, d) is the function $f \circ g: \mathbf{R}^2 \rightarrow \mathbf{R}^2$ where

$$f \circ g(x, y) = f(g(x, y)) = f(x + c, y + d) = (x + c + a, y + d + b).$$

This is a translation in the direction of $(c + a, d + b)$. It can be easily verified that the set of all translations in $\mathbf{R}^2$, forms an Abelian group, $(\mathrm{T}(2), \circ)$, under composition. The identity is the identity transformation $e: \mathbf{R}^2 \rightarrow \mathbf{R}^2$ where $e(x, y) = (x, y)$. The inverse of the translation in the direction (a, b) is the translation in the opposite direction $(-a, -b)$.

3.02 DEFINITION. A *permutation* or *symmetry* of a set X is a bijection from X to itself.

In order to show that the set of permutations of X forms a group, we look at some of the properties of bijections.

The function $f: X \rightarrow Y$ is called *injective* or *one-to-one* if $f(x_1) = f(x_2)$ implies that $x_1 = x_2$. In other words, an injective function never takes two different points to the same point. The function $f: X \rightarrow Y$ is called *surjective* or *onto* if, for any $y \in Y$, there exists $x \in X$ with $y = f(x)$; that is, if the image $f(X)$ is the whole set Y. A *bijective* function or *one-to-one correspondence* is a function that is both injective and surjective.

3.03 LEMMA.   If $f: X \to Y$ and $g: Y \to Z$ are two functions, then

(i)   if f and g are injective, $g \circ f$ is injective.
(ii)   if f and g are surjective, $g \circ f$ is surjective.
(iii)   if f and g are bijective, $g \circ f$ is bijective.

PROOF.   (i)   Suppose that $g \circ f(x_1) = g \circ f(x_2)$. Then $g(f(x_1)) = g(f(x_2))$, and, since g is injective, $f(x_1) = f(x_2)$. Since f is also injective, $x_1 = x_2$. This proves that $g \circ f$ is injective.

(ii)   Let $z \in Z$. Since g is surjective, there exists $y \in Y$ with $g(y) = z$, and since f is also surjective, there exists $x \in X$ with $f(x) = y$. Hence $g \circ f(x) = g(f(x)) = g(y) = z$, and $g \circ f$ is surjective.

(iii)   follows from parts (i) and (ii).   □

The function $h: Y \to X$ is called the *inverse* to the function $f: X \to Y$ if $h \circ f = 1_X$, the identity function on X, and $f \circ h = 1_Y$, the identity function on Y. The following theorem gives a necessary and sufficient condition for a function to have an inverse.

3.04 INVERSION THEOREM.   The function $f: X \to Y$ has an inverse if and only if f is bijective.

PROOF.   Suppose that $h: Y \to X$ is an inverse to f. The function f is injective because, if $f(x_1) = f(x_2)$, it follows that $h \circ f(x_1) = h \circ f(x_2)$, and so $x_1 = x_2$. The function f is surjective because, if y is any element of Y and $x = h(y)$, it follows that $f(x) = f(h(y)) = y$. Therefore, f is bijective.

Conversely, suppose f is bijective. We define the function $h: Y \to X$ as follows. For any $y \in Y$, there exists $x \in X$ with $y = f(x)$. Since f is injective, there is only one such element x. Define $h(y) = x$. This function h is an inverse to f because $f(h(y)) = f(x) = y$, and $h(f(x)) = h(y) = x$.   □

3.05 THEOREM.   If $S(X)$ is the set of bijections from any set X to itself, then $(S(X), \circ)$ is a group under composition. This group is called the *symmetric group* or *permutation group* of X.

PROOF.   It follows from Lemma 3.03 that the composition of two bijections is a bijection; thus $S(X)$ is closed under composition. The composition of functions is always associative, and the identity of $S(X)$ is the function $1_X: X \to X$ defined by $1_X(x) = x$. The Inversion Theorem 3.04 proves that any bijective function $f \in S(X)$ has an inverse $f^{-1} \in S(X)$. Therefore, $(S(X), \circ)$ satisfies all the axioms for a group.   □

For example, if $X = \{a, b\}$ is a two-element set, the only bijections from X to itself are the identity 1_X and the symmetry $f: X \to X$, defined by $f(a) = b$, $f(b) = a$, that interchanges the two elements. The use of the term "symmetry" to describe the bijection f agrees with one of our everyday uses of the word. In the phrase "the Boolean expression $(a \wedge b) \vee (a' \wedge b')$ is symmetrical in a and b" we mean that the expression is unchanged when we interchange a and b. The symmetric group of X, $\mathbb{S}(X) = \{1_X, f\}$ and its group table is given in Table 3.2. The composition $f \circ f$ interchanges the two elements a and b twice; thus it is the identity.

Table 3.2 The symmetry group of $\{a, b\}$

$\circ$	1_X	f
1_X	1_X	f
f	f	1_X

Since the composition of functions is not generally commutative, $\mathbb{S}(X)$ is not usually an Abelian group. Consider the elements f and g in the permutation group of $\{1, 2, 3\}$ where $f(1) = 2$, $f(2) = 3$, $f(3) = 1$ and $g(1) = 1$, $g(2) = 3$, $g(3) = 2$. Then $f \circ g(1) = 2$, $f \circ g(2) = 1$, $f \circ g(3) = 3$, while $g \circ f(1) = 3$, $g \circ f(2) = 2$, $g \circ f(3) = 1$; hence $f \circ g \neq g \circ f$, and $\mathbb{S}(\{1, 2, 3\})$ is not Abelian.

A nonsingular linear transformation of the plane is a bijective function of the form $f: \mathbf{R}^2 \to \mathbf{R}^2$ where $f(x, y) = (a_{11}x + a_{12}y, a_{21}x + a_{22}y)$ with $a_{11}a_{22} - a_{12}a_{21} \neq 0$. It can be verified that the composition of two such linear transformations is another linear transformation. The set of all nonsingular linear transformations, L, forms a non-Abelian group $(L, \circ)$.

Besides talking about the symmetries of a distinct set of elements, we often refer, in everyday language, to a geometric object or figure as being symmetrical. We now make this notion more mathematically precise.

3.06 DEFINITION. If F is a figure in the plane or in space, a *symmetry of the figure F* or *isometry* of F is a bijection $f: F \to F$, which preserves distances; that is, for all points $p, q \in F$, the distance from $f(p)$ to $f(q)$ must be the same as the distance from p to q.

One can visualize this operation by imagining F to be a solid object that can be picked up and turned in some manner so that it assumes a configuration identical to the one it had in its original position. For example, the design on the left of Figure 3.01 has two symmetries, the identity and a half turn about a vertical axis, called an axis of symmetry.

Figure 3.01.   Symmetrical designs.

The design on the right of Figure 3.01 has three symmetries, the identity and rotations of one-third and two-thirds of a revolution about its center.

The set of all symmetries of a geometric figure forms a group under composition because the composition and inverse of two distance-preserving functions is distance preserving.

3.07 EXAMPLE.   Write down the table for the group of symmetries of a rectangle with unequal sides.

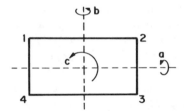

Figure 3.02.   Symmetries of a rectangle.

SOLUTION.   Label the corners of the rectangle 1, 2, 3, and 4 as in Figure 3.02. Any symmetry of the rectangle will send corner points to corner points and so 'will permute the corners among themselves. Denote the symmetry obtained by reflecting the rectangle in the horizontal axis through the center, by a; then $a(1)=4$, $a(2)=3$, $a(3)=2$, and $a(4)=1$. This symmetry can also be considered as a rotation of the rectangle through half a revolution about this horizontal axis. There is a similar symmetry, b, about the vertical axis through the center. A third symmetry, c, is obtained by rotating the rectangle in its plane through half a revolution about its center. Finally, the identity map, e, is a symmetry. These are the only symmetries because it can be verified that any other bijection between the corners will not preserve distances.

The group of symmetries of the rectangle is $(\{e,a,b,c\}, \circ)$, and its table, as shown in Table 3.3, can be calculated as follows. The symmetries a, b, and c are all half turns, so $a \circ a$, $b \circ b$, and $c \circ c$ are full turns and are therefore equal to the identity. The function $a \circ b$ acts on the corner points by $a \circ b(1) = a(b(1)) = a(2) = 3$, $a \circ b(2) = 4$, $a \circ b(3) = 1$, and $a \circ b(4) = 2$. Therefore, $a \circ b = c$. The other products can be calculated similarly. $\square$

Table 3.3. Symmetry group of a rectangle

$\circ$	e	a	b	c
e	e	a	b	c
a	a	e	c	b
b	b	c	e	a
c	c	b	a	e

This group of symmetries of a rectangle is sometimes called the *Klein 4-group*, after the German geometer Felix Klein (1849–1925). The rotations a and b cannot be performed without going outside the plane of the rectangle; such rotations are called *improper*. However, the rotation c can be performed in the plane; rotations of this type are called *proper*.

We have seen that the group operation can be denoted by various symbols, the most common being multiplication, composition, and addition. It is conventional to use addition only for Abelian groups. Furthermore, the identity under addition is usually denoted by 0 and the inverse of a by $-a$. Hence expressions of the form $a \cdot b^{-1}$ and $a^n = a \cdots a$, in multiplicative notation, would be written as $a - b$ and $na = a + \cdots + a$, respectively, in additive notation.

In propositions and theorems concerning groups in general, it is conventional to use multiplicative notation and also to omit the dot in writing a product; therefore, $a \cdot b$ would just be written as ab.

Whenever the operation in a group is clearly understood, we denote the group just by its underlying set. Therefore, the groups $(\mathbf{Z}, +)$, $(\mathbf{Q}, +)$, $(\mathbf{R}, +)$, and $(\mathbf{C}, +)$ are usually denoted just by $\mathbf{Z}$, $\mathbf{Q}$, $\mathbf{R}$, and $\mathbf{C}$, respectively. This should cause no confusion because $\mathbf{Z}$, $\mathbf{Q}$, $\mathbf{R}$, and $\mathbf{C}$ are not groups under multiplication (since the element 0 has no multiplicative inverse). The symmetric group of X is denoted just by $\mathbb{S}(X)$, the operation of composition being understood. Moreover, if we refer to a group G without explicitly defining the group or the operation, it can be assumed that the operation in G is multiplication.

We now prove two propositions that will enable us to manipulate the elements of a group more easily. Recall from Proposition 2.12 that the identity of any binary operation is unique. We first show that the inverse of any element of a group is unique.

3.08 PROPOSITION. Let $\star$ be an associative binary operation on a set S that has identity e. Then, if an element a has an inverse, this inverse is unique.

PROOF.   Suppose that b and c are both inverses of a; thus $a\star b = b\star a = e$, and $a\star c = c\star a = e$. Now

$$
\begin{aligned}
b &= b\star e & \text{since } e \text{ is the identity} \\
&= b\star(a\star c) = (b\star a)\star c & \text{since } \star \text{ is associative} \\
&= e\star c = c & \text{since } e \text{ is the identity.}
\end{aligned}
$$

Hence the inverse of a is unique. $\square$

3.09 PROPOSITION.   If a, b, and c are elements of a group G, then

(i) $(a^{-1})^{-1} = a$

(ii) $(ab)^{-1} = b^{-1}a^{-1}$

(iii) $ab = ac$ or $ba = ca$ implies $b = c$.   (Cancellation Law)

PROOF.   (i)   The inverse of a^{-1} is an element b such that $a^{-1}b = ba^{-1} = e$. But a is such an element, and by Proposition 3.08 we know that the inverse is unique. Hence $(a^{-1})^{-1} = a$.

(ii)   We have $(ab)(b^{-1}a^{-1}) = a((bb^{-1})a^{-1})$ by associativity
$$= a(ea^{-1}) = aa^{-1} = e.$$

Hence $b^{-1}a^{-1}$ is the unique inverse of ab.

(iii)   Suppose $ab = ac$. Then $a^{-1}(ab) = a^{-1}(ac)$ and $(a^{-1}a)b = (a^{-1}a)c$. That is, $eb = ec$ and $b = c$. Similarly, $ba = ca$ implies $b = c$. $\square$

Notice in part (ii) that the order of multiplication is reversed. This should be familiar from the particular case of the group of nonsingular $n \times n$ matrices under multiplication. A more everyday example is the operation of putting on socks and shoes. To reverse the procedure, the shoes are taken off first and then the socks.

SUBGROUPS

It often happens that some subset of a group will also form a group under the same operation. Such a group is called a subgroup. For example, $(\mathbf{R}, +)$ is a subgroup of $(\mathbf{C}, +)$.

3.10 DEFINITION. If $(G, \cdot)$ is a group and H is a nonempty subset of G, then $(H, \cdot)$ is called a *subgroup* of $(G, \cdot)$ if

(i) $a \cdot b \in H$ for all $a, b \in H$ (Closure)

and

(ii) $a^{-1} \in H$ for all $a \in H$.   (Existence of inverses)

3.11 PROPOSITION. If H is a subgroup of $(G, \cdot)$, then $(H, \cdot)$ is also a group.

PROOF. If H is a subgroup of $(G, \cdot)$, we show that $(H, \cdot)$ satisfies all the group axioms. The definition above implies that H is closed under the operation; that is, $\cdot$ is a binary operation on H. If $a, b, c \in H$, then $(a \cdot b) \cdot c = a \cdot (b \cdot c)$ in $(G, \cdot)$ and hence also in $(H, \cdot)$. Since H is nonempty, it contains at least one element, say h. Now $h^{-1} \in H$ and $h \cdot h^{-1}$, which is the identity, is in H. The definition of subgroup implies that $(H, \cdot)$ contains inverses. Therefore, $(H, \cdot)$ satisfies all the axioms of a group. □

Conditions (i) and (ii) are equivalent to the single condition:

(iii) $a \cdot b^{-1} \in H$ for all $a, b \in H$.

However, when H is finite, the following result shows that it is sufficient just to check condition (i).

3.12 PROPOSITION. If H is a nonempty *finite* subset of a group G and $ab \in H$ for all $a, b \in H$, then H is a subgroup of G.

PROOF. We have to show that for each element $a \in H$, its inverse is also in H. All the elements, $a, a^2 = aa, a^3 = aaa, a^4, \ldots$ belong to H, and, since H is finite, these cannot all be distinct. Therefore $a^i = a^j$ for some $1 \leqslant i < j$. By Proposition 3.09(iii), we can cancel a^i from each side to obtain $e = a^{j-i}$, where $j - i > 0$. Therefore, $e \in H$ and this equation can be written as $e = a(a^{j-i-1}) = (a^{j-i-1})a$. Hence $a^{-1} = a^{j-i-1}$, which belongs to H, since $j - i - 1 \geqslant 0$. □

In the group $(\{1, -1, i, -i\}, \cdot)$, the subset $\{1, -1\}$ forms a subgroup because this subset is closed under multiplication. In the group of translations of the plane, the set of translations in the horizontal direction only forms a subgroup, because the composition and inverses of horizontal translations are still horizontal translations.

The group $\mathbf{Z}$ is a subgroup of $\mathbf{Q}$, $\mathbf{Q}$ is a subgroup of $\mathbf{R}$, and $\mathbf{R}$ is a subgroup of $\mathbf{C}$. (Remember that addition is the operation in all these groups.)

However, the set of nonnegative integers $\mathbf{N} = \{0, 1, 2, \ldots\}$ is a subset of $\mathbf{Z}$ but *not* a subgroup, because the inverse of 1, namely, -1, is not in $\mathbf{N}$. This example shows that Proposition 3.12 is false if we drop the condition that H be finite.

The relation of being a subgroup is transitive, and, in fact, for any group G, the inclusion relation between the subgroups of G is a partial order relation.

3.13 EXAMPLE.   Draw the poset diagram of the subgroups of the group of symmetries of a rectangle.

SOLUTION.   By looking at the table of this group in Table 3.3, we see that $\circ$ is a binary operation on $\{e, a\}$; thus $\{e, a\}$ is a subgroup. Also $\{e, b\}$ and $\{e, c\}$ are subgroups. If a subgroup contains a and b, it must contain $a \circ b = c$ and so it is the whole group. Similarly, subgroups containing a and c or b and c must be the whole group. The poset diagram of subgroups is given in Figure 3.03. $\square$

Figure 3.03.   **Subgroups of the group of symmetries of a rectangle.**

CYCLIC GROUPS AND DIHEDRAL GROUPS

3.14 DEFINITION: The number of elements in a group G is written $\# G$ and is called the *order of the group*. G is called a *finite group* if $\# G$ is finite and is called an *infinite group* otherwise.

An important class of groups consists of those for which every element can be written as a power (positive or negative) of some fixed element. These are called cyclic groups.

3.15 DEFINITION.   A group $(G, \cdot)$ is called *cyclic* if there exists an element $g \in G$ such that $G = \{ g^n \mid n \in \mathbf{Z}\}$. The element g is called a *generator* of the cyclic group.

Every cyclic group is Abelian because $g^r \cdot g^s = g^{r+s} = g^s \cdot g^r$.

The group $(\{1, -1, i, -i\}, \cdot)$ is a cyclic group of order 4 generated by i because $i^0 = 1$, $i^1 = i$, $i^2 = -1$, $i^3 = -i$, $i^4 = 1$, $i^5 = i$, etc. Hence the group can be written as $(\{1, i, i^2, i^3\}, \cdot)$.

In additive notation, the group $(G, +)$ is cyclic if $G = \{ng \mid n \in \mathbf{Z}\}$ for some $g \in G$. The group $(\mathbf{Z}, +)$ is an infinite cyclic group with generator 1 (or -1).

3.16 DEFINITION. The *order of an element* g in a group $(G, \cdot)$ is the least positive integer, r, such that $g^r = e$. If no such r exists, the order of the element is said to be infinite.

Note the difference between the order of an *element* and the order of a *group*. (Definitions 3.16 and 3.14). We find connections between these two orders and later prove Lagrange's Theorem, which implies that, in a finite group, the order of an element divides the order of the group.

For example, in $(\{1, -1, i, -i\}, \cdot)$, the identity 1 has order $1, -1$ has order 2 because $(-1)^2 = 1$, whereas i and $-i$ both have order 4. The group has order 4.

Let $\mathbf{Q}^* = \mathbf{Q} - \{0\}$ be the set of nonzero rational numbers. Then $(\mathbf{Q}^*, \cdot)$ is a group under multiplication. The order of the identity element 1 is 1, and the order of -1 is 2. The order of every other element is infinite, because the only solutions to $q^r = 1$ with $q \in \mathbf{Q}^*$, $r \geqslant 1$ are $q = \pm 1$. The group has infinite order. However, it is not cyclic, because there is no rational number r such that every nonzero rational can be written as r^n, for some $n \in \mathbf{Z}$.

For any element g in a group $(G, \cdot)$ we can look at all the powers of this element, namely, $\{g^r \mid r \in \mathbf{Z}\}$. This may not be the whole group, but it will be a subgroup.

3.17 PROPOSITION. If g is any element of order k in a group $(G, \cdot)$, then $H = \{g^r \mid r \in \mathbf{Z}\}$ is a subgroup of order k in $(G, \cdot)$.

This is called the *cyclic subgroup generated by* g.

PROOF. We first check that H is a subgroup of $(G, \cdot)$. This follows from the fact that $g^r \cdot g^s = g^{r+s} \in H$ and $(g^r)^{-1} = g^{-r} \in H$ for all $r, s \in \mathbf{Z}$.

If the order of the element g is infinite, we show that the elements g^r are all distinct. Suppose $g^r = g^s$ where $r > s$. Then $g^{r-s} = e$ with $r - s > 0$, which contradicts the fact that g has infinite order. In this case, $\# H$ is infinite.

If the order of the element g is k, which is finite, we show that $H = \{g^0 = e, g^1, g^2, \dots, g^{k-1}\}$. Suppose $g^r = g^s$ where $0 \leqslant s < r \leqslant k - 1$. Multiply both sides by g^{-s} so that $g^{r-s} = e$ with $0 < r - s < k$. This contradicts the fact that k is the order of g. Hence the elements $g^0, g^1, g^2, \dots, g^{k-1}$ are all distinct. For any other element, g^t, we can write $t = qk + r$, where $0 \leqslant r < k$, and so

$$g^t = g^{qk+r} = (g^k)^q (g^r) = (e^q)(g^r) = g^r.$$

Hence $H = \{g^0, g^1, g^2, \dots, g^{k-1}\}$ and $\# H = k$. $\square$

For example, in $(\mathbf{Z}, +)$, the subgroup generated by 3 is $\{\ldots, -3, 0, 3, 6, 9, \ldots\}$, an infinite subgroup that we write as $3\mathbf{Z} = \{3r \mid r \in \mathbf{Z}\}$.

3.18 THEOREM.   If the finite group G is of order n and has an element g of order n, then G is a cyclic group generated by g.

PROOF.   From the previous proposition we know that H, the subgroup of G generated by g has order n. Therefore, H is a subset of the finite set G with the same number of elements. Hence $G = H$ and G is a cyclic group generated by g. $\square$

3.19 EXAMPLE.   Show that the Klein 4-group of symmetries of a rectangle, described in Example 3.07, is not cyclic.

SOLUTION.   In the Klein 4-group, the identity has order 1, whereas all the other elements have order 2. As it has no element of order 4, it cannot be cyclic. $\square$

All the elements of the Klein 4-group can be written in terms of a and b. We therefore say that this group can be *generated* by the two elements a and b.

3.20 EXAMPLE.   Show that the group of proper rotations of a regular n-gon in the plane is a cyclic group of order n generated by a rotation of $2\pi / n$ radians. This group is denoted by $\mathcal{C}_n$.

SOLUTION.   This is the group of those symmetries of the regular n-gon that can be performed in the plane; that is, without turning the n-gon over.
Label the vertices 1 through n as in Figure 3.04. Under any symmetry, the center must be fixed, and the vertex 1 can be taken to any of the n vertices. The image of 1 determines the rotation; hence the group is of order n.

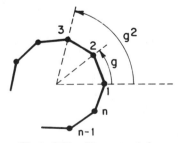

Figure 3.04.   Elements of $\mathcal{C}_n$.

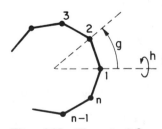

Figure 3.05.   Elements of $\mathcal{D}_n$.

Let g be the counterclockwise rotation of the n-gon through $2\pi/n$. Then g has order n, and, by Theorem 3.18, the group is cyclic of order n. Hence $\mathcal{C}_n = \{e, g, g^2, \ldots, g^{n-1}\}$. □

Let us now consider the group of *all* symmetries (both proper and improper rotations) of the regular n-gon. We call this group the *dihedral group* and denote it by $\mathcal{D}_n$.

3.21 EXAMPLE. Show that the dihedral group, $\mathcal{D}_n$, is of order $2n$ and is not cyclic.

SOLUTION. Label the vertices 1 to n in a counterclockwise direction around the n-gon. Let g be a counterclockwise rotation through $2\pi/n$, and let h be the improper rotation of the n-gon about an axis through the center and vertex 1, as indicated in Figure 3.05. The element g generates the group $\mathcal{C}_n$, which is a cyclic subgroup of $\mathcal{D}_n$. The element h has order 2 and generates a subgroup $\{e, h\}$.

Any symmetry will fix the origin and is determined by the image of two adjacent vertices, say 1 and 2. The vertex 1 can be taken to any of the n vertices, and then 2 must be taken to one of the two vertices adjacent to the image of 1. Hence $\mathcal{D}_n$ has order $2n$.

If the image of 1 is $r+1$, the image of 2 must be $r+2$ or r. If the image of 2 is $r+2$, the symmetry is g^r. If the image of 2 is r, the symmetry is $g^r h$. Figure 3.06 shows that the symmetries $g^r h$ and hg^{-r} have the same effect and therefore imply the relation $g^r h = hg^{-r} = hg^{n-r}$.

Hence the dihedral group is

$$\mathcal{D}_n = \{e, g, g^2, \ldots, g^{n-1}, h, gh, g^2 h, \ldots, g^{n-1}h\}.$$

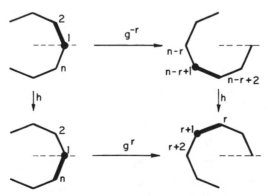

Figure 3.06.   The relation $g^r h = hg^{-r}$ in $\mathcal{D}_n$.

Note that, if $n \geqslant 3$, then $gh \neq hg$; thus $\mathcal{D}_n$ is a *noncommutative* group. Therefore this group cannot be cyclic. $\square$

$\mathcal{D}_2$ can be defined as the symmetries of the figure in Figure 3.07. Hence $\mathcal{D}_2 = \{e, g, h, gh\}$, and each nonidentity element has order 2.

Figure 3.07. The symmetries of a 2-gon.

3.22 EXAMPLE.   Draw the group table for $\mathcal{C}_4$ and $\mathcal{D}_4$.

SOLUTION. $\mathcal{D}_4$ is the group of symmetries of the square, and its table, which is calculated using the relation $g^r h = hg^{4-r}$, is given in Table 3.4. For example, $(g^2 h)(gh) = g^2(hg)h = g^2(g^3 h)h = g^5 h^2 = g$. Since $\mathcal{C}_4$ is a subgroup of $\mathcal{D}_4$, the table for $\mathcal{C}_4$ appears inside the dotted lines in the top left corner. $\square$

Table 3.4.   The group $\mathcal{D}_4$

$\cdot$	e	g	g^2	g^3	h	gh	g^2h	g^3h
e	e	g	g^2	g^3	h	gh	g^2h	g^3h
g	g	g^2	g^3	e	gh	g^2h	g^3h	h
g^2	g^2	g^3	e	g	g^2h	g^3h	h	gh
g^3	g^3	e	g	g^2	g^3h	h	gh	g^2h
h	h	g^3h	g^2h	gh	e	g^3	g^2	g
gh	gh	h	g^3h	g^2h	g	e	g^3	g^2
g^2h	g^2h	gh	h	g^3h	g^2	g	e	g^3
g^3h	g^3h	g^2h	gh	h	g^3	g^2	g	e

Note that the order of each of the elements h, gh, g^2h, g^3h is 2. In general, the element $g^r h$ in $\mathcal{D}_n$ is a reflection in the line through the center of the n-gon bisecting the angle between vertices 1 and $r+1$. Therefore, $g^r h$ always has order 2.

Morphisms

Recall that a morphism between two algebraic structures is a function that preserves their operations. For instance, in Example 3.07, each element of the group K of symmetries of the rectangle induces a permutation of the vertices $1, 2, 3, 4$. This defines a function $f: K \to S(\{1, 2, 3, 4\})$ with the property that the composition of two symmetries of the rectangle corresponds to the composition of permutations of the set $\{1, 2, 3, 4\}$. Since this function preserves the operations, it is a morphism of groups.

Two groups are isomorphic if their structures are essentially the same. For example, the group tables of the cyclic group C_4 and $(\{1, -1, i, -i\}, \cdot)$ would be identical if we replaced a rotation through $n\pi/2$ by i^n. We would therefore say that $(C_4, \circ)$ and $(\{1, -1, i, -1\}, \cdot)$ are isomorphic.

3.23 DEFINITION. If $(G, \cdot)$ and $(H, \cdot)$ are two groups, the function $f: G \to H$ is called a *group morphism* if

$$f(a \cdot b) = f(a) \cdot f(b) \quad \text{for all} \quad a, b \in G.$$

If the groups have different operations, say they are $(G, \cdot)$ and $(H, \star)$, the condition would be written as

$$f(a \cdot b) = f(a) \star f(b).$$

We often use the notation $f: (G, \cdot) \to (H, \star)$ for such a morphism. Many authors use the term "homomorphism" instead of "morphism."

A *group isomorphism* is a bijective group morphism. If there is an isomorphism between the groups $(G, \cdot)$ and $(H, \star)$, we say $(G, \cdot)$ and $(H, \star)$ are *isomorphic* and write $(G, \cdot) \cong (H, \star)$.

If G and H are any two groups, the trivial function that maps every element of G to the identity of H is always a morphism. If $i: \mathbf{Z} \to \mathbf{Q}$ is the inclusion map, i is a group morphism from $(\mathbf{Z}, +)$ to $(\mathbf{Q}, +)$. In fact, if H is a subgroup of G, the inclusion map is always a group morphism.

Let $f: \mathbf{Z} \to \{1, -1\}$ be the function defined by $f(n) = 1$, if n is even, and $f(n) = -1$, if n is odd. Then it can be verified that $f(m + n) = f(m) \cdot f(n)$ for any $m, n \in \mathbf{Z}$, and so this defines a group morphism $f: (\mathbf{Z}, +) \to (\{1, -1\}, \cdot)$.

Let $GL(2, \mathbf{R})$ be the set of 2×2 nonsingular real matrices. The one-to-one correspondence between the set, L, of nonsingular linear transformations of the plane and the 2×2 coefficient matrices is an isomorphism between the groups $(L, \circ)$ and $(GL(2, \mathbf{R}), \cdot)$.

Isomorphic groups have exactly the same properties as each other, and we sometimes identify the groups via the isomorphism and give them the same name. If $f: G \to H$ is an isomorphism between finite groups, the group table of H is the same as that of G, when each element $g \in G$ is replaced by $f(g) \in H$.

Besides preserving the operations of a group, the following result shows that morphisms also preserve the identity and inverses.

3.24 PROPOSITION. If $f: G \to H$ is a group morphism,

(i) $f(e_G) = e_H$, where e_G and e_H are the identities of G and H, respectively.
(ii) $f(a^{-1}) = f(a)^{-1}$ for all $a \in G$.

PROOF. (i) $f(e_G) = f(e_G \cdot e_G) = f(e_G) \cdot f(e_G)$, since f is a morphism. Multiplying each side by $f(e_G)^{-1}$, we have $f(e_G)^{-1} \cdot f(e_G) = f(e_G)^{-1} \cdot f(e_G) \cdot f(e_G)$ $= e_H \cdot f(e_G) = f(e_G)$. Hence $f(e_G) = f(e_G)^{-1} \cdot f(e_G) = e_H$.

(ii) $f(a) \cdot f(a^{-1}) = f(a \cdot a^{-1}) = f(e_G) = e_H$. Hence $f(a^{-1})$ is the unique inverse of $f(a)$, and $f(a^{-1}) = f(a)^{-1}$. $\square$

3.25 THEOREM. Cyclic groups of the same order are isomorphic.

PROOF. Let G and H be cyclic groups generated by g and h, respectively. If G and H are of infinite order, define $f: G \to H$ by $f(g^r) = h^r$ for $r \in \mathbf{Z}$. Then f is a bijection between G and H. Also

$$f(g^r \cdot g^s) = f(g^{r+s}) = h^{r+s} = h^r \cdot h^s = f(g^r) \cdot f(g^s),$$

so f is a group isomorphism.

If G and H have finite order n, $G = \{e, g, g^2, \ldots, g^{n-1}\}$ and $H = \{e, h, h^2, \ldots, h^{n-1}\}$. Define the bijection $f: G \to H$ by $f(g^r) = h^r$ for $r = 0, 1, \ldots, n-1$. If r and s lie between 0 and $n-1$, let $r + s = kn + l$, where l lies between 0 and $n-1$. Then

$$f(g^r \cdot g^s) = f(g^{r+s}) = f(g^{kn+l}) = f\big((g^n)^k \cdot g^l\big) = f(e^k \cdot g^l) = f(g^l) = h^l$$

and

$$f(g^r) \cdot f(g^s) = h^r \cdot h^s = h^{r+s} = h^{kn+l} = (h^n)^k \cdot h^l = e^k \cdot h^l = h^l$$

and so f is an isomorphism. $\square$

Hence every cyclic group is either isomorphic to $(\mathbf{Z}, +)$ or $(\mathcal{C}_n, \cdot)$ for some n. In the next chapter, we see that another important class of cyclic groups consists of the integers modulo n, $(\mathbf{Z}_n, +)$. Of course, the above theorem implies that $(\mathbf{Z}_n, +) \cong (\mathcal{C}_n, \cdot)$.

Any morphism, $f: G \to H$, from a *cyclic* group G to any group H is determined just by the image of a generator. If g generates G and $f(g) = h$, it follows from the definition of a morphism that $f(g^r) = f(g)^r = h^r$ for all $r \in \mathbf{Z}$.

3.26 PROPOSITION. Corresponding elements under a group isomorphism have the same order.

PROOF. Let $f: G \to H$ be an isomorphism, and let $f(g) = h$. Suppose that g and h have orders m and n, respectively, where m is finite. Then $h^m = f(g)^m = f(g^m) = f(e) = e$. So, n, the order of h, is also finite, and $n \leqslant m$, since n is the *least* positive integer with the property $h^n = e$.

Now, if n is finite, $f(g^n) = f(g)^n = h^n = e = f(e)$. Since f is bijective, $g^n = e$, and hence m is finite and $m \leqslant n$.

Therefore, either m and n are both finite, and $m = n$ or m and n are both infinite. $\square$

3.27 EXAMPLE. Is $\mathcal{D}_2$ isomorphic to $\mathcal{C}_4$ or the Klein 4-group of symmetries of a rectangle?

SOLUTION. Compare the orders of the elements given in Table 3.5.

Table 3.5

$\mathcal{D}_2$		$\mathcal{C}_4$		Klein 4-group	
element	order	element	order	element	order
e	1	e	1	e	1
g	2	g	4	a	2
h	2	g^2	2	b	2
gh	2	g^3	4	c	2

We see that $\mathcal{D}_2$ cannot be isomorphic to $\mathcal{C}_4$, but could possibly be isomorphic to the Klein 4-group.

In the Klein 4-group we can write $c = a \circ b$ and we can obtain a bijection, f, from $\mathcal{D}_4$ to the Klein 4-group, by defining $f(g) = a$ and $f(h) = b$. Table 3.6 for the two groups show that this is an isomorphism. $\square$

Table 3.6. Isomorphic groups

The group $\mathcal{D}_2$

·	e	g	h	gh
e	e	g	h	gh
g	g	e	gh	h
h	h	gh	e	g
gh	gh	h	g	e

The Klein 4-group

∘	e	a	b	c
e	e	a	b	c
a	a	e	c	b
b	b	c	e	a
c	c	b	a	e

PERMUTATION GROUPS

A permutation of n elements is a bijective function from the set of the n elements to itself. The permutation groups of two sets, with the same number of elements, are isomorphic. We denote the permutation group of $X = \{1,2,3,\ldots,n\}$ by $(\mathbb{S}_n, \circ)$ and call it the *symmetric group on* n *elements*. Hence $\mathbb{S}_n \cong \mathbb{S}(Y)$ for any n element set Y.

3.28 PROPOSITION. $\# \mathbb{S}_n = n!$

PROOF.   The order of $\mathbb{S}_n$ is the number of bijections from $\{1,2,\ldots,n\}$ to itself. There are n possible choices for the image of 1 under a bijection. Once the image of 1 has been chosen, there are $n-1$ choices for the image of 2. Then there are $n-2$ choices for the image of 3. Continuing in this way, we see that $\# \mathbb{S}_n = n(n-1)(n-2)\cdots 2\cdot 1 = n!$ $\square$

If $\pi : \{1,2,\ldots,n\} \to \{1,2,\ldots,n\}$ is a permutation, we denote it by

$$\begin{pmatrix} 1 & 2 & \cdots & n \\ \pi(1) & \pi(2) & \cdots & \pi(n) \end{pmatrix}.$$

For example, the permutation of $\{1,2,3\}$ that interchanges 1 and 3 is written $\begin{pmatrix} 1 & 2 & 3 \\ 3 & 2 & 1 \end{pmatrix}$. We think of this as

$$\begin{bmatrix} 1 & 2 & 3 \\ \downarrow & \downarrow & \downarrow \\ 3 & 2 & 1 \end{bmatrix}.$$

We can write $\mathbb{S}_2 = \left\{ \begin{pmatrix} 1 & 2 \\ 1 & 2 \end{pmatrix}, \begin{pmatrix} 1 & 2 \\ 2 & 1 \end{pmatrix} \right\}$, which has 2 elements and

$$\mathbb{S}_3 = \left\{ \begin{pmatrix} 1 & 2 & 3 \\ 1 & 2 & 3 \end{pmatrix}, \begin{pmatrix} 1 & 2 & 3 \\ 3 & 1 & 2 \end{pmatrix}, \begin{pmatrix} 1 & 2 & 3 \\ 2 & 3 & 1 \end{pmatrix}, \right.$$

$$\left. \begin{pmatrix} 1 & 2 & 3 \\ 1 & 3 & 2 \end{pmatrix}, \begin{pmatrix} 1 & 2 & 3 \\ 2 & 1 & 3 \end{pmatrix}, \begin{pmatrix} 1 & 2 & 3 \\ 3 & 2 & 1 \end{pmatrix} \right\},$$

which is of order $3! = 6$.

If $\pi, \rho \in \mathbb{S}_n$ are two permutations, their product $\pi \circ \rho$ is the permutation obtained by applying ρ first and then π. This agrees with our notion of composition of functions because $(\pi \circ \rho)(x) = \pi(\rho(x))$. (However, the reader should be aware that some authors use the opposite convention in which π is applied first and then ρ.)

3.29 EXAMPLE. If $\pi = \begin{pmatrix} 1 & 2 & 3 \\ 3 & 1 & 2 \end{pmatrix}$ and $\rho = \begin{pmatrix} 1 & 2 & 3 \\ 3 & 2 & 1 \end{pmatrix}$ are two elements of $\mathbb{S}_3$, calculate $\pi \circ \rho$ and $\rho \circ \pi$.

SOLUTION. $\pi \circ \rho = \begin{pmatrix} 1 & 2 & 3 \\ 3 & 1 & 2 \end{pmatrix} \circ \begin{pmatrix} 1 & 2 & 3 \\ 3 & 2 & 1 \end{pmatrix}$. To calculate this, we start at the right and trace the image of each element under the composition.

$$\begin{bmatrix} 1 & 2 & 3 \\ & \downarrow & \\ 3 & 1 & 2 \end{bmatrix} \circ \begin{bmatrix} 1 & 2 & 3 \\ & & \downarrow \\ 3 & 2 & 1 \end{bmatrix} = \begin{bmatrix} 1 & 2 & 3 \\ \downarrow & & \\ 2 & & \end{bmatrix}$$

Under ρ, 1 is mapped to 3, and under π, 3 is mapped to 2; thus under $\pi \circ \rho$, 1 is mapped to 2. Tracing the images of 2 and 3, we see that

$$\pi \circ \rho = \begin{pmatrix} 1 & 2 & 3 \\ 3 & 1 & 2 \end{pmatrix} \circ \begin{pmatrix} 1 & 2 & 3 \\ 3 & 2 & 1 \end{pmatrix} = \begin{pmatrix} 1 & 2 & 3 \\ 2 & 1 & 3 \end{pmatrix}.$$

In a similar way we can show that

$$\rho \circ \pi = \begin{pmatrix} 1 & 2 & 3 \\ 3 & 2 & 1 \end{pmatrix} \circ \begin{pmatrix} 1 & 2 & 3 \\ 3 & 1 & 2 \end{pmatrix} = \begin{pmatrix} 1 & 2 & 3 \\ 1 & 3 & 2 \end{pmatrix}.$$

Note that $\pi \circ \rho \neq \rho \circ \pi$ and that $\mathbb{S}_3$ is not commutative. $\square$

The permutation $\pi = \begin{pmatrix} 1 & 2 & 3 \\ 3 & 1 & 2 \end{pmatrix}$ has the effect of moving the elements around in a cycle. This is called a *cycle of length 3*, and we write this as $(1 \quad 3 \quad 2)$. We think of this as $(1 \rightarrow 3 \rightarrow 2)$. The permutation π could also be written as $(3 \quad 2 \quad 1)$ or $(2 \quad 1 \quad 3)$ in cycle notation.

3.30 DEFINITION. If $a_1, a_2, \ldots, a_r$ are distinct elements of $\{1, 2, 3, \ldots, n\}$, the permutation $\pi \in \mathbb{S}_n$, defined by

$$\pi(a_1) = a_2, \pi(a_2) = a_3, \ldots, \pi(a_{r-1}) = a_r, \pi(a_r) = a_1$$

and $\pi(x) = x$ if $x \notin \{a_1, a_2, \ldots, a_r\}$, is called a *cycle of length* r or an *r-cycle*. We denote it by $(a_1 a_2 \ldots a_r)$.

Note that the value of n does not appear in the cycle notation. For example, $\begin{pmatrix} 1 & 2 & 3 & 4 \\ 3 & 1 & 4 & 2 \end{pmatrix} = (1 \quad 3 \quad 4 \quad 2)$, is a 4-cycle in $\mathbb{S}_4$, whereas $\begin{pmatrix} 1 & 2 & 3 & 4 & 5 & 6 \\ 3 & 1 & 4 & 2 & 5 & 6 \end{pmatrix} = (1 \quad 3 \quad 4 \quad 2)$ is a 4-cycle in $\mathbb{S}_6$, and $\begin{pmatrix} 1 & 2 & 3 & 4 & 5 & 6 \\ 1 & 5 & 3 & 2 & 4 & 6 \end{pmatrix} = (2 \quad 5 \quad 4)$, is a 3-cycle in $\mathbb{S}_6$.

3.31 PROPOSITION. An *r-cycle* in $\mathbb{S}_n$ has order r.

PROOF. If $\pi = (a_1 a_2 \ldots a_r)$ is an *r*-cycle in $\mathbb{S}_n$, then $\pi(a_1) = a_2$, $\pi^2(a_1) = a_3$, $\pi^3(a_1) = a_4, \ldots$ and $\pi^r(a_1) = a_1$. Similarly, $\pi^r(a_i) = a_i$ for $i = 1, 2, \ldots, r$. None of the permutations $\pi, \pi^2, \ldots, \pi^{r-1}$ is equal to the identity permutation because they all move the element a_1. Hence the order of π is greater than $r-1$. Since π^r fixes $a_1, \ldots, a_r$ and π^r leaves the other elements alone, it follows that π^r is the identity and that π has order r. $\square$

3.32 EXAMPLE. Write down $\pi = (1 \quad 3 \quad 4 \quad 2)$, $\rho = (1 \quad 3)$, and $\sigma = (1 \quad 2) \circ (3 \quad 4)$ as permutations in $\mathbb{S}_4$. Calculate $\pi \circ \rho \circ \sigma$.

SOLUTION.

$$(1 \quad 3 \quad 4 \quad 2) = \begin{pmatrix} 1 & 2 & 3 & 4 \\ 3 & 1 & 4 & 2 \end{pmatrix}, \quad (1 \quad 3) = \begin{pmatrix} 1 & 2 & 3 & 4 \\ 3 & 2 & 1 & 4 \end{pmatrix},$$

and $(1 \quad 2) \circ (3 \quad 4) = \begin{pmatrix} 1 & 2 & 3 & 4 \\ 2 & 1 & 4 & 3 \end{pmatrix}$. We can either calculate a product of cycles from the permutation representation or we can use the cycle representation directly. Let us calculate $\pi \circ \rho \circ \sigma$ from their cycles. Remember that a cycle in $\mathbb{S}_4$ is a bijection from $\{1,2,3,4\}$ to itself, and a product of cycles is a composition of functions. In calculating such a composition, we begin at the right and work our way left. Consider the effect of $\pi \circ \rho \circ \sigma$ on each of the elements 1, 2, 3, and 4.

$$\pi \circ \rho \circ \sigma = (1 \quad 3 \quad 4 \quad 2) \circ (1 \quad 3) \circ (1 \quad 2) \circ (3 \quad 4)$$

$$
\begin{array}{ccccccccc}
1 & \longleftarrow & 2 & \longleftarrow & 2 & \longleftarrow & 1 & \longleftarrow & 1 \\
4 & \longleftarrow & 3 & \longleftarrow & 1 & \longleftarrow & 2 & \longleftarrow & 2 \\
2 & \longleftarrow & 4 & \longleftarrow & 4 & \longleftarrow & 4 & \longleftarrow & 3 \\
3 & \longleftarrow & 1 & \longleftarrow & 3 & \longleftarrow & 3 & \longleftarrow & 4
\end{array}
$$

For example, 2 is left unchanged by $(3 \quad 4)$; then 2 is sent to 1 under $(1 \quad 2)$, 1 is sent to 3 under $(1 \quad 3)$, and finally 3 is sent to 4 under $(1 \quad 3 \quad 4 \quad 2)$. Hence $\pi \circ \rho \circ \sigma$ sends 2 to 4. The permutation $\pi \circ \rho \circ \sigma$ also sends 4 to 3, 3 to 2, and fixes 1. Therefore, $\pi \circ \rho \circ \sigma = (2 \quad 4 \quad 3)$. $\square$

Permutations that are not cycles can be split up into two or more cycles as follows. If π is a permutation in $\mathbb{S}_n$ and $a \in \{1,2,3,\ldots,n\}$, the *orbit* of a under π consists of the distinct elements $a, \pi(a), \pi^2(a), \pi^3(a), \ldots$. We can split a permutation up into its different orbits, and each orbit will give rise to a cycle.

Consider the permutation $\pi = \begin{pmatrix} 1 & 2 & 3 & 4 & 5 & 6 & 7 & 8 \\ 3 & 2 & 8 & 1 & 5 & 7 & 6 & 4 \end{pmatrix} \in \mathbb{S}_8$. Here $\pi(1) = 3$, $\pi^2(1) = \pi(3) = 8$, $\pi^3(1) = 4$, and $\pi^4(1) = 1$; thus the orbit of 1 is $\{1, 3, 8, 4\}$. This is also the orbit of 3, 4 and 8. This orbit gives rise to the cycle $(1 \quad 3 \quad 8 \quad 4)$. Since π leaves 2 and 5 fixed, their orbits are just

themselves. The orbit of 6 and 7 is $\{6,7\}$, which gives rise to the 2-cycle
$(6 \quad 7)$. We can picture the orbits and their corresponding cycles as in
Figure 3.08.

It can be verified that $\pi = (1 \quad 3 \quad 8 \quad 4) \circ (2) \circ (5) \circ (6 \quad 7)$. Since no
number is in two different cycles, these cycles are called *disjoint*. If a
permutation is written as a product of disjoint cycles, it does not matter in which order we write the cycles. We could write $\pi =$
$(5) \circ (6 \quad 7) \circ (2) \circ (1 \quad 3 \quad 8 \quad 4)$. When writing down a product of cycles,
we often omit the 1-cycles and write $\pi = (1 \quad 3 \quad 8 \quad 4) \circ (6 \quad 7)$. The
identity permutation is usually just written as (1).

Figure 3.08.   Disjoint cycle decomposition.

3.33 PROPOSITION. Every permutation can be written as a product of
disjoint cycles.

PROOF. Let π be a permutation and let $\gamma_1, \ldots, \gamma_k$ be the cycles obtained
in the above way from the orbits of π. Let a_1 be any number in the domain
of π, and let $\pi(a_1) = a_2$. If γ_i is the cycle containing a_1, we can write
$\gamma_i = (a_1 a_2 \ldots a_r)$; the other cycles will not contain any of the elements
$a_1, a_2, \ldots, a_r$ and hence will leave them all fixed. Therefore, the product
$\gamma_1 \circ \gamma_2 \circ \cdots \circ \gamma_k$ will map a_1 to a_2, because the only cycle to move a_1 or a_2 is
γ_i. Hence $\pi = \gamma_1 \circ \gamma_2 \circ \cdots \circ \gamma_k$, because they both have the same effect on
all the numbers in the domain of π. $\square$

3.34 COROLLARY. The order of a permutation is the least common
multiple of the lengths of its disjoint cycles.

PROOF. If π is written in terms of disjoint cycles as $\gamma_1 \circ \gamma_2 \circ \cdots \circ \gamma_k$, the
order of the cycles can be changed because they are disjoint. Therefore, for
any integer m, $\gamma^m = \gamma_1^m \circ \gamma_2^m \circ \cdots \circ \gamma_k^m$. Because the cycles are disjoint, this
is the identity if and only if γ_i^m is the identity for each i. The least such
integer is the least common multiple of the orders of the cycles. $\square$

3.35 EXAMPLE. Find the order of the permutation

$$\pi = \begin{pmatrix} 1 & 2 & 3 & 4 & 5 & 6 & 7 & 8 \\ 3 & 5 & 8 & 7 & 1 & 4 & 6 & 2 \end{pmatrix}.$$

SOLUTION.   We can write this permutation in terms of disjoint cycles as

$$\pi = (1 \quad 3 \quad 8 \quad 2 \quad 5) \circ (4 \quad 7 \quad 6).$$

Hence the order of π is LCM$(5,3) = 15$. Of course, we could calculate $\pi^2, \pi^3, \pi^4, \ldots$ until we obtained the identity, but this would take much longer. $\square$

3.36 EXAMPLE.   Show that $\mathcal{D}_3$ is isomorphic to $\mathcal{S}_3$ and write out the table for the latter group.

SOLUTION. $\mathcal{D}_3$ is the group of symmetries of an equilateral triangle, and any symmetry induces a permutation of the vertices. This defines a function $f: \mathcal{D}_3 \to \mathcal{S}_3$. If $\sigma, \tau \in \mathcal{D}_3$, then $f(\sigma \circ \tau)$ is the induced permutation on the vertices, which is the same as $f(\sigma) \circ f(\tau)$. Hence f is a morphism. Figure 3.09 illustrates the six elements of $\mathcal{D}_3$ and their corresponding permutations. We shade the underside of the triangle and block in the corner near vertex 1 to illustrate how the triangle moves. To visualize this, imagine a triangular jigsaw puzzle piece and consider all the possible ways of fitting this piece into a triangular hole. Any proper rotation will leave the white side uppermost, whereas an improper rotation will leave the shaded side uppermost.

The six permutations are all distinct; thus f is a bijection and an isomorphism between $\mathcal{D}_3$ and $\mathcal{S}_3$. The group table for $\mathcal{S}_3$ is given in Table 3.7. $\square$

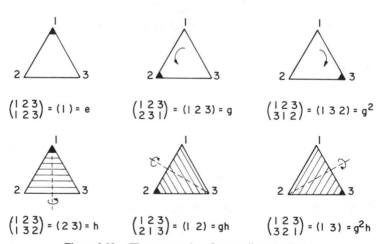

Figure 3.09.   The symmetries of an equilateral triangle.

Table 3.7.   The group $\mathbb{S}_3$

∘	(1)	(123)	(132)	(23)	(12)	(13)
(1)	(1)	(123)	(132)	(23)	(12)	(13)
(123)	(123)	(132)	(1)	(12)	(13)	(23)
(132)	(132)	(1)	(123)	(13)	(23)	(12)
(23)	(23)	(13)	(12)	(1)	(132)	(123)
(12)	(12)	(23)	(13)	(123)	(1)	(132)
(13)	(13)	(12)	(23)	(132)	(123)	(1)

Even and Odd Permutations

We now show that every permutation can be given a parity, either even or odd.

Consider the permutation $\pi = \begin{pmatrix} 1 & 2 & 3 & 4 & 5 \\ 3 & 1 & 5 & 4 & 2 \end{pmatrix} \in \mathbb{S}_5$. In π, the second row is a rearrangement or permutation of the first row. We count the number of pairs in the second row that are out of their natural order; this is called the number of inversions that have occurred. For example, 1 precedes 3 in the natural order, but 3 prededes 1 in the permutation π; thus 1 and 3 are inverted. The pair 1 and 4, however, remain in their natural order. The number of such inversions of a permutaton will determine its parity.

In Figure 3.10, a line is drawn connecting the number i in the top row to its image $\pi(i)$ in the bottom row. If two lines cross, an inversion occurs between the pair of numbers in the bottom row. Draw the lines so that no three go through one point, and let the number of crossover points be $I(\pi)$. In this case $I(\pi) = 5$, corresponding to inversions of the five pairs $\{1,3\}$, $\{2,3\}$, $\{2,4\}$, $\{2,5\}$, and $\{4,5\}$.

An alternative way of counting the number of inversions is to look at the second row and, below each number k, write a tick for each number less than k that occurs after it in the permutation.

$$\begin{pmatrix} 1 & 2 & 3 & 4 & 5 \\ 3 & 1 & 5 & 4 & 2 \end{pmatrix}$$
$$\quad \vee \vee \qquad \vee \vee \quad \vee \qquad \leftarrow\text{Total number of inversions} = 5$$

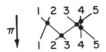

Figure 3.10.   Number of inversions of a permutation.

In our example, we put two ticks below 3, because 1 and 2 follow it; we put two ticks below 5, because 4 and 2 follow it; and we put one tick below 4 because 2 follows it. The total number of ticks is the number of inversions.

3.37 DEFINITION. Let $\pi \in \mathfrak{S}_n$ and $1 \leqslant i < j \leqslant n$. Define

$$\phi(i,j) = \begin{cases} 0 & \text{if} \quad \pi(i) < \pi(j) \\ 1 & \text{if} \quad \pi(i) > \pi(j) \end{cases}.$$

Then $I(\pi) = \Sigma_{1 \leqslant i < j \leqslant n}\phi(i,j)$ is called the *number of inversions* of the permutation π. The permutation π is called *even* if $I(\pi)$ is even, and *odd* if $I(\pi)$ is odd.

For example, the number of inversions of the n-cycle $(1234\ldots n)$ is $n - 1$. This follows from Figure 3.11 or by counting the number of tick marks.

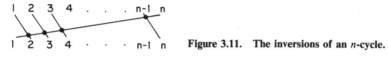

Figure 3.11. The inversions of an n-cycle.

$$\begin{pmatrix} 1 & 2 & 3 & 4 & \ldots & n-1 & n \\ 2 & 3 & 4 & 5 & \ldots & n & 1 \\ \vee & \vee & \vee & \vee & \ldots & \vee & \end{pmatrix}$$

3.38 PROPOSITION. If $\pi, \rho \in \mathfrak{S}_n$, then $(-1)^{I(\pi \circ \rho)} = (-1)^{I(\pi)} \cdot (-1)^{I(\rho)}$. In other words

> (even permutation) $\circ$ (even permutation) = even permutation;
> (even permutation) $\circ$ (odd permutation) = odd permutation;
> (odd permutation) $\circ$ (even permutation) = odd permutation;
> (odd permutation) $\circ$ (odd permutation) = even permutation.

PROOF. Let $1 \leqslant i < j \leqslant n$ and consider the effect of $\pi \circ \rho$ on i and j. For each pair of numbers i and j, we have one of the four cases illustrated in Figure 3.12.

Let m_q denote the number of times the qth case occurs as i and j vary with the restriction $1 \leqslant i < j \leqslant n$. Now $I(\rho) = m_3 + m_4$, because $\rho(i)$ and $\rho(j)$ cross over under ρ only in cases 3 and 4. $I(\pi) = m_2 + m_4$, because $\pi(\rho(i))$ and $\pi(\rho(j))$ cross over under π only in cases 2 and 4. Since $I(\pi \circ \rho) = m_2 + m_3$, it follows that $I(\pi \circ \rho) = I(\pi) + I(\rho) - 2m_4$, and $(-1)^{I(\pi \circ \rho)} = (-1)^{I(\pi) + I(\rho)} = (-1)^{I(\pi)} \cdot (-1)^{I(\rho)}$. $\square$

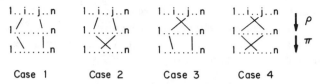

Figure 3.12. Inversions under composition.

The *determinant* of an $n \times n$ matrix (a_{ij}) is often defined in terms of odd and even permutations by

$$\det(a_{ij}) = \sum_{\pi \in S_n} (-1)^{I(\pi)} a_{1\pi 1} a_{2\pi 2} \cdots a_{n\pi n}.$$

For example, the 3×3 determinant is given by

$$\det(a_{ij}) = a_{11}a_{22}a_{33} + a_{12}a_{23}a_{31} + a_{13}a_{21}a_{32} - a_{11}a_{23}a_{32} - a_{12}a_{21}a_{33} - a_{13}a_{22}a_{31}.$$

3.39 PROPOSITION. Every permutation can be written as a product of 2-cycles (not necessarily disjoint), and the parity of the permutation is the same as the parity of the number of 2-cycles.

A 2-cycle is often called a *transposition* because it transposes two elements.

PROOF. We can write any n-cycle as a product of $n-1$ transpositions

$$(a_1 a_2 \ldots a_n) = (a_1 a_2) \circ (a_2 a_3) \circ \cdots \circ (a_{n-1} a_n).$$

Hence, by Proposition 3.33, we can write every permutation as a product of 2-cycles.

A 2-cycle is always odd because, if $i < j$, it follows from Figure 3.13 that $I(ij) = 2(j - i) - 1$. Hence, by Proposition 3.38, a permutation is even if it can be written as an even product of transpositions, and it is odd otherwise. $\square$

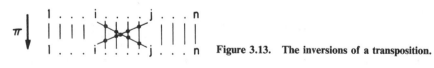

Figure 3.13. The inversions of a transposition.

3.40 COROLLARY. An n-cycle is an even permutation if n is odd and an odd permutation if n is even.

PROOF. This follows from the proof of the above proposition. $\square$

It is rather tedious to determine the parity of a permutation by counting the number of inversions. However, Proposition 3.38 and Corollary 3.40 give us a much simpler method for finding the parity of a permutation by looking at its cycle decomposition.

3.41 EXAMPLE.   Write the permutation

$$\pi = \begin{pmatrix} 1 & 2 & 3 & 4 & 5 & 6 & 7 & 8 \\ 4 & 1 & 8 & 2 & 7 & 3 & 6 & 5 \end{pmatrix}$$

as a product of disjoint cycles and determine its order and parity.

SOLUTION.   As disjoint cycles $\pi = (1 \quad 4 \quad 2) \circ (3 \quad 8 \quad 5 \quad 7 \quad 6)$. Hence the order of π is LCM$(3,5) = 15$. The parity of the 3-cycle $(1 \quad 4 \quad 2)$ is even, and the parity of the 5-cycle $(3 \quad 8 \quad 5 \quad 7 \quad 6)$ is even; therefore, the parity of π is (even) $\circ$ (even) = even. $\square$

3.42 DEFINITION.   Denote the set of even permutations of n elements by $\mathcal{Q}_n$. It follows from Proposition 3.38 that $\mathcal{Q}_n$ is a subgroup of $\mathcal{S}_n$. This subgroup is called the *alternating group on* n *elements*.
   For example, $\mathcal{Q}_3 = \{(1),(123),(132)\}$, which is a cyclic group of order 3. $\mathcal{Q}_4 = \{(1), (123), (124), (134), (234), (132), (142), (143), (243), (12) \circ (34), (13) \circ (24), (14) \circ (23)\}$, a group of order 12. We show in the next chapter that $\# \mathcal{Q}_n = n!/2$.

3.43 PROPOSITION.   Every even permutation can be written as a product of 3-cycles (not necessarily disjoint).

PROOF.   By Proposition 3.39, an even permutation can be written as a product of an even number of transpositions. Consider the product of two transpositions. If these two transpositions are identical, their product is the identity. If the two transpositions have one element in common, say (ab) and (bc), their product $(ab) \circ (bc) = (abc)$, a 3-cycle. If the two transpositions have no elements in common, say (ab) and (cb), we can write their product as

$$(ab) \circ (cd) = (ab) \circ (bc) \circ (bc) \circ (cd) = (abc) \circ (bcd),$$

a product of two 3-cycles. Hence any even permutation can be written as a product of 3-cycles. $\square$

   Propositions 3.39 and 3.43 imply that $\mathcal{S}_n$ is generated by 2-cycles, where $\mathcal{Q}_n$ is generated by 3-cycles.

CAYLEY'S REPRESENTATION THEOREM

At the beginning of the nineteenth century, groups appeared only in a very concrete form, such as symmetry groups or permutation groups. Arthur Cayley (1821–1895) was the first mathematician to deal with groups abstractly in terms of axioms, but he showed that any abstract group can be considered as a subgroup of a symmetric group. Hence, in some sense, if you know all about symmetry groups and permutation groups, you know all about group theory. This result is analogous to Stone's Representation Theorem for Boolean algebras, which proves that any abstract Boolean algebra can be considered as an algebra of subsets of a set.

3.44 CAYLEY'S THEOREM. Every group $(G, \cdot)$ is isomorphic to a subgroup of its symmetric group $(\mathbb{S}(G), \circ)$.

PROOF. For each element $g \in G$, define $\pi_g : G \to G$ by $\pi_g(x) = g \cdot x$. We show that π_g is a bijection. It is surjective because, for any $y \in G, \pi_g(g^{-1} \cdot y) = g \cdot (g^{-1} \cdot y) = y$. It is injective because $\pi_g(x) = \pi_g(y)$ implies $g \cdot x = g \cdot y$, and so, by Proposition 3.09, $x = y$. Hence $\pi_g \in \mathbb{S}(G)$.

Let $H = \{\pi_g \in \mathbb{S}(G) | g \in G\}$. We show that $(H, \circ)$ is a subgroup of $(\mathbb{S}(G), \circ)$ isomorphic to $(G, \cdot)$.

Define the function $\psi : G \to H$ by $\psi(g) = \pi_g$. This is clearly surjective. It is also injective because $\psi(g) = \psi(h)$ implies $\pi_g = \pi_h$, and $\pi_g(e) = \pi_h(e)$ implies that $g = h$.

We now have to show that ψ preserves the group operation. If $g, h \in G, \pi_{g \cdot h}(x) = (g \cdot h)(x) = g \cdot (h \cdot x) = \pi_g(h \cdot x) = (\pi_g \circ \pi_h)(x)$ and $\pi_{g \cdot h} = \pi_g \circ \pi_h$. Also, $\pi_{h^{-1}} \circ \pi_h = \pi_{h^{-1} \cdot h} = \pi_e$; thus $(\pi_h)^{-1} = \pi_{h^{-1}} \in H$. Hence H is a subgroup of $\mathbb{S}(G)$, and $\psi(g \cdot h) = \psi(g) \circ \psi(h)$.

Therefore, we have shown that $\psi : (G, \cdot) \to (H, \circ)$ is an isomorphism of groups. $\square$

3.45 COROLLARY. If G is a finite group of order n, then G is isomorphic to a subgroup of $\mathbb{S}_n$.

PROOF. This follows because $\mathbb{S}(G)$ is isomorphic to $\mathbb{S}_n$. $\square$

This is not of very much practical value, however, because $\mathbb{S}_n$ has order $n!$, which is much larger than the order of G, in general.

Exercises

1–4. Construct tables for the following groups.

1. $\mathcal{C}_5$.
2. $\mathcal{D}_4$.
3. $(\mathcal{P}(X), \Delta)$ where $X = \{a, b, c\}$.
4. $\mathcal{Q}_4$.

5-16. Which of the following are groups and which are Abelian groups? Give reasons for your answers.

5. $(\mathfrak{M}(m \times n; \mathbf{R}), +)$.
6. $(\{1, 2, 3, 4, 6, 12\}, \text{GCD})$.

7. $(\{a + b\sqrt{2} \mid a, b \in \mathbf{Q}\}, +)$.
8. $(\{a/b \in \mathbf{Q} \mid a, b \in \mathbf{Z}, b \text{ odd}\}, +)$.

9. $(\{z \in \mathbf{C} \mid |z| = 1\}, +)$.
10. $(\{z \in \mathbf{C} \mid |z| = 1\}, \cdot)$.

11. $\left(\left\{ \begin{pmatrix} 1 & 0 \\ 0 & 1 \end{pmatrix}, \begin{pmatrix} -1 & 0 \\ 0 & 1 \end{pmatrix}, \begin{pmatrix} 1 & 0 \\ 0 & -1 \end{pmatrix}, \begin{pmatrix} -1 & 0 \\ 0 & -1 \end{pmatrix} \right\}, \cdot \right)$.

12. $(\mathfrak{M}(n \times n; \mathbf{R}) - \{0\}, \cdot)$ where 0 is the $n \times n$ zero matrix.

13. $(\{1, \zeta, \zeta^2, \zeta^3, \ldots, \zeta^{n-1}\}, \cdot)$ where ζ is a complex nth root of 1.

14. $(\{e, a\}, \star)$ where $e \star e = e$ and $e \star a = a \star e = a \star a = a$.

15. $(\mathbf{R}^*, \sim)$ where $\mathbf{R}^* = \mathbf{R} - \{0\}$ and $x \sim y$ is xy, if $x > 0$, and x/y, if $x < 0$.

16. $(\mathbf{Z}, \star)$ where $m \star n$ is $m + n$, if m is even, and $m - n$, if m is odd.

17. Prove that in any group $(G, \cdot)$, $(a_1 \cdots a_n)^{-1} = a_n^{-1} \cdots a_1^{-1}$.

18. If k is an integer (positive or negative), prove that $(a^{-1}ba)^k = a^{-1}b^k a$ in any group $(G, \cdot)$.

19. If G is a group in which $a^2 = e$, the identity for all $a \in G$, show that G is Abelian.

20. Prove that G is Abelian if and only if $(ab)^2 = a^2 b^2$ for all $a, b \in G$.

21. If a is not the identity in a group and $a^4 b = ba^5$, prove that $ab \neq ba$.

22. Prove that the order of the element g^{-1} is the same as the order of g.

23. Prove that the order of the element ab is the same as the order of ba.

24. Prove that every subgroup of a cyclic group is cyclic.

25. The *Gaussian integers* is the set of complex numbers $\mathbf{Z}[i] = \{a + ib \mid a, b \in \mathbf{Z}\}$. Is the group $(\mathbf{Z}[i], +)$ a cyclic group?

26–33. Describe the symmetry groups of the figures in Figure 3.14.

26. A   27. H   28. R   29. ✳

30. N   31.   32. 卐   33. O

Figure 3.14

34–35. Describe the symmetry groups of the frieze patterns in Figure 3.15. These patterns are repeated indefinitely in both directions.

Figure 3.15

36. Prove that the relation of being a subgroup is a partial order on the set of subgroups of a group G.

37. Draw the poset diagram of the subgroups of $\mathcal{C}_6$.

38. Draw the poset diagram of the subgroups of $\mathcal{S}_3$.

39. If H and K are subgroups of a group G, prove that $H \cap K$ is also a subgroup of G. Is $H \cup K$ necessarily a subgroup of G?

40. If $f: G \to H$ and $g: H \to K$ are group morphisms, show that $g \circ f: G \to K$ is also a group morphism.

41. Find all the group morphisms from $(\mathbf{Z}, +)$ to $(\mathbf{Q}, +)$.

42. Show that the set $\{f_i: \mathbf{R} - \{0,1\} \to \mathbf{R} - \{0,1\}\}$, of functions under composition, is isomorphic to $\mathcal{S}_3$ where

$$f_1(x) = x, f_2(x) = 1 - x, f_3(x) = \frac{1}{x}, f_4(x) = 1 - \frac{1}{x}, f_5(x) = \frac{1}{1-x}$$

$$\text{and} \quad f_6(x) = \frac{x}{x-1}.$$

43. Is $(\mathbf{Z}, +)$ isomorphic to $(\mathbf{Q}^*, \cdot)$, where $\mathbf{Q}^* = \mathbf{Q} - \{0\}$? Give reasons.

44. Is $(\mathbf{R}, +)$ isomorphic to $(\mathbf{R}_{>0}, \cdot)$ where $\mathbf{R}_{>0} = \{x \in \mathbf{R} | x > 0\}$? Give reasons.

45. Find the orders of all the elements in $\mathcal{C}_4$.

46. Is $\mathcal{C}_4 \cong \mathcal{D}_6$? Give reasons.

47. Draw the table and find the order of all the elements in the group $(\{\pm 1, \pm i, \pm j, \pm k\}, \cdot)$ where $i^2 = j^2 = k^2 = -1, ij = k = -ji, jk = i = -kj$ and $ki = j = -ik$. This is called the *quaternion group* $\mathcal{Q}$ of order 8.

48. Let G be the group generated by the matrices $\begin{pmatrix} 0 & 1 \\ -1 & 0 \end{pmatrix}$ and $\begin{pmatrix} 0 & 1 \\ 1 & 0 \end{pmatrix}$ under matrix multiplication. Show that G is a non-Abelian group of order 8. Is it isomorphic to $\mathcal{D}_4$ or the quaternion group $\mathcal{Q}$?

49. Show that $\mathcal{D}_k$ is isomorphic to the group generated by $\begin{pmatrix} 0 & 1 \\ 1 & 0 \end{pmatrix}$ and $\begin{pmatrix} \zeta & 0 \\ 0 & \zeta^{-1} \end{pmatrix}$ under matrix multiplication, where $\zeta = \exp(2\pi i/k)$, a complex kth root of unity.

50. Construct the table for the group generated by g and h where g and h satisfy the relations $g^3 = h^2 = e$ and $gh = hg^2$.

51. Prove that a group of even order always has at least one element of order 2.

52. Find a subgroup of $\mathcal{S}_7$ of order 10.

53. Find a subgroup of $\mathcal{S}_5$ of order 3.

54. Find a subgroup of $\mathcal{Q}_4$ isomorphic to the Klein 4-group.

55–58. Multiply out the following permutations

55. $\begin{pmatrix} 1 & 2 & 3 & 4 \\ 2 & 4 & 3 & 1 \end{pmatrix} \circ \begin{pmatrix} 1 & 2 & 3 & 4 \\ 4 & 3 & 2 & 1 \end{pmatrix}$.

56. $\begin{pmatrix} 1 & 2 & 3 & 4 & 5 & 6 \\ 4 & 5 & 2 & 6 & 3 & 1 \end{pmatrix}^3$.

57. $(1 \quad 2 \quad 3 \quad 4 \quad 5) \circ (2 \quad 3 \quad 4)$.

58. $(3 \quad 6 \quad 2) \circ (1 \quad 5) \circ (4 \quad 2)$.

59–62. Write the following permutations as a product of disjoint cycles. Find the order of each permutation and state whether the permutation is even or odd.

59. $\begin{pmatrix} 1 & 2 & 3 & 4 & 5 & 6 \\ 6 & 1 & 2 & 3 & 4 & 5 \end{pmatrix}$.   **60.** $\begin{pmatrix} 1 & 2 & 3 & 4 & 5 & 6 & 7 \\ 2 & 4 & 6 & 1 & 5 & 7 & 3 \end{pmatrix}$.

61. $\begin{pmatrix} 1 & 2 & 3 & 4 & 5 & 6 \\ 5 & 6 & 4 & 3 & 2 & 1 \end{pmatrix}$.

62. $\begin{pmatrix} 1 & 2 & 3 & 4 & 5 & 6 & 7 & 8 & 9 \\ 8 & 9 & 4 & 2 & 7 & 3 & 5 & 1 & 6 \end{pmatrix}$.

63–66. Find the following permutations.

63. $\begin{pmatrix} 1 & 2 & 3 & 4 & 5 \\ 5 & 1 & 2, & 3 & 4 \end{pmatrix}^{-1}$.   **64.** $\begin{pmatrix} 1 & 2 & 3 & 4 & 5 & 6 \\ 2 & 1 & 6 & 5 & 3 & 4 \end{pmatrix}^{-1}$.

65. $(1 \quad 2 \quad 3)^{-1}$.   **66.** $(1 \quad 2 \quad 4 \quad 6 \quad 5 \quad 7)^{-2}$.

67–69. For each polynomial, find the permutations of the subscripts that leave the value of the polynomial unchanged. These will form subgroups of $\mathcal{S}_4$, called the symmetry groups of the polynomials.

67. $(x_1 + x_2)(x_3 + x_4)$.   **68.** $(x_1 - x_2)(x_3 - x_4)$.

69. $(x_1 - x_2)^2 + (x_2 - x_3)^2 + (x_3 - x_4)^2 + (x_4 - x_1)^2$.

70. Describe the group of proper rotations of the tetrahedron with vertices $(0,0,0)$, $(1,0,0)$, $(0,1,0)$, and $(0,0,1)$ in $\mathbf{R}^3$.

71. What is the number of generators of the cyclic group $\mathcal{C}_n$?

72. Express $(123) \circ (456)$ as the power of a single cycle in $\mathcal{S}_6$. Can you generalize this result?

73. A perfect interlacing shuffle of a deck of $2n$ cards is the permutation $\begin{pmatrix} 1 & 2 & 3 & \ldots & n & n+1 & n+2 & \ldots & 2n \\ 2 & 4 & 6 & \ldots & 2n & 1 & 3 & \ldots & 2n-1 \end{pmatrix}$. What is the least number of perfect shuffles that have to be performed on a deck of 52 cards before the cards are back in their original position? If there were 50 cards, what would be the least number?

74. The *center* of a group G is the set $Z(G)=\{x\in G|xg=gx$ for all $g\in G\}$. Show that $Z(G)$ is an Abelian subgroup of G.

75. Find the center of $\mathcal{D}_3$.

76. Find the center of $\mathcal{D}_4$.

77. Prove that $\mathcal{S}_n$ is generated by the elements $(12),(23),(34),\ldots,(n-1\,n)$.

78. Prove that $\mathcal{S}_n$ is generated by the elements $(1\,2\,3\ldots n)$ and (12).

79. Prove that $\mathcal{A}_n$ is generated by the set $\{(12r)|r=3,4,\ldots,n\}$.

80. The well-known 15-puzzle consists of a shallow box filled with 16 small squares in a 4×4 array. The bottom right corner square is removed, and the other squares are labeled as in Figure 3.16. By sliding the squares around (without lifting them up), show that the set of possible permutations that can be obtained with the bottom right square blank is precisely $\mathcal{A}_{15}$. (There is no known easy proof that all elements in $\mathcal{A}_{15}$ must occur.)

1	2	3	4
5	6	7	8
9	10	11	12
13	14	15	

Initial position

81.

4	3	2	1
5	6	7	8
12	11	10	9
13	14	15	

82.

10	9	8	7
11	2	1	6
12	3	4	5
13	14	15	

83.

8	14	11	3
12	2	15	9
6	4	13	1
	7	10	5

Figure 3.16. The 15-puzzle.

81–83. Which of the positions of the 15-puzzle shown in Figure 3.16 can be achieved?

84. An *automorphism* of a group G is an isomorphism from G to itself. Prove that the set of all automorphisms of G forms a group under composition.

85. Find the automorphism group of the Klein 4-group.

86. Find the automorphism group of $\mathcal{C}_3$.

87. Find the automorphism group of $\mathcal{C}_4$.

88. Find the automorphism group of $\mathcal{S}_3$.

89. A word on $\{x,y\}$ is a finite string of the symbols x,x^{-1},y,y^{-1} where x and x^{-1} cannot be adjacent and y and y^{-1} cannot be adjacent; for example, $xxy^{-1}x$ and $x^{-1}x^{-1}yxy$ are words. Let F be the set of such words together with the empty word, which is denoted by 1. The operation of concatenation places one word after another. Show that F is a group under concatenation, where any strings of the form $xx^{-1},x^{-1}x,yy^{-1},y^{-1}y$ are deleted in a concatenated word. F is called the *free group on two generators*. Is F Abelian? What is the inverse of $x^{-1}x^{-1}yxy$?

Quotient Groups 4

Certain techniques are fundamental to the study of algebra. One such technique is the construction of the quotient set of an algebraic object by means of an equivalence relation on the underlying set. For example, if the object is the group of integers $(\mathbf{Z}, +)$, the congruence relation modulo n on $\mathbf{Z}$ will define the quotient group of integers modulo n.

This quotient construction can be applied to numerous algebraic structures, including groups, Boolean algebras, and vector spaces.

In this chapter, we introduce the concept of an equivalence relation and go on to apply this to groups. We obtain Lagrange's Theorem, which states that the order of a subgroup divides the order of the group, and we also obtain the Morphism Theorem for Groups. We study the implications of these two theorems and classify the groups of low order.

EQUIVALENCE RELATIONS

Relations are one of the basic building blocks of mathematics (as well as of the rest of the world). A *relation* R from a set S to a set T is a subset of

$S \times T$. We say that a is related to b under R if the pair (a,b) belongs to the subset, and we write this as aRb. If (a,b) does not belong to the subset, we say that a is not related to b, and write $a\not\!Rb$. This definition even covers many relations in everyday life, such as "is the father of", "is richer than" and "goes to the same school as" as well as mathematical relations such as "is equal to", "is a member of" and "is similar to". A relation R from S to T has the property that, for any elements a in S, and b in T, either aRb or $a\not\!Rb$.

Any function $f: S \to T$ gives rise to a relation R from S to T by taking aRb to mean $f(a) = b$. The subset R of $S \times T$ is the graph of the function. However, relations are much more general than functions. One element can be related to many elements, or to no elements at all.

A relation from a set S to itself is called a *relation on* S. Any partial order on a set, such as " $\leqslant$ " on the real numbers, or "is a subset of" on a power set $\mathcal{P}(X)$, is a relation on that set. "Equals" is a relation on any set S and is defined by the subset $\{(a,a) | a \in S\}$ of $S \times S$. An equivalence relation is a relation that has the most important properties of the "equals" relation.

4.01 DEFINITION. A relation E on a set S is called an *equivalence relation* if the following conditions hold.

(i) aEa for all $a \in S$.   (Reflexive Condition)
(ii)   If aEb then bEa.   (Symmetric Condition)
(iii)   If aEb and bEc, then aEc.   (Transitive Condition)

If E is an equivalence relation on S and $a \in S$, then $[a] = \{x \in S | xEa\}$ is called the *equivalence class* containing a. The set of all equivalence classes is called the *quotient set* of S by E and is denoted by S/E. Hence

$$S/E = \{[a] | a \in S\}.$$

4.02 PROPOSITION. If E is an equivalence relation on a set S, then

(i)   if aEb, then $[a] = [b]$.
(ii)   if $a\not\!Eb$, then $[a] \cap [b] = \varnothing$.
(iii) S can be written as the disjoint union of all the distinct equivalence classes.

PROOF. (i)   If aEb, let x be any element of $[a]$. Then xEa and, from the transitivity, xEb. Hence $x \in [b]$ and $[a] \subseteq [b]$.

The symmetry of E implies that bEa and an argument similar to the above shows that $[b] \subseteq [a]$. This proves $[a] = [b]$.

(ii)   Suppose $a\not\!E b$. If there was an element $x \in [a] \cap [b]$, then xEa, xEb, and, by symmetry and transitivity, aEb. Hence $[a] \cap [b] = \varnothing$.

(iii)   Parts (i) and (ii) show that two equivalence classes are either the same or disjoint. The reflexivity of E implies that each element $a \in S$ is in the equivalence class $[a]$. Hence S is the disjoint union of all the equivalence classes.   ☐

A collection of nonempty subsets is said to *partition* a set S if the union of the subsets is S and any two subsets are disjoint. The previous proposition shows that any equivalence relation partitions the set into its equivalence classes. Each element of the set belongs to one and only one equivalence class.

It can also be shown that every partition of a set gives rise to an equivalence relation whose classes are precisely the subsets in the partition.

4.03 EXAMPLE.   Let n be a fixed positive integer and a and b any two integers. We say that *a is congruent to b modulo n* if n divides $a - b$. We denote this by

$$a \equiv b \bmod n.$$

Show that this congruence relation modulo n is an equivalence relation on **Z**.

The set of equivalence classes is called the *set of integers modulo* n and is denoted by $\mathbf{Z}_n$.

SOLUTION.   Write "$n|m$" for "n divides m", which means that there is some integer k such that $m = nk$. Hence $a \equiv b \bmod n$ if and only if $n|(a - b)$.

(i)   For all $a \in \mathbf{Z}$, $n|(a - a)$, so $a \equiv a \bmod n$ and the relation is reflexive.

(ii)   If $a \equiv b \bmod n$, then $n|(a - b)$, and so $n|-(a - b)$. Hence $n|(b - a)$ and $b \equiv a \bmod n$.

(iii)   If $a \equiv b \bmod n$ and $b \equiv c \bmod n$, then $n|(a - b)$ and $n|(b - c)$, and so $n|(a - b) + (b - c)$. Therefore, $n|(a - c)$ and $a \equiv c \bmod n$.

Hence congruence modulo n is an equivalence relation on **Z**.   ☐

In the congruence relation modulo 3, we have the following equivalence classes.

$$[0] = \{\ldots, -3, 0, 3, 6, 9, \ldots\} \qquad [3] = \{\ldots, 0, 3, 6, 9, 12, \ldots\} = [0]$$

$$[1] = \{\ldots, -2, 1, 4, 7, 10, \ldots\}$$

$$[2] = \{\ldots, -1, 2, 5, 8, 11, \ldots\}$$

Any equivalence class must be one of [0], [1], or [2], so $\mathbf{Z}_3 = \{[0],[1],[2]\}$.

In general, $\mathbf{Z}_n = \{[0],[1],[2],\ldots,[n-1]\}$, since any integer is congruent modulo n to its remainder when divided by n.

One set of equivalence classes that is introduced in elementary school is the set of rational numbers. Students soon become used to the fact that $1/2$ and $3/6$ represent the same rational number. We need to use the concept of equivalence class to define a rational number precisely. Define the relation E on $\mathbf{Z} \times \mathbf{Z}^*$ (where $\mathbf{Z}^* = \mathbf{Z} - \{0\}$) by $(a,b)E(c,d)$ if and only if $ad = bc$. This is an equivalence relation on $\mathbf{Z} \times \mathbf{Z}^*$, and the equivalence classes are called *rational numbers*. We denote the equivalence class $[(a,b)]$ by a/b. Therefore, since $(1,2)E(3,6)$, it follows that $1/2 = 3/6$.

Two series-parallel circuits involving the switches $A_1, A_2, \ldots, A_n$ are said to be equivalent if they both are open or both are closed for any position of the n switches. This is an equivalence relation, and the equivalence classes are the 2^{2^n} distinct types of circuits controlled by n switches.

Any permutation π on a set S induces an equivalence relation, $\sim$, on S where $a \sim b$ if and only if $b = \pi^r(a)$, for some $r \in \mathbf{Z}$. The equivalence classes are the *orbits* of π. In the decomposition of the permutation π into disjoint cycles, the elements in each cycle constitute one orbit.

Cosets and Lagrange's Theorem

The congruence relation modulo n on $\mathbf{Z}$ can be defined by $a \equiv b \bmod n$ if and only if $a - b \in n\mathbf{Z}$, where $n\mathbf{Z}$ is the subgroup of $\mathbf{Z}$ consisting of all multiples of n. We now generalize this notion and define congruence in any group modulo one of its subgroups. We are interested in the equivalence classes, which we call cosets.

4.04 DEFINITION. Let $(G, \cdot)$ be a group with subgroup H. For $a, b \in G$, we say that *a is congruent to b modulo H*, and write $a \equiv b \bmod H$, if and only if $ab^{-1} \in H$.

4.05 PROPOSITION. The relation $a \equiv b \bmod H$ is an equivalence relation on G. The equivalence class containing a can be written in the form $Ha = \{ha \mid h \in H\}$, and it is called a *right coset* of H in G. The element a is called a *representative* of the coset Ha.

PROOF.

(i) For all $a \in G, aa^{-1} = e \in H$; thus the relation is reflexive.

(ii) If $a \equiv b \bmod H$, then $ab^{-1} \in H$; thus $(ab^{-1})^{-1} = ba^{-1} \in H$. Hence

$b \equiv a \bmod H$, and the relation is symmetric.

(iii) If $a \equiv b$ and $b \equiv c \bmod H$, then ab^{-1} and $bc^{-1} \in H$. Hence $(ab^{-1})(bc^{-1}) = ac^{-1} \in H$ and $a \equiv c \bmod H$. The relation is transitive and is therefore an equivalence relation.

The equivalence class containing a is

$$\{x \in G \mid x \equiv a \bmod H\} = \{x \in G \mid xa^{-1} = h \in H\}$$

$$= \{x \in G \mid x = ha, \text{ where } h \in H\}$$

$$= \{ha \mid h \in H\}$$

which we denote by Ha. $\square$

4.06 EXAMPLE. Find the right cosets of $\mathcal{C}_3$ in $\mathcal{S}_3$.

SOLUTION. One coset is the subgroup itself $\mathcal{C}_3 = \{(1), (123), (132)\}$. Take any element not in the subgroup, say (12). Then another coset is

$$\mathcal{C}_3(12) = \{(12), (123) \circ (12), (132) \circ (12)\} = \{(12), (13), (23)\}.$$

Since the right cosets form a partition of $\mathcal{S}_3$ and the two above cosets contain all the elements of $\mathcal{S}_3$, it follows that these are the only two cosets. In fact, $\mathcal{C}_3 = \mathcal{C}_3(123) = \mathcal{C}_3(132)$ and $\mathcal{C}_3(12) = \mathcal{C}_3(13) = \mathcal{C}_3(23)$. $\square$

4.07 EXAMPLE. Find the right cosets of $H = \{e, g^4, g^8\}$ in $\mathcal{C}_{12} = \{e, g, g^2, \ldots, g^{11}\}$.

SOLUTION. H itself is one coset. Another is $Hg = \{g, g^5, g^9\}$. These two cosets have not exhausted all the elements of $\mathcal{C}_{12}$, so pick an element, say g^2, which is not in H or Hg. A third coset is $Hg^2 = \{g^2, g^6, g^{10}\}$ and a fourth is $Hg^3 = \{g^3, g^7, g^{11}\}$.

Since $\mathcal{C}_{12} = H \cup Hg \cup Hg^2 \cup Hg^3$, these are all the cosets. $\square$

As the above examples suggest, every coset contains the same number of elements. We use this result to prove the famous theorem of Joseph Lagrange (1736–1813).

4.08 LEMMA. There is a bijection between any two right cosets of H in G.

PROOF. Let Ha be a right coset of H in G. We produce a bijection between Ha and H, from which it follows that there is a bijection between any two right cosets.

Define $\psi : H \rightarrow Ha$ by $\psi(h) = ha$. Then ψ is clearly surjective. Now suppose $\psi(h_1) = \psi(h_2)$ so that $h_1 a = h_2 a$. Multiplying each side by a^{-1} on the right, we obtain $h_1 = h_2$. Hence ψ is a bijection. $\square$

4.09 LAGRANGE'S THEOREM. If G is a finite group and H is a subgroup of G, then $\#H$ divides $\#G$.

PROOF.   The right cosets of H in G form a partition of G, so G can be written as a disjoint union

$G = Ha_1 \cup Ha_2 \cup \ldots \cup Ha_k$ for a finite set of elements $a_1, a_2, \ldots, a_k \in G$.

By Lemma 4.08, the number of elements in each coset is $\#H$. Hence, adding up all the elements in the above disjoint union, we see that $\#G = k(\#H)$. Therefore, $\#H$ divides $\#G$. $\square$

4.10 DEFINITION.   If H is a subgroup of G, the number of distinct right cosets of H in G is called the *index* of H in G and is written $\#(G:H)$.

The following is a direct consequence of the proof of Lagrange's Theorem.

4.11 COROLLARY.   If G is a finite group with subgroup H, then

$$\#(G:H) = \#G / \#H.$$
$\square$

4.12 COROLLARY.   If a is an element of a finite group G, then the order of a divides the order of G.

PROOF.   Let $H = \{a^r | r \in \mathbb{Z}\}$ be the cyclic subgroup generated by a. By Proposition 3.17, the order of the subgroup H is the same as the order of the element a. Hence, by Lagrange's Theorem, the order of a divides the order of G. $\square$

4.13 COROLLARY.   If a is an element of the finite group G, then $a^{\#G} = e$.

PROOF.   If m is the order of a, then $\#G = mk$ for some integer k. Hence $a^{\#G} = a^{mk} = (a^m)^k = e^k = e$. $\square$

4.14 COROLLARY.   If G is a group of prime order, then G is cyclic.

PROOF.   Let $\#G = p$, a prime number. By Corollary 4.12, every element has order 1 or p. But the only element of order 1 is the identity. Therefore, all the other elements have order p and there is at least one, because $\#G \geqslant 2$. Hence by Theorem 3.18, G is a cyclic group. $\square$

The converse of Lagrange's Theorem is false, as the following example shows. That is, if k is a divisor of the order of G, it does not necessarily follow that G has a subgroup of order k.

4.15 EXAMPLE. $\mathcal{C}_4$ is a group of order 12 having no subgroup of order 6.

SOLUTION. $\mathcal{C}_4$ contains one identity element, eight 3-cycles of the form (abc) and three pairs of transpositions of the form $(ab) \circ (cd)$, where a, b, c, and d are distinct elements of $\{1,2,3,4\}$. If a subgroup contains a 3-cycle (abc), it must also contain its inverse (acb). If a subgroup of order 6 exists, it must contain the identity and a product of two transpositions, because the odd number of nonidentity elements cannot be made up of 3-cycles and their inverses. A subgroup of order 6 must also contain at least two 3-cycles because $\mathcal{C}_4$ only contains four elements that are not 3-cycles.

Without loss of generality, suppose that a subgroup of order 6 contains the elements (abc) and $(ab) \circ (cd)$. Then it must also contain the elements $(abc)^{-1} = (acb), (abc) \circ (ab) \circ (cd) = (acd), (ab) \circ (cd) \circ (abc) = (bdc), (acd)^{-1} = (adc)$ which, together with the identity, gives more than six elements. Hence $\mathcal{C}_4$ contains no subgroup of order 6. $\square$

NORMAL SUBGROUPS AND QUOTIENT GROUPS

Let G be a group with subgroup H. The right cosets of H in G are equivalence classes under the relation $a \equiv b \bmod H$, defined by $ab^{-1} \in H$. We can also define the relation L on G so that aLb if and only if $b^{-1}a \in H$. This relation, L, is an equivalence relation, and the equivalence class containing a is the *left coset* $aH = \{ah \mid h \in H\}$. As the following example shows, the left coset of an element does not necessarily equal the right coset.

4.16 EXAMPLE. Find the left and right cosets of $H = \mathcal{C}_3$ and $K = \{(1),(12)\}$ in $\mathcal{S}_3$.

SOLUTION. We calculated the right cosets of $H = \mathcal{C}_3$ in Example 4.06.

Right Cosets	Left Cosets
$H \quad = \{(1),(123),(132)\}$	$H = \{(1),(123),(132)\}$
$H(12) = \{(12),(13),(23)\}$	$(12)H = \{(12),(23),(13)\}$

In this case, the left and right cosets of H are the same.

However, the left and right cosets of K are not all the same.

	Right Cosets		Left Cosets
K	$= \{(1),(12)\}$	$K = \{(1),(12)\}$	
$K(13)$	$= \{(13),(132)\}$	$(13)K = \{(13),(123)\}$	
$K(23)$	$= \{(23),(123)\}$	$(23)K = \{(23),(132)\}$	$\square$

Since $a \equiv b \bmod H$ is an equivalence relation for any subgroup H of a group G and the quotient set is the set of right cosets $\{Ha | a \in G\}$, it is natural to ask whether this quotient *set* is also a *group* with a multiplication induced by the multiplication in G. We show that this is the case if and only if the right cosets of H equal the left cosets.

4.17 DEFINITION. A subgroup H of a group G is called a *normal subgroup* of G if $g^{-1}hg \in H$ for all $g \in G$ and $h \in H$.

4.18 PROPOSITION. $Hg = gH$, for all $g \in G$, if and only if H is a normal subgroup of G.

PROOF. Suppose $Hg = gH$. Then, for any element $h \in H, hg \in Hg = gH$. Hence $hg = gh_1$ for some $h_1 \in H$ and $g^{-1}hg = g^{-1}gh_1 = h_1 \in H$. Therefore, H is a normal subgroup.

Conversely, if H is normal, let $hg \in Hg$ and $g^{-1}hg = h_1 \in H$. Then $hg = gh_1 \in gH$ and $Hg \subseteq gH$. Also, $ghg^{-1} = (g^{-1})^{-1}hg^{-1} = h_2 \in H$, since H is normal, and so $gh = h_2 g \in Hg$. Hence, $gH \subseteq Hg$ and $Hg = gH$. $\square$

Therefore, $\mathcal{C}_3$ is a normal subgroup of $\mathcal{S}_3$, whereas $\{(1),(12)\}$ is not.

4.19 PROPOSITION. Any subgroup of an Abelian group is normal.

PROOF. If H is a subgroup of an Abelian group, G, then $g^{-1}hg = hg^{-1}g = h \in H$ for all $g \in G, h \in H$. Hence H is normal. $\square$

If N is a normal subgroup of a group G, the left cosets of N in G are the same as the right cosets of N in G and so there will be no ambiguity in just talking about the cosets of N in G.

4.20 THEOREM. If N is a normal subgroup of $(G, \cdot)$, the set of cosets $G/N = \{Ng | g \in G\}$ forms a group $(G/N, \cdot)$ where the operation is defined by $(Ng_1) \cdot (Ng_2) = N(g_1 \cdot g_2)$.

4.21 DEFINITION. This group is called the *quotient group* or *factor group* of G by N.

PROOF. The operation of multiplying two cosets, Ng_1 and Ng_2, is defined in terms of particular elements, g_1 and g_2, of the cosets. For this operation to make sense, we have to verify that, if we choose different elements, h_1 and h_2, in the same cosets, the product coset $N(h_1 \cdot h_2)$ is the same as $N(g_1 \cdot g_2)$. In other words, we have to show that multiplication of cosets is well defined.

Since h_1 is in the same coset as g_1, we have $h_1 \equiv g_1 \bmod N$. Similarly, $h_2 \equiv g_2 \bmod N$. We show that $Nh_1 h_2 = Ng_1 g_2$. We have $h_1 g_1^{-1} = n_1 \in N$ and $h_2 g_2^{-1} = n_2 \in N$ and so $h_1 h_2 (g_1 g_2)^{-1} = h_1 h_2 g_2^{-1} g_1^{-1} = n_1 g_1 n_2 g_2 g_2^{-1} g_1^{-1} = n_1 g_1 n_2 g_1^{-1}$. Now N is a normal subgroup, so $g_1 n_2 g_1^{-1} \in N$ and $n_1 g_1 n_2 g_1^{-1} \in N$. Hence $h_1 h_2 \equiv g_1 g_2 \bmod N$ and $Nh_1 h_2 = Ng_1 g_2$. Therefore, the operation is well defined.

The operation is associative because $(Ng_1 \cdot Ng_2) \cdot Ng_3 = N(g_1 g_2) \cdot Ng_3 = N(g_1 g_2) g_3$ and also $Ng_1 \cdot (Ng_2 \cdot Ng_3) = Ng_1 \cdot N(g_2 g_3) = Ng_1 (g_2 g_3) = N(g_1 g_2) g_3$.

Since $Ng \cdot Ne = Nge = Ng$ and $Ne \cdot Ng = Ng$, the identity is $Ne = N$. The inverse of Ng is Ng^{-1} because $Ng \cdot Ng^{-1} = N(g \cdot g^{-1}) = Ne = N$ and also $Ng^{-1} \cdot Ng = N$.

Hence $(G/N, \cdot)$ is a group. $\square$

The order of G/N is the number of cosets of N in G. Hence

$$\#(G/N) = \#(G:N) = \#G/\#N.$$

We have seen in Example 4.16 that $\mathcal{C}_3$ is a normal subgroup of $\mathcal{S}_3$; therefore $\mathcal{S}_3 / \mathcal{C}_3$ is a quotient group. If $H = \mathcal{C}_3$, the elements of this group are the cosets H and $H(12)$, and its multiplication table is given in Table 4.1.

Table 4.1. The quotient group $\mathcal{S}_3 / \mathcal{C}_3$

∘	H	$H(12)$
H	H	$H(12)$
$H(12)$	$H(12)$	H

4.22 EXAMPLE. $(\mathbf{Z}_n, +)$ is the quotient group of $(\mathbf{Z}, +)$ by the subgroup $n\mathbf{Z} = \{nz \mid z \in \mathbf{Z}\}$.

SOLUTION. Since $(\mathbf{Z}, +)$ is Abelian, every subgroup is normal. The set $n\mathbf{Z}$ can be verified to be a subgroup, and the relationship $a \equiv b \bmod n\mathbf{Z}$ is equivalent to $a - b \in n\mathbf{Z}$ and to $n \mid a - b$. Hence $a \equiv b \bmod n\mathbf{Z}$ is the same relation as $a \equiv b \bmod n$. Therefore, $\mathbf{Z}_n$ is the quotient group $\mathbf{Z}/n\mathbf{Z}$, where the operation on congruence classes is defined by $[a] + [b] = [a + b]$. $\square$

$(\mathbf{Z}_n, +)$ is a cyclic group with 1 as a generator, and therefore, by Theorem 3.25, is isomorphic to $\mathcal{C}_n$. The group $(\mathbf{Z}_5, +)$ is shown in Table 4.2.

Table 4.2.   The group $(\mathbf{Z}_5, +)$

+	[0]	[1]	[2]	[3]	[4]
[0]	[0]	[1]	[2]	[3]	[4]
[1]	[1]	[2]	[3]	[4]	[0]
[2]	[2]	[3]	[4]	[0]	[1]
[3]	[3]	[4]	[0]	[1]	[2]
[4]	[4]	[0]	[1]	[2]	[3]

When there is no confusion, we write the elements of $\mathbf{Z}_n$ as $0, 1, 2, 3, \ldots, n-1$ instead of $[0], [1], [2], [3], \ldots, [n-1]$.

4.23 PROPOSITION. If H is a subgroup of index 2 in G, so that $\#(G:H) = 2$, then H is a normal subgroup of G, and G/H is cyclic group of order 2.

PROOF.   Since $\#(G:H) = 2$, there are only two right cosets of H in G. One must be H and the other can be written as Hg, where g is any element of G that is not in H. To show that H is a normal subgroup of G, we need to show that $g^{-1}hg \in H$ for all $g \in G$ and $h \in H$. If g is an element of H, it is clear that $g^{-1}hg \in H$ for all $h \in H$. If g is not an element of H, suppose that $g^{-1}hg \notin H$. In this case, $g^{-1}hg$ must be an element of the other right coset Hg, and we can write $g^{-1}hg = h_1 g$, for some $h_1 \in H$. It follows that $g = hh_1^{-1} \in H$, which contradicts the fact that $g \notin H$. Hence $g^{-1}hg \in H$ for all $g \in G$ and $h \in H$; in other words, H is normal in G. $\square$

4.24 THEOREM. If G is a finite Abelian group and the prime p divides the order of G, then G contains an element of order p and hence a subgroup of order p.

PROOF.   We prove this result by induction on the order of G. For a particular prime p, suppose that all Abelian groups of order less than k, whose order is divisible by p, contain an element of order p. The result is vacuously true for groups of order 1. Now suppose G is a group of order k. If p divides k, choose any nonidentity element $g \in G$. Let t be the order of the element g.

Case 1. If p divides t, say $t = pr$, then g^r is an element of order p. This follows because g^r is not the identity, but $(g^r)^p = g^t = e$, and p is a prime.

Case 2.   On the other hand, if p does not divide t, let K be the subgroup generated by g. Since G is Abelian, K is normal, and the quotient group G/K has order $\# G/t$, which is divisible by p. Therefore, by the induction hypothesis, G/K has an element of order p, say Kh. If u is the order of h in G, then $h^u = e$ and $(Kh)^u = Kh^u = K$. Since Kh has order p in G/K, u is a multiple of p, and we are back to Case 1.

The result now follows from the Principle of Mathematical Induction.

$\square$

This result is a partial converse to Lagrange's Theorem. It is a special case of some important results, in more advanced group theory, known as the Sylow Theorems. These theorems give information on the subgroups of prime power order, and they can be found in books such as Herstein [4] or Hall [21].

4.25 EXAMPLE.   Show that $\mathcal{C}_5$ has no proper normal subgroups.

SOLUTION.   It follows from Corollary 3.40 that $\mathcal{C}_5$ contains three types of nonidentity elements, namely, 3-cycles, 5-cycles, and pairs of disjoint transpositions. Suppose N is a normal subgroup of $\mathcal{C}_5$ that contains more than one element.

Case 1.   Suppose N contains the 3-cycle (abc). From the definition of normal subgroup, $g^{-1} \circ (abc) \circ g \in N$ for all $g \in \mathcal{C}_5$. If we take $g = (ab) \circ (cd)$, we obtain

$$(ab) \circ (cd) \circ (abc) \circ (ab) \circ (cd) = (adb) \in N$$

and also $(adb)^{-1} = (abd) \in N$. In a similar way, we can show that N contains every 3-cycle. Therefore, by Proposition 3.43, N must be the whole alternating group.

Case 2.   Suppose N contains the 5-cycle $(abcde)$. Then

$$(abc)^{-1} \circ (abcde) \circ (abc) = (acb) \circ (abcde) \circ (abc) = (abdec) \in N$$

$$\text{and } (abcde) \circ (abdec)^{-1} = (abcde) \circ (acedb) = (adc) \in N.$$

We are now back to Case 1, and hence $N = \mathcal{C}_5$.

Case 3.   Suppose N contains the pair of disjoint transpositions $(ab) \circ (cd)$. Then, if e is the element of $\{1, 2, 3, 4, 5\}$ not appearing in these transpositions, we have

$$(abe)^{-1} \circ (ab) \circ (cd) \circ (abe) = (aeb) \circ (ab) \circ (cd) \circ (abe) = (ae) \circ (cd) \in N.$$

Also $(ab) \circ (cd) \circ (ae) \circ (cd) = (aeb) \in N$, and again we are back to Case 1.

We have shown that any normal subgroup of $\mathcal{Q}_5$ containing more than one element must be $\mathcal{Q}_5$ itself. $\square$

A group without any proper normal subgroups is called a *simple* group. The term "simple" must be understood in the technical sense that it cannot be broken down, because it cannot have any nontrivial quotient groups. This is analogous to a prime number, which has no nontrivial quotients. Apart from the cyclic groups of prime order, which have no proper subgroups of any kind, simple groups are comparatively rare.

$\mathcal{Q}_5$ is of great interest to mathematicians because it is used in Galois Theory to show that there is an equation of the fifth degree that cannot be solved by any algebraic formula.

MORPHISM THEOREM

One of the basic results of group theory is the Morphism Theorem, which describes the relationship between morphisms, normal subgroups, and quotient groups. There is an analogous result for most algebraic systems, including rings and vector spaces.

4.26 DEFINITION. If $f: G \to H$ is a group morphism, then the *kernel* of f, denoted by $\mathrm{Ker} f$, is the set of elements of G that are mapped by f to the identity of H. That is, $\mathrm{Ker} f = \{ g \in G \mid f(g) = e_H \}$.

4.27 PROPOSITION. Let $f: G \to H$ be a group morphism. Then

        (i) $\mathrm{Ker} f$ is a normal subgroup of G

and   (ii) f is injective if and only if $\mathrm{Ker} f = \{ e_G \}$.

PROOF. (i) We first show that $\mathrm{Ker} f$ is a subgroup of G. Let $a, b \in \mathrm{Ker} f$ so that $f(a) = f(b) = e_H$. Then

$$f(ab) = f(a)f(b) = e_H e_H = e_H \text{ and so } ab \in \mathrm{Ker} f$$

and

$$f(a^{-1}) = f(a)^{-1} = e_H^{-1} = e_H \text{ and so } a^{-1} \in \mathrm{Ker} f.$$

Therefore, $\mathrm{Ker} f$ is a subgroup of G.

If $a \in \mathrm{Ker} f$ and $g \in G$, then

$$f(g^{-1}ag) = f(g^{-1})f(a)f(g) = f(g)^{-1}e_H f(g) = f(g)^{-1}f(g) = e_H.$$

Hence $g^{-1}ag \in \mathrm{Ker} f$, and $\mathrm{Ker} f$ is a normal subgroup of G.

(ii)   If f is injective, only one element maps to the identity of H. Hence $\mathrm{Ker} f = \{e_G\}$. Conversely, if $\mathrm{Ker} f = \{e_G\}$, suppose that $f(g_1) = f(g_2)$. Then $f(g_1 g_2^{-1}) = f(g_1) f(g_2)^{-1} = e_H$ and $g_1 g_2^{-1} \in \mathrm{Ker} f = \{e_G\}$. Hence $g_1 = g_2$, and f is injective. $\square$

4.28 PROPOSITION.   For any group morphism $f: G \to H$, the *image* of f, $\mathrm{Im} f = \{f(g) | g \in G\}$, is a subgroup of H (although not necessarily normal).

PROOF.   Let $f(g_1), f(g_2) \in \mathrm{Im} f$. Then $f(g_1) f(g_2) = f(g_1 g_2) \in \mathrm{Im} f$, and $f(g_1)^{-1} = f(g_1^{-1}) \in \mathrm{Im} f$, so $\mathrm{Im} f$ is a subgroup of H. $\square$

4.29 MORPHISM THEOREM FOR GROUPS.   Let K be the kernel of the group morphism $f: G \to H$. Then G/K is isomorphic to the image of f, and the isomorphism $\psi: G/K \to \mathrm{Im} f$ is defined by $\psi(Kg) = f(g)$.

This result is also known as the First Isomorphism Theorem; the Second and Third Isomorphism Theorems are given in Exercises 43 and 44.

PROOF.   The function ψ is defined on a coset by using one particular element in the coset, so we have to check that ψ is well defined, and it does not matter which element we use. If $Kg = Kg'$, then $g' \equiv g \bmod K$ and $g'g^{-1} = k \in K = \mathrm{Ker} f$. Hence

$$f(g') = f(kg) = f(k)f(g) = e_H f(g) = f(g)$$

and ψ is well defined on cosets.

The function ψ is a morphism because

$$\psi(Kg_1 Kg_2) = \psi(Kg_1 g_2) = f(g_1 g_2) = f(g_1) f(g_2) = \psi(Kg_1) \psi(Kg_2).$$

If $\psi(Kg) = e_H$, then $f(g) = e_H$ and $g \in K$. Hence the only element in the kernel of ψ is the identity coset K, and ψ is injective. Finally, $\mathrm{Im} \psi = \mathrm{Im} f$, by the definition of ψ. Therefore, ψ is the required isomorphism between G/K and $\mathrm{Im} f$. $\square$

Conversely, it can be shown that, if K is any normal subgroup of G, there is a morphism from G to G/K, whose kernel is precisely K.

By taking f to be the identity morphism from G to itself, the Morphism Theorem implies that $G/\{e\} \cong G$.

The function $f: \mathbf{Z} \to \mathbf{Z}_n$, defined by $f(x) = [x]$, has $n\mathbf{Z}$ as its kernel, and therefore the Morphism Theorem yields the fact that $\mathbf{Z}/n\mathbf{Z} \cong \mathbf{Z}_n$.

If a and b are the generators of the cyclic groups $\mathcal{C}_{12}$ and $\mathcal{C}_6$, respectively, define the morphism $f: \mathcal{C}_{12} \to \mathcal{C}_6$ by $f(a^r) = b^{2r}$. The kernel is $K =$

$\{e,a^3,a^6,a^9\}$, and so $\mathcal{C}_{12}/K\cong\mathrm{Im}f=\{e,b^2,b^4\}$. This isomorphism is obtained by mapping the coset Ka^r to b^{2r}.

4.30 PROPOSITION. $\mathcal{C}_n$ is a normal subgroup of $\mathbb{S}_n$ and $\#\mathcal{C}_n=n!/2$.

PROOF. If $I(\pi)$ is the number of inversions of the permutation $\pi\in\mathbb{S}_n$, define $f:(\mathbb{S}_n,\circ)\to(\mathbb{Z}_2,+)$ by $f(\pi)=[I(\pi)]$. Now, by Proposition 3.38,

$$f(\pi\circ\rho)=[I(\pi\circ\rho)]=[I(\pi)]+[I(\rho)]=f(\pi)+f(\rho).$$

Hence f is a group morphism, and its kernel is the alternating group $\mathcal{C}_n$ of even permutations. By Proposition 4.27, $\mathcal{C}_n$ is a normal subgroup of $\mathbb{S}_n$, and, by the Morphism Theorem, $\mathbb{S}_n/\mathcal{C}_n\cong\mathbb{Z}_2$. By Corollary 4.11, $\#\mathcal{C}_n=\#\mathbb{S}_n/2=n!/2$. $\square$

4.31 EXAMPLE. Show that the quotient group $\mathbf{R}/\mathbf{Z}$, of real numbers modulo 1, is isomorphic to the circle group $W=\{e^{i\theta}\in\mathbf{C}|\theta\in\mathbf{R}\}$.

SOLUTION. The set of points on the circle of complex numbers of unit modulus, W, form a group under multiplication. Define the function $f:\mathbf{R}\to W$ by $f(x)=e^{2\pi ix}$. This is a morphism from $(\mathbf{R},+)$ to $(W,\cdot)$ because

$$f(x+y)=e^{2\pi i(x+y)}=e^{2\pi ix}\cdot e^{2\pi iy}=f(x)\cdot f(y).$$

This function can be visualized in Figure 4.01 as wrapping the real line round and round the circle.

The morphism f is clearly surjective, and its kernel is $\{x\in\mathbf{R}|e^{2\pi ix}=1\}=\mathbf{Z}$. Therefore, the Morphism Theorem implies that $\mathbf{R}/\mathbf{Z}\cong W$. The quotient space $\mathbf{R}/\mathbf{Z}$ is the set of equivalence classes of $\mathbf{R}$ under the relation defined by $x\equiv y\bmod\mathbf{Z}$ if and only if the real numbers x and y differ by an integer. This quotient space is called the group of real numbers modulo 1. $\square$

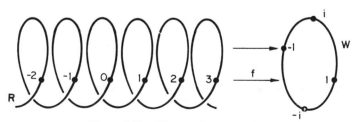

Figure 4.01. The morphism $f:\mathbf{R}\to W$.

4.32 PROPOSITION. If G and H are finite groups whose orders are relatively prime, there is only one morphism from G to H.

PROOF. Let K be the kernel of a morphism f from G to H. Then $G/K \cong \text{Im}f$, a subgroup of H. Now $\#(G/K) = \#G/\#K$, which is a divisor of $\#G$. But, by Lagrange's Theorem, $\#\text{Im}f$ is a divisor of $\#H$. Since $\#G$ and $\#H$ are relatively prime, we must have $\#(G/K) = \#\text{Im}f = 1$. Therefore, $f: G \to H$ is the trivial morphism defined by $f(g) = e_H$ for all $g \in G$. $\square$

4.33 EXAMPLE. Find all the subgroups and quotient groups of $\mathcal{D}_4$, the symmetry group of a square, and draw the poset diagram of its subgroups.

SOLUTION. Any symmetry of the square induces a permutation of its vertices. Thus, as in Example 3.36, this defines a group morphism $f: \mathcal{D}_4 \to \mathcal{S}_4$. However, unlike the case of the symmetries of an equilateral triangle, this is not an isomorphism because $\#\mathcal{D}_4 = 8$, whereas $\#\mathcal{S}_4 = 24$. The kernel of f consists of symmetries fixing the vertices and so consists of the identity only. Therefore, by the Morphism Theorem, $\mathcal{D}_4$ is isomorphic to the image of f in $\mathcal{S}_4$. We equate an element of $\mathcal{D}_4$ with its image in $\mathcal{S}_4$. All the elements of $\mathcal{D}_4$ are shown in Figure 4.02. The corner by the vertex 1 is blocked in, and the reverse side of the square is shaded to illustrate the effect of the symmetries. The order of each symmetry is given in Table 4.3.

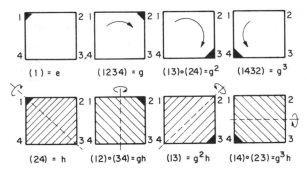

Figure 4.02.   **Symmetries of the square.**

Table 4.3.   **The orders of the symmetries of the square**

Elements of $\mathcal{D}_4$:	e	g	g^2	g^3	h	gh	g^2h	g^3h
Order of element:	1	4	2	4	2	2	2	2

The cyclic subgroups generated by the elements are $\{e\}$, $\mathcal{C}_4$ $= \{e,g,g^2,g^3\}$, $\{e,g^2\}$, $\{e,h\}$, $\{e,gh\}$, $\{e,g^2h\}$, and $\{e,g^3h\}$.

By Lagrange's Theorem, any proper subgroup must have order 2 or 4. Since any group of order 2 is cyclic, the only proper subgroups that are not cyclic are of order 4 and contain elements of order 1 and 2. There are two such subgroups, $K_1 = \{e,g^2,h,g^2h\}$ and $K_2 = \{e,g^2,gh,g^3h\}$. All the subgroups are illustrated in Figure 4.03.

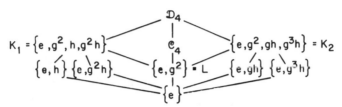

Figure 4.03.   The poset diagram of subgroups of $\mathcal{D}_4$.

To find all the quotient groups, we must determine which subgroups are normal.

The trivial group $\{e\}$ and the whole group $\mathcal{D}_4$ are normal subgroups. Since $\mathcal{C}_4$, K_1, and K_2 have index 2 in $\mathcal{D}_4$, they are normal by Proposition 4.23, and their quotient groups are cyclic of order 2.

Subgroup H	Left Coset gH	Right Coset Hg
$\{e,h\}$	$\{g,gh\}$	$\{g,hg\} = \{g,g^3h\}$
$\{e,g^2h\}$	$\{g,g^3h\}$	$\{g,g^2hg\} = \{g,gh\}$
$\{e,gh\}$	$\{g,g^2h\}$	$\{g,ghg\} = \{g,h\}$
$\{e,g^3h\}$	$\{g,h\}$	$\{g,g^3hg\} = \{g,g^2h\}$

For each of the above subgroups, the left and right cosets containing g are different; therefore, none of the subgroups are normal.

Left Cosets of L	Right Cosets of L
$L = \{e,g^2\}$	$L = \{e,g^2\}$
$gL = \{g,g^3\}$	$Lg = \{g,g^3\}$
$hL = \{h,hg^2\} = \{h,g^2h\}$	$Lh = \{h,g^2h\}$
$ghL = \{gh,ghg^2\} = \{gh,g^3h\}$	$Lgh = \{gh,g^3h\}$

The above table shows that $L = \{e,g^2\}$ is a normal subgroup. The multiplication table for $\mathcal{D}_4/L$ given in Table 4.4 shows that it is isomorphic to the Klein 4-group. $\square$

Table 4.4.   The group $\mathcal{D}_4/L$

	L	Lh	Lg	Lgh
L	L	Lh	Lg	Lgh
Lh	Lh	L	Lgh	Lg
Lg	Lg	Lgh	L	Lh
Lgh	Lgh	Lg	Lh	L

DIRECT PRODUCTS

Given two sets, S and T, we can form their Cartesian product, $S \times T = \{(s,t) | s \in S, t \in T\}$, whose elements are ordered pairs. For example, the product of the real line, **R**, with itself is the plane, $\mathbf{R} \times \mathbf{R} = \mathbf{R}^2$. We now show how to define the product of any two groups; the underlying set of the product is the Cartesian product of the underlying sets of the original groups.

4.34 PROPOSITION.   If $(G, \circ)$ and $(H, \star)$ are two groups, then $(G \times H, \cdot)$ is a group under the operation $\cdot$ defined by

$$(g_1, h_1) \cdot (g_2, h_2) = (g_1 \circ g_2, h_1 \star h_2).$$

The group $(G \times H, \cdot)$ is called the *direct product* of the groups $(G, \circ)$ and $(H, \star)$.

PROOF.   All the group axioms follow from the axioms for $(G, \circ)$ and $(H, \star)$. The identity of $G \times H$ is (e_G, e_H), and the inverse of (g, h) is (g^{-1}, h^{-1}).   □

This construction can be iterated any finite number of times to obtain the direct product of n groups.

Sometimes the direct product of two groups G and H is called the *direct sum* and is denoted by $G \oplus H$. (The direct sum of a finite number of groups is the same as the direct product. It is possible to define a direct sum and direct product of an infinite number of groups; these are different. An element of the direct product is obtained by taking one element from each group, while an element of the direct sum is obtained by taking one element from each group, but with only a finite number different from the identity.)

4.35 EXAMPLE.   Write down the table for the direct product of $\mathcal{C}_2$ with itself.

SOLUTION.   Let $C_2 = \{e, g\}$ so that $C_2 \times C_2 = \{(e,e),(e,g),(g,e),(g,g)\}$. Its table is given in Table 4.5.

Table 4.5.   The group $C_2 \times C_2$

·	(e,e)	(e,g)	(g,e)	(g,g)
(e,e)	(e,e)	(e,g)	(g,e)	(g,g)
(e,g)	(e,g)	(e,e)	(g,g)	(g,e)
(g,e)	(g,e)	(g,g)	(e,e)	(e,g)
(g,g)	(g,g)	(g,e)	(e,g)	(e,e)

We see that this group $C_2 \times C_2$ is isomorphic to the Klein 4-group of symmetries of a rectangle.   □

4.36 THEOREM.   If $GCD(m,n) = 1$, then $C_{mn} \cong C_m \times C_n$.

PROOF.   Let g, h, and k be the generators of C_{mn}, C_m, and C_n, respectively. Define

$$f : C_{mn} \to C_m \times C_n$$

by $f(g^r) = (h^r, k^r)$ for $r \in \mathbf{Z}$. This is well defined for all integers r because, if $g^r = g^{r'}$, then $r - r'$ is a multiple of mn, and so $r - r'$ is a multiple of m and of n. Hence $h^r = h^{r'}$ and $k^r = k^{r'}$.

Now

$$f(g^r \cdot g^s) = f(g^{r+s}) = (h^{r+s}, k^{r+s}) = (h^r \cdot h^s, k^r \cdot k^s) = (h^r, k^r) \cdot (h^s, k^s)$$

$$= f(g^r) \cdot f(g^s).$$

Hence f is a group morphism.

If $g^r \in \mathrm{Ker} f$, then $h^r = e$ and $k^r = e$. Therefore, r is divisible by m and n and, since $GCD(m,n) = 1$, r is divisible by mn. Hence $\mathrm{Ker} f = \{e\}$, and C_{mn} is isomorphic to the image of f. However, $\# C_{mn} = mn$ and $\#(C_m \times C_n) = \# C_m \cdot \# C_n = mn$; hence $\mathrm{Im} f = C_m \times C_n$, and f is an isomorphism.   □

The following is an easy consequence of this result.

4.37 COROLLARY.   Let $n = p_1^{\alpha_1} \cdot p_2^{\alpha_2} \cdots p_r^{\alpha_r}$ be a decomposition of the integer n into powers of distinct primes $p_1, p_2, \ldots, p_r$. Then $C_n \cong C_{p_1^{\alpha_1}} \times C_{p_2^{\alpha_2}} \times \cdots \times C_{p_r^{\alpha_r}}$.   □

If m and n are not coprime, then $\mathcal{C}_{mn}$ is never isomorphic to $\mathcal{C}_m \times \mathcal{C}_n$. For example, $\mathcal{C}_2 \times \mathcal{C}_2$ is not isomorphic to $\mathcal{C}_4$ because the direct product contains no element of order 4. In general, the order of the element (h, k) in $H \times K$ is the least common multiple of the orders of h and k, because $(h, k)^r = (h^r, k^r) = (e, e)$ if and only if $h^r = e$ and $k^r = e$. Hence, if $\mathrm{GCD}(m, n) > 1$, the order of (h, k) in $\mathcal{C}_m \times \mathcal{C}_n$ is always less than mn.

Direct products can be used to classify all finite Abelian groups. It can be shown that *any finite Abelian group* is isomorphic to a *direct product of cyclic groups*. For example, see Baumslag and Chandler [19]. The above results can be used to sort out those products of cyclic groups that are isomorphic to each other. For example, there are three nonisomorphic Abelian groups with 24 elements, namely,

$$\mathcal{C}_8 \times \mathcal{C}_3 \cong \mathcal{C}_{24}$$

$$\mathcal{C}_2 \times \mathcal{C}_4 \times \mathcal{C}_3 \cong \mathcal{C}_6 \times \mathcal{C}_4 \cong \mathcal{C}_2 \times \mathcal{C}_{12}$$

and

$$\mathcal{C}_2 \times \mathcal{C}_2 \times \mathcal{C}_2 \times \mathcal{C}_3 \cong \mathcal{C}_6 \times \mathcal{C}_2 \times \mathcal{C}_2.$$

4.38 THEOREM. If $(G, \cdot)$ is a finite group for which every element $g \in G$ satisfies $g^2 = e$, then $\# G = 2^n$, and G is isomorphic to the n-fold direct product $\mathcal{C}_2^n$.

PROOF. Every element in G has order 1 or 2, and the identity is the only element of order 1. Therefore, every element of G is its own inverse. The group G is Abelian because, for any $g, h \in G, gh = (gh)^{-1} = h^{-1}g^{-1} = hg$.

Choose the elements $a_1, a_2, \ldots, a_n \in G$ so that $a_1 \neq e$ and a_i cannot be written in terms of $a_1, \ldots, a_{i-1}$. Furthermore, choose n maximal, so that every element can be written in terms of the $a_i s$. We show that the function

$$f: \mathcal{C}_2^n \to G, \text{ défined by } f(g^{r_1}, g^{r_2}, \ldots, g^{r_n}) = a_1^{r_1} a_2^{r_2} \cdots a_n^{r_n}$$

is an isomorphism. It is well defined for all integers r_i, because, if $g^{r_i} = g^{q_i}$, then $a_i^{r_i} = a_i^{q_i}$. Now

$$f((g^{r_1}, \ldots, g^{r_n}) \cdot (g^{s_1}, \ldots, g^{s_n})) = f(g^{r_1 + s_1}, \ldots, g^{r_n + s_n}) = a_1^{r_1 + s_1} \cdots a_n^{r_n + s_n}$$

$$= a_1^{r_1} \cdots a_n^{r_n} \cdot a_1^{s_1} \cdots a_n^{s_n}, \text{ because } G \text{ is Abelian}$$

$$= f(g^{r_1}, \ldots, g^{r_n}) \cdot f(g^{s_1}, \ldots, g^{s_n}).$$

Hence f is a group morphism.

Let $(g^{r_1},\ldots,g^{r_n})\in\mathrm{Ker}f$. Suppose that r_i is the last odd exponent, so that $r_{i+1},r_{i+2},\ldots,r_n$ are all even. Then $a_1^{r_1}\cdots a_{i-1}^{r_{i-1}}a_i=e$ and

$$a_i=a_i^{-1}=a_1^{r_1}\cdots a_{i-1}^{r_{i-1}}$$

which is a contradiction. Therefore, all the exponents are even, and f is injective. The choice of the elements a_i guarantees that f is surjective. Hence f is the required isomorphism. $\square$

4.39 EXAMPLE. Describe all the group morphisms from $\mathcal{C}_{10}$ to $\mathcal{C}_2\times\mathcal{C}_5$. Which of these are isomorphisms?

SOLUTION. Since $\mathcal{C}_{10}$ is a cyclic group, generated by g, for example, a morphism from $\mathcal{C}_{10}$ is determined by the image of g. Let h and k be generators of $\mathcal{C}_2$ and $\mathcal{C}_5$, respectively, and consider the function $f_{r,s}:\mathcal{C}_{10}\to\mathcal{C}_2\times\mathcal{C}_5$ which maps g to the element $(h^r,k^s)\in\mathcal{C}_2\times\mathcal{C}_5$. Then, if $f_{r,s}$ is a morphism, $f_{r,s}(g^n)=(h^{rn},k^{sn})$ for $0\leqslant n\leqslant 9$. However, this would also be true for all integers n, because, if $g^n=g^m$, then $10|n-m$. Hence $2|n-m$ and $5|n-m$ and $h^{rn}=h^{rm}$ and $k^{sn}=k^{sm}$.

We now verify that $f_{r,s}$ is a morphism for any r and s. We have

$$f_{r,s}(g^ag^b)=f_{r,s}(g^{a+b})=(h^{(a+b)r},k^{(a+b)s})=(h^{ar},k^{as})(h^{br},k^{bs})$$

$$=f_{r,s}(g^a)f_{r,s}(g^b).$$

Therefore, there are ten morphisms, $f_{r,s}$, from $\mathcal{C}_{10}$ to $\mathcal{C}_2\times\mathcal{C}_5$ corresponding to the ten elements (h^r,k^s) of $\mathcal{C}_2\times\mathcal{C}_5$.
Now

$$\mathrm{Ker}f_{r,s}=\{\,g^n|(h^{rn},k^{sn})=(e,e)\}=\{\,g^n|rn\equiv 0\bmod 2\text{ and }sn\equiv 0\bmod 5\}.$$

Hence $\mathrm{Ker}f_{r,s}=\{e\}$ if $(r,s)=(1,1)$, $(1,2)$, $(1,3)$, or $(1,4)$, while $\mathrm{Ker}f_{0,0}=\mathcal{C}_{10}$, $\mathrm{Ker}f_{1,0}=\{e,g^2,g^4,g^6,g^8\}$, and $\mathrm{Ker}f_{0,s}=\{e,g^5\}$, if $s=1$, 2, 3, or 4. If $\mathrm{Ker}f_{r,s}$ contains more than one element, $f_{r,s}$ is not an injection and cannot be an isomorphism. By the Morphism Theorem,

$$\#\mathcal{C}_{10}/\#\mathrm{Ker}f_{r,s}=\#\mathrm{Im}f_{r,s}$$

and if $\mathrm{Ker}f_{r,s}=\{e\}$, then $\#\mathrm{Im}f_{r,s}=10$, so $f_{r,s}$ is surjective also. Therefore, the isomorphisms are $f_{1,1}$, $f_{1,2}$, $f_{1,3}$, and $f_{1,4}$. $\square$

GROUPS OF LOW ORDER

We find all the possible isomorphism classes of groups with eight or fewer elements.

4.40 LEMMA. If g is an element of order r, then g^t and g^{-t} have order $r/\text{GCD}(r,t) = \text{LCM}(r,t)/t$.

PROOF. Suppose $(g^t)^s = e$, so that $g^{ts} = e$. This happens if $r|ts$, or equivalently if $r/\text{GCD}(r,t)|ts/\text{GCD}(r,t)$. Since $r/\text{GCD}(r,t)$ and $t/\text{GCD}(r,t)$ are coprime, this happens if and only if $r/\text{GCD}(r,t)|s$. Therefore, the order of g^t, which is the least value of s, is $r/\text{GCD}(r,t)$. Using the relationship $rt = \text{GCD}(r,t) \cdot \text{LCM}(r,t)$, the order of g^t can be written as $\text{LCM}(r,t)/t$.

The inverse g^{-t} has the same order because $(g^{-t})^s = e$ if and only if $g^{ts} = e$. $\square$

4.41 LEMMA. If g and h are elements of coprime orders r and s in an Abelian group, then gh has order rs.

PROOF. Suppose $(gh)^u = e$ so that $g^u h^u = e$ and $g^u = h^{-u}$. The previous lemma implies that $r/\text{GCD}(r,u) = s/\text{GCD}(s,u)$. Since r and s have no common factors, we must have $\text{GCD}(r,u) = r$ and $\text{GCD}(s,u) = s$. Therefore, u must be a multiple of r and s, and the order of gh is rs. $\square$

Using these and previous results we can describe the groups of order eight or less.

Order 1.   Every trivial group is isomorphic to $\{e\}$.

Order 2.   By Corollary 4.14, every group of order 2 is cyclic.

Order 3.   By Corollary 4.14, every group of order 3 is cyclic.

Order 4.   Each element has order 1, 2, or 4.

Case (i).   If there is an element of order 4, the group is cyclic.

Case (ii).   If not, every element has order 1 or 2 and, by Theorem 4.38, the group is isomorphic to $\mathcal{C}_2 \times \mathcal{C}_2$.

Order 5.   By Corollary 4.14, every group of order 5 is cyclic.

Order 6.   Each element has order 1, 2, 3, or 6.

Case (i).   If there is an element of order 6, the group is cyclic.

Case (ii). If not, the elements have orders 1, 2, or 3. By Theorem 4.38, all the elements in a group of order 6 cannot have orders 1 and 2. Hence there is an element, say a, of order 3. The subgroup $H = \{e, a, a^2\}$ has index 2, and if $b \notin H$, the underlying set of the group is then $H \cup Hb = \{e, a, a^2, b, ab, a^2b\}$. By Proposition 4.23, H is normal, and the quotient group of H is cyclic of order 2. Hence

$$b^r \in Hb^r = (Hb)^r = \begin{cases} H & \text{if } r \text{ is even} \\ Hb & \text{if } r \text{ is odd} \end{cases}.$$

Therefore, b has even order. It cannot be 6, so it must be 2. As H is normal, $bab^{-1} \in H$. We cannot have $bab^{-1} = e$, because $a \neq e$. If $bab^{-1} = a$, then $ba = ab$, and we can prove that the whole group is Abelian. This cannot happen because, by Lemma 4.41, ab would have order 6. Therefore, $bab^{-1} = a^2$, and the group is generated by a and b with relations $a^3 = b^2 = e$ and $ba = a^2b$. This group is isomorphic to $\mathcal{D}_3$ and $\mathcal{S}_3$.

Order 7. Every group of order 7 is cyclic.

Order 8. Each element has order 1, 2, 4, or 8.

Case (i). If there is an element of order 8, the group is cyclic.

Case (ii). If all elements have order 1 or 2 the group is isomorphic to $\mathcal{C}_2 \times \mathcal{C}_2 \times \mathcal{C}_2$.

Case (iii). Otherwise, there is an element of order 4, say a. The subgroup $H = \{e, a, a^2, a^3\}$ is of index 2 and therefore normal. If $b \notin H$, the underlying set of the group is $H \cup Hb = \{e, a, a^2, a^3, b, ab, a^2b, a^3b\}$. Now $b^2 \in H$, but b^2 cannot have order 4, otherwise b would have order 8. Therefore, $b^2 = e$ or a^2. As H is normal, $bab^{-1} \in H$ and has the same order as a because $(bab^{-1})^k = ba^k b^{-1}$.

Case (iiia). If $bab^{-1} = a$, then $ba = ab$, and the whole group can be proved to be Abelian. If $b^2 = e$, each element can be written uniquely in the form $a^r b^s$ where $0 \leqslant r \leqslant 3$, and $0 \leqslant s \leqslant 1$. Hence the group is isomorphic to $\mathcal{C}_4 \times \mathcal{C}_2$ by mapping $a^r b^s$ to (a^r, b^s). If $b^2 = a^2$, let $c = ab$ so that $c^2 = a^2 b^2 = a^4 = e$. Each element of the group can now be written uniquely in the form $a^r c^s$ where $0 \leqslant r \leqslant 3, 0 \leqslant s \leqslant 1$, and the group is still isomorphic to $\mathcal{C}_4 \times \mathcal{C}_2$.

Case (iiib). If $bab^{-1} = a^3$ and $b^2 = e$, the group is generated by a and b with the relations $a^4 = b^2 = e, ba = a^3b$. This is isomorphic to the dihedral group $\mathcal{D}_4$.

Case (iiic). If $bab^{-1} = a^3$ and $b^2 = a^2$, then the group is isomorphic to the quaternion group $\mathcal{Q}$, described in Exercise 47 at the end of Chapter 3. The isomorphism maps $a^r b^s$ to $i^r j^s$.

Any group with eight or fewer elements is isomorphic to exactly one group in Table 4.6.

Table 4.6. Groups of low order

Order	1	2	3	4	5	6	7	8
Abelian Groups	$\{e\}$	C_2	C_3	C_4 $\quad$ $C_2 \times C_2$	C_5	C_6	C_7	C_8 $\quad$ $C_4 \times C_2$ $\quad$ $C_2 \times C_2 \times C_2$
Non-Abelian Groups						S_3		$\mathcal{D}_4$ $\quad$ $\mathcal{Q}$

ACTION OF A GROUP ON A SET

The concept of a group acting on a set X is a slight generalization of the group of symmetries of X. It is equivalent to considering a subgroup of $S(X)$. This concept is useful for determining the order of the symmetry groups of solids in three dimensions, and it is indispensable in Chapter 6, when we look at the Pólya–Burnside method of enumerating sets with symmetries.

4.42 DEFINITION. The group $(G, \cdot)$ *acts on the set* X if there is a function

$$\psi : G \times X \to X$$

such that, when we write $g(x)$ for $\psi(g, x)$, we have

    (i) $(g_1 g_2)(x) = g_1(g_2(x))$ for all $g_1, g_2 \in G, x \in X$

and   (ii) $e(x) = x$ if e is the identity of G and $x \in X$.

4.43 PROPOSITION. If g is an element of a group G acting on the set X, then the function $g : X \to X$, which maps x to $g(x)$, is a bijection. This defines a morphism

$$\chi : G \to S(X)$$

from G to the group of symmetries of X.

PROOF. The function $g : X \to X$ is injective because, if $g(x) = g(y)$, then $g^{-1}g(x) = g^{-1}g(y)$, and $e(x) = e(y)$ or $x = y$. It is surjective because, if $z \in X$,

$$g\big(g^{-1}(z)\big) = gg^{-1}(z) = e(z) = z.$$

Hence g is bijective, and g can be considered as an element of $\mathcal{S}(X)$, the group of symmetries of X.

The function $\chi : G \rightarrow \mathcal{S}(X)$, which takes the element $g \in G$ to the bijection $g : X \rightarrow X$, is a group morphism because $\chi(g_1 g_2)$ is the function from X to X defined by $\chi(g_1 g_2)(x) = (g_1 g_2)(x) = (g_1(g_2(x)) = \chi(g_1) \circ \chi(g_2)(x)$; thus $\chi(g_1 g_2) = \chi(g_1) \circ \chi(g_2)$. $\square$

If $\operatorname{Ker} \chi = \{e\}$, then χ is injective, and the group G is said to *act faithfully on the set* X. G acts faithfully on X if the only element of G, which fixes every element of X, is the identity $e \in G$. In this case, we identify G with $\operatorname{Im} \chi$ and regard G as a subgroup of $\mathcal{S}(X)$.

For example, consider the cyclic group of order 2, $\mathcal{C}_2 = \{e, h\}$, acting on the regular hexagon in Figure 4.04, where h reflects the hexagon about the line joining vertex 3 to vertex 6. Then $\mathcal{C}_2$ acts faithfully and can be identified with the subgroup $\{(1), (15) \circ (24)\}$ of $\mathcal{D}_6$.

The cyclic group $\mathcal{C}_3 = \{e, g, g^2\}$ acts faithfully on the cube in Figure 4.05, where g rotates the cube through one-third of a revolution about a line joining two opposite vertices. This group action can be considered as the subgroup $\{(1), (163) \circ (457), (136) \circ (475)\}$ of the symmetry group of the cube.

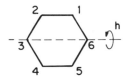

Figure 4.04. $\mathcal{C}_2$ acting on a hexagon.

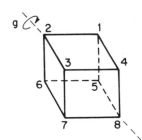

Figure 4.05. $\mathcal{C}_3$ acting on a cube.

4.44 PROPOSITION.   If G acts on a set X and $x \in X$, then

$$\operatorname{Stab} x = \{ g \in G \mid g(x) = x \}$$

is a subgroup of G, called the *stabilizer* of x. It is the set of elements of G that fix x.

PROOF.   Stab x is a subgroup because

(i)   if $g_1, g_2 \in$ Stab x, then $(g_1 g_2)(x) = g_1(g_2(x)) = g_1(x) = x$, and so $g_1 g_2 \in$ Stab x;

(ii)   if $g \in$ Stab x, then $g^{-1}(x) = x$, and so $g^{-1} \in$ Stab x. $\square$

4.45 DEFINITION.   The set of all images of an element $x \in X$ under the action of a group G is called the *orbit* of x under G and is denoted by

$$\text{Orb } x = \{ g(x) | g \in G \}.$$

The orbit of x is the equivalence class of x under the equivalence relation on X in which x is equivalent to y if and only if $y = g(x)$ for some $g \in G$. If π is a permutation in S_n, the subgroup generated by π acts on the set $\{1, 2, \ldots, n\}$, and this definition of orbit agrees with our previous one.

A graphic illustration of orbits can be obtained by looking at the group of matrices

$$\text{SO}(2) = \left\{ \begin{pmatrix} \cos\theta & -\sin\theta \\ \sin\theta & \cos\theta \end{pmatrix} \middle| \theta \in \mathbf{R} \right\}$$

under matrix multiplication. This group is called the special orthogonal group and is isomorphic to the circle group W. SO(2) acts on $\mathbf{R}^2$ as follows. The matrix $M \in$ SO(2) takes the vector $\mathbf{x} \in \mathbf{R}^2$ to the vector $M\mathbf{x}$. The orbit of any element $\mathbf{x} \in \mathbf{R}^2$ is the circle through $\mathbf{x}$ with center at the origin. Since the origin is the only fixed point for any of the nonidentity transformations, the stabilizer of the origin is the whole group, whereas the stabilizer of any other element is the subgroup consisting of the identity matrix only.

The orbits of the cyclic group $\mathcal{C}_2$ acting on the hexagon in Figure 4.04 are $\{1, 5\}$, $\{2, 4\}$, $\{3\}$, and $\{6\}$.

There is an important connection between the number of elements in the orbit of a point x and the stabilizer of that point.

4.46 LEMMA.   If G acts on X, then for each $x \in X$

$$\#(G : \text{Stab } x) = \# \text{Orb } x.$$

PROOF.   Let $H = $ Stab x and define the function

$$\xi : G/H \rightarrow \text{Orb } x$$

by $\xi(Hg) = g^{-1}(x)$. This is well defined on cosets because, if $Hg = Hk$, then

$k = hg$ for some $h \in H$, and so $k^{-1}(x) = (hg)^{-1}(x) = g^{-1}h^{-1}(x) = g^{-1}(x)$, since $h^{-1} \in H = \operatorname{Stab} x$.

The function ξ is surjective by the definition of the orbit of x. It is also injective, because $\xi(Hg_1) = \xi(Hg_2)$ implies that $g_1^{-1}(x) = g_2^{-1}(x)$, and so $g_2 g_1^{-1}(x) = x$ and $g_2 g_1^{-1} \in \operatorname{Stab} x = H$. Therefore, ξ is a bijection, and the result follows. $\square$

Note that ξ is *not* a morphism. $G/\operatorname{Stab} x$ is just a set of cosets because $\operatorname{Stab} x$ is not necessarily normal. Furthermore, we have placed no group structure on $\operatorname{Orb} x$.

4.47 THEOREM. If the finite group G acts on a set X, then, for each $x \in X$,
$$\# G = (\# \operatorname{Stab} x)(\# \operatorname{Orb} x).$$

PROOF. This follows from the above lemma and Corollary 4.11. $\square$

4.48 EXAMPLE. Find the number of proper rotations of a cube.

SOLUTION. Let G be the group of proper rotations of a cube; that is, rotations that can be carried out in three dimensions. The stabilizer of the vertex 1 in Figure 4.06 is $\operatorname{Stab} 1 = \{(1), (245) \circ (386), (254) \circ (368)\}$. The orbit of 1 is the set of all the vertices, because there is an element of G that will take 1 to any other vertex. Therefore, by Theorem 4.47
$$\# G = (\# \operatorname{Stab} 1)(\# \operatorname{Orb} 1) = 3 \cdot 8 = 24. \quad \square$$

The full symmetry group of the cube would include improper rotations such as the reflection in the plane as shown in Figure 4.07. This induces

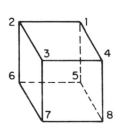

Figure 4.06.   The cube.

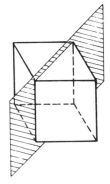

Figure 4.07.   Reflection in a plane.

the permutation $(24) \circ (68)$ on the vertices, and it cannot be obtained by physically rotating the cube in three dimensions. Under this group

$$\text{Stab } 1 = \{(1), (245) \circ (368), (254) \circ (386), (24) \circ (68), (25) \circ (38), (45) \circ (36)\}$$

and so the order of the full symmetry group of the cube is

$$(\#\text{Stab } 1)(\#\text{Orb } 1) = 6 \cdot 8 = 48.$$

Therefore, there are 24 proper and 24 improper rotations of the cube.

The article by Shapiro [29] contains many applications, mainly to group theory, of the actions of a group on a set.

Exercises

1–4. Which of the following relations are equivalence relations? Describe the equivalence classes of those relations which are equivalence relations.

1. The relation $\sim$ on $\mathbf{P} \times \mathbf{P}$ defined by $(a,b) \sim (c,d)$ if and only if $a + d = b + c$.

2. The relation T on the set of continuous functions from $\mathbf{R}$ to $\mathbf{R}$, where fTg if and only if $f(3) = g(3)$.

3. The inclusion relation on the power set $\mathcal{P}(X)$.

4. The relation C on a group G, where aCb if and only if $ab = ba$.

5. Find the left and right cosets of $H = \{(1), (12), (34), (12) \circ (34)\}$ in $\mathcal{S}_4$.

6. Let H be the subgroup of $\mathcal{C}_4$ that fixes 1. Find the left and right cosets of H in $\mathcal{C}_4$. Is H normal? Describe the left cosets in terms of their effect on the element 1. Can you find a similar description for the right cosets?

7–12. Verify that each of the following functions is well defined. Determine which are group morphisms, and find the kernels and images of all the morphisms. The element of $\mathbf{Z}_n$ containing x is denoted by $[x]_n$.

7. $f : \mathbf{Z}_{12} \to \mathbf{Z}_{12}$ where $f([x]_{12}) = [x+1]_{12}$.

8. $f : \mathcal{C}_{12} \to \mathcal{C}_{12}$ where $f(g) = g^3$.

9. $f : \mathbf{Z} \to \mathbf{Z}_2 \times \mathbf{Z}_4$ where $f(x) = ([x]_2, [x]_4)$.

10. $f : \mathbf{Z}_8 \to \mathbf{Z}_2$ where $f([x]_8) = [x]_2$.

11. $f : \mathcal{C}_2 \times \mathcal{C}_3 \to \mathcal{S}_3$ where $f(h^r, k^s) = (12)^r \circ (123)^s$.

12. $f : \mathcal{S}_n \to \mathcal{S}_{n+1}$ where $f(\pi)$ is the permutation on $\{1, 2, \ldots, n+1\}$ defined by $f(\pi)(i) = \pi(i)$, if $i \leqslant n$, and $f(\pi)(n+1) = n+1$.

13. If H is a subgroup of an Abelian group G, prove that the quotient group G/H is Abelian.

14. If H is a subgroup of G, show that $g^{-1}Hg = \{g^{-1}hg \mid h \in H\}$ is a subgroup for each $g \in G$.

15. Prove that the subgroup H is normal in G if and only if $g^{-1}Hg = H$ for all $g \in G$.

16. If H is the only subgroup of a given order in a group G, prove that H is normal in G.

17. Let H be any subgroup of a group G. Prove that there is a one-to-one correspondence between the set of left cosets of H in G and the set of right cosets of H in G.

18. Is the cyclic subgroup $\{(1),(123),(132)\}$ normal in $\mathcal{S}_4$?

19. Is the cyclic subgroup $\{(1),(123),(132)\}$ normal in $\mathcal{Q}_4$?

20. Is $\{(1),(1234),(13)\circ(24),(1432),(13),(24),(14)\circ(23),(12)\circ(34)\}$ normal in $\mathcal{S}_4$?

21. Find all the group morphisms from $\mathcal{C}_3$ to $\mathcal{C}_4$.

22. Find all the group morphisms from $\mathbf{Z}$ to $\mathbf{Z}_4$.

23. Find all the group morphisms from $\mathcal{C}_6$ to $\mathcal{C}_6$.

24. Find all the group morphisms from $\mathbf{Z}$ to $\mathcal{D}_4$.

25–29.   Which of the following pairs of groups are isomorphic? Give reasons.

25. $\mathcal{C}_{60}$ and $\mathcal{C}_{10} \times \mathcal{C}_6$.

26. $(\mathcal{P}\{a,b,c\}, \Delta)$ and $\mathcal{C}_2 \times \mathcal{C}_2 \times \mathcal{C}_2$.

27. $\mathcal{D}_n$ and $\mathcal{C}_n \times \mathcal{C}_2$.   **28.** $\mathcal{D}_6$ and $\mathcal{Q}_4$.

29. $\mathbf{Z}_4 \times \mathbf{Z}_2$ and $(\{\pm 1, \pm i, \pm(1+i)/\sqrt{2}, \pm(1-i)/\sqrt{2}\}, \cdot)$.

30. If $G \times H$ is cyclic prove that G and H are cyclic.

31. If π is an r-cycle in $\mathcal{S}_n$, prove that $\rho^{-1}\circ\pi\circ\rho$ is also an r-cycle for each $\rho \in \mathcal{S}_n$.

32. Find four different subgroups of $\mathcal{S}_4$ that are isomorphic to $\mathcal{S}_3$.

33. Find all the isomorphism classes of groups of order 10.

34. Find all the ten subgroups of $\mathcal{Q}_4$ and draw the poset diagram under inclusion. Which of the subgroups are normal?

35. For any groups G and H, prove that $G \times H/G' \cong H$ and $G \times H/H' \cong G$, where $G' = \{(g,e) \in G \times H \mid g \in G\}$ and $H' = \{(e,h) \in G \times H \mid h \in H\}$.

36. Show that $\mathbf{Q}/\mathbf{Z}$ is an infinite group but that every element has finite order.

37. If G is a subgroup of $\mathcal{S}_n$ and G contains an odd permutation, prove that G contains a normal subgroup of index 2.

38. In any group $(G, \cdot)$ the element $a^{-1}b^{-1}ab$ is called the *commutator* of *a* and *b*. Let G' be the subset of G consisting of all finite products of commutators. Show that G' is a normal subgroup of G. This is called the *commutator subgroup*. Also prove that G/G' is Abelian.

39. Let $\mathbf{C}^*$ be the group of nonzero complex numbers under multiplication and W be the multiplicative group of complex numbers of unit modulus. Describe $\mathbf{C}^*/W$.

40. Show that $K = \{(1), (12) \circ (34), (13) \circ (24), (14) \circ (23)\}$ is a subgroup of $\mathbb{S}_4$ isomorphic to the Klein 4-group. Prove that K is normal and that $\mathbb{S}_4/K \cong \mathbb{S}_3$.

41. If K is the group given in the previous exercise, prove that K is normal in $\mathcal{Q}_4$ and that $\mathcal{Q}_4/K \cong \mathcal{C}_3$. This shows that $\mathcal{Q}_4$ is not a simple group.

42. The *cross-ratio* of the four distinct real numbers x_1, x_2, x_3, x_4, in that order, is the ratio $\lambda = (x_2 - x_4)(x_3 - x_1)/(x_2 - x_1)(x_3 - x_4)$. Find the subgroup, K of $\mathbb{S}_4$, of all those permutations of the four numbers that preserve the value of the cross-ratio. Show that, if λ is the cross-ratio of four numbers taken in a certain order, the cross-ratio of these numbers in any other order must belong to the set

$$\left\{ \lambda, 1 - \lambda, \frac{1}{\lambda}, 1 - \frac{1}{\lambda}, \frac{1}{1 - \lambda}, \frac{\lambda}{\lambda - 1} \right\}.$$

Furthermore, show that all permutations in the same coset of K in $\mathbb{S}_4$ give rise to the same cross-ratio. In other words, prove that the quotient group $\mathbb{S}_4/K$ is isomorphic to the group of functions given in Exercise 42 of Chapter 3. The cross-ratio is very useful in projective geometry because it is preserved under projective transformations.

43. *(Second Isomorphism Theorem)* Let N be a normal subgroup of G, and H be any subgroup of G. Show that $HN = \{hn \mid h \in H, n \in N\}$ is a subgroup of G and that $H \cap N$ is a normal subgroup of H. Also prove that

$$H/(H \cap N) \cong HN/N.$$

44. *(Third Isomorphism Theorem)* Let M and N be normal subgroups of G, and N be a normal subgroup of M. Show that $\phi : G/N \to G/M$ is a well-defined morphism if $\phi(Ng) = Mg$, and prove that

$$(G/N)/(M/N) \cong G/M.$$

45. If a finite group contains no nontrivial subgroups, prove that it is either trivial or cyclic of prime order.

46. If d is a divisor of the order of a finite cyclic group G, prove that G contains a subgroup of order d.

47. If G is a finite Abelian group and p is a prime such that $g^p = e$, for all $g \in G$, prove that G is isomorphic to $\mathbf{Z}_p^n$, for some integer n.

48. What is the symmetry group of a rectangular box with sides of length 2, 3, and 4 cm?

49. Let

$$G_p = \left\{ \begin{pmatrix} a & b \\ c & d \end{pmatrix} \in \mathfrak{M}(2 \times 2; \mathbf{Z}_p) \mid ad - bc = 1 \text{ in } \mathbf{Z}_p \right\}.$$

If p is prime, show that $(G_p, \cdot)$ is a group of order $p(p^2 - 1)$, and find a group isomorphic to G_2.

50. Show that $(\mathbf{R}^*, \cdot)$ acts on $\mathbf{R}^{n+1}$ by scalar multiplication. What are the orbits under this action? The set of orbits, excluding the origin, form the n-dimensional real *projective space*.

51–58. Let $\mathbf{Z}_m^* = \{[x] \in \mathbf{Z}_m \mid \mathrm{GCD}(x, m) = 1\}$. The number of elements in this set is denoted by $\phi(m)$ and is called the Euler ϕ-function. *For example, $\phi(4) = 2$, $\phi(6) = 2$, and $\phi(8) = 4$.*

51. Show that $\phi(p^r) = p^r - p^{r-1}$, if p is a prime.

52. Show that $\phi(mn) = \phi(m)\phi(n)$ if $\mathrm{GCD}(m, n) = 1$.

53. Prove that $(\mathbf{Z}_m^*, \cdot)$ is an Abelian group.

54. Write out the multiplication table for $(\mathbf{Z}_8^*, \cdot)$.

55. Prove that $(\mathbf{Z}_6^*, \cdot)$ and $(\mathbf{Z}_{17}^*, \cdot)$ are cyclic and find their generators.

56. Find groups in Table 4.6 that are isomorphic to $(\mathbf{Z}_8^*, \cdot)$, $(\mathbf{Z}_9^*, \cdot)$, $(\mathbf{Z}_{10}^*, \cdot)$, and $(\mathbf{Z}_{15}^*, \cdot)$ and describe the isomorphisms.

57. Prove that, if $\mathrm{GCD}(a, m) = 1$, then $a^{\phi(m)} \equiv 1 \bmod m$. [This result was known to Leonhard Euler (1707–1783).]

58. Prove that if p is a prime, then for any integer $a, a^p \equiv a \bmod p$. [This result was known to Pierre de Fermat (1601–1665).]

59. If G is a group of order 35 acting on a set with 13 elements, show that G must have a fixed point; that is, a point $x \in S$ such that $g(x) = x$ for all $g \in G$.

60. If G is a group of order p^r acting on a set with m elements, show that G has a fixed point if p does not divide m.

Symmetry Groups in Three Dimensions

5

In this chapter, we determine the symmetry groups that can be realized in two- and three-dimensional space. We rely heavily on geometric intuition, not only to simplify arguments, but also to give geometric flavor to the group theory. Because we live in a three-dimensional world, these symmetry groups play a crucial role in the application of modern algebra to physics and chemistry.

We first show how the group of isometries of $\mathbf{R}^n$ can be broken down into translations and orthogonal transformations fixing the origin. Since the orthogonal transformations can be represented as a group of matrices, we look at the properties of matrix groups. We then use these matrix groups to determine all the finite rotation groups in two and three dimensions, and we find polyhedra that realize these symmetry groups.

TRANSLATIONS AND THE EUCLIDEAN GROUP

Euclidean geometry in n dimensions is concerned with those properties that are preserved under isometries (rigid motions) of Euclidean n-space, $\mathbf{R}^n$. The group of all isometries of $\mathbf{R}^n$ is called the *Euclidean group* in n

dimensions and is denoted by $E(n)$. The set of translations in $\mathbf{R}^n$ forms a subgroup that we denote by $T(n)$. We now proceed to show that $T(n)$ is a normal subgroup of $E(n)$ whose quotient group, $E(n)/T(n)$, is isomorphic to $O(n)$, the *orthogonal group* of dimension n. The group $O(n)$ consists of all the $n \times n$ real orthogonal matrices under multiplication. An orthogonal matrix corresponds to a linear transformation of $\mathbf{R}^n$ that preserves inner products and distances. The real $n \times n$ matrix A is orthogonal if $AA^T = I$, the identity matrix, where A^T is the transpose of the matrix A.

5.01 PROPOSITION.   The group of isometries in $\mathbf{R}^n$ that fix the origin is isomorphic to the orthogonal group $O(n)$.

PROOF.   We first show that an isometry α in $\mathbf{R}^n$, which fixes the origin, is a linear transformation. If $\mathbf{x}$ and $\mathbf{y}$ are two vectors in $\mathbf{R}^n$, consider the parallelogram with vertices, $\mathbf{0}$, $\mathbf{x}$, $\mathbf{y}$, and $\mathbf{x} + \mathbf{y}$. This is transformed, under α, to the parallelogram with vertices $\mathbf{0}$, $\alpha(\mathbf{x})$, $\alpha(\mathbf{y})$, and $\alpha(\mathbf{x} + \mathbf{y})$. Hence $\alpha(\mathbf{x} + \mathbf{y}) = \alpha(\mathbf{x}) + \alpha(\mathbf{y})$. The line joining $\mathbf{0}$ to $\mathbf{x}$ passes through $\lambda \mathbf{x}$, for each real number λ, and then is transformed, under α, to the line joining $\mathbf{0}$ to $\alpha(\mathbf{x})$, which passes through $\alpha(\lambda \mathbf{x})$. Also $\alpha(\lambda \mathbf{x}) = \lambda \alpha(\mathbf{x})$, and so α is a linear transformation.

If A is the matrix of the linear transformation α with respect to some coordinate system, then A must be an orthogonal matrix. Conversely, any orthogonal matrix defines a linear transformation that preserves distances and so is an isometry fixing the origin. This correspondence between matrices and linear transformations yields an isomorphism between $O(n)$ and the isometries fixing the origin. $\square$

Henceforth we identify the orthogonal group $O(n)$ with the isometries of $\mathbf{R}^n$ that fix the origin.

5.02 LEMMA.   If τ is a translation in $\mathbf{R}^n$ taking $\mathbf{x}$ to $\mathbf{y}$ and α is any isometry of $\mathbf{R}^n$, then $\alpha \circ \tau \circ \alpha^{-1}$ is a translation that takes $\alpha(\mathbf{x})$ to $\alpha(\mathbf{y})$.

PROOF.   Choose $n - 1$ points $\mathbf{z}_1, \mathbf{z}_2, \ldots, \mathbf{z}_{n-1}$ so that $\mathbf{y} - \mathbf{x}, \mathbf{z}_1 - \mathbf{x}, \ldots,$ $\mathbf{z}_{n-1} - \mathbf{x}$ form an independent set of vectors in $\mathbf{R}^n$. Any isometry is determined by the images of the $n + 1$ points $\mathbf{x}, \mathbf{y}, \mathbf{z}_1, \ldots, \mathbf{z}_{n-1}$.

We now consider the situation in $\mathbf{R}^2$ that is illustrated in Figure 5.01; the case of $\mathbf{R}^n$ is similar, except that there are more points involved. Let Δ be the triangle with vertices $\alpha(\mathbf{x})$, $\alpha(\mathbf{y})$, and $\alpha(\mathbf{z}_1)$. Then Figure 5.01 shows that the effect of $\alpha \circ \tau \circ \alpha^{-1}$ on the triangle Δ is to translate it and that the vertex $\alpha(\mathbf{x})$ is taken to $\alpha(\mathbf{y})$. Since the triangle Δ is nondegenerate, the effect of $\alpha \circ \tau \circ \alpha^{-1}$ on the whole plane must be a translation that takes $\alpha(\mathbf{x})$ to $\alpha(\mathbf{y})$. $\square$

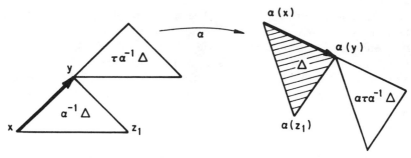

Figure 5.01.   The effect of the isometry $\alpha \circ \tau \circ \alpha^{-1}$.

This lemma shows that the group of translations, $T(n)$, is a *normal* subgroup of the Euclidean group, $E(n)$.

5.03 THEOREM.   There is a surjective morphism $\phi : E(n) \rightarrow O(n)$, which takes the isometry $\alpha \in E(n)$ to $\tau \circ \alpha$, where τ is the translation taking $\alpha(0)$ to **0**. The kernel of ϕ is the normal subgroup $T(n)$.

PROOF.   Since $\tau(\alpha(0)) = 0$, $\tau \circ \alpha$ is an element of $O(n)$. We now show that ϕ is a morphism. Let $\alpha, \beta \in E(n)$ and let τ, σ, and ρ be translations such that $\tau(\alpha(0)) = 0$, $\sigma(\beta(0)) = 0$, and $\rho(\alpha \circ \beta(0)) = 0$, so that $\phi(\alpha) = \tau \circ \alpha$, $\phi(\beta) = \sigma \circ \beta$, and $\phi(\alpha \circ \beta) = \rho \circ \alpha \circ \beta$. Now

$$\phi(\alpha) \circ \phi(\beta) = \tau \circ \alpha \circ \sigma \circ \beta = \tau \circ \left(\alpha \circ \sigma \circ \alpha^{-1}\right) \circ \alpha \circ \beta.$$

By Lemma 5.02, $\alpha \circ \sigma \circ \alpha^{-1}$ is the translation taking $\alpha(\beta(0))$ to $\alpha(0)$. Hence $\tau \circ (\alpha \circ \sigma \circ \alpha^{-1})$ is the translation taking $\alpha(\beta(0))$ to **0** and it must equal ρ. Therefore

$$\phi(\alpha) \circ \phi(\beta) = \rho \circ \alpha \circ \beta = \phi(\alpha \circ \beta),$$

and ϕ is a group morphism. This morphism is surjective because each isometry that fixes the origin maps to itself.

Any translation τ belongs to the kernel of ϕ because $\phi(\tau) = \tau^{-1} \circ \tau$, which is the identity. Conversely, if $\alpha \in \mathrm{Ker}\,\phi$, and τ is the translation taking $\alpha(0)$ to **0**, then $\tau \circ \alpha$ is the identity, and $\alpha = \tau^{-1}$. Therefore, $\mathrm{Ker}\,\phi = T(n)$, the group of translations. $\square$

Every isometry, α, can be written as an orthogonal transformation, $\tau \circ \alpha$, followed by a translation, τ^{-1}.

5.04 PROPOSITION.   Every finite subgroup of isometries of n dimensional space fixes at least one point.

PROOF.   Let G be a finite subgroup of isometries, and let **x** be any point of n dimensional space. The orbit of **x** consists of a finite number of points that are permuted among themselves by any element of G. Since all the elements of G are rigid motions. the centroid of Orb **x** must be always be sent to itself. Therefore, the centroid is a fixed point under G. $\square$

If the fixed point of any finite subgroup G of isometries is taken as the origin, then G is a subgroup of O(n), and all its elements can be written as orthogonal matrices. We now look at the structure of groups whose elements can be written as matrices.

MATRIX GROUPS

In physical sciences and in mathematical theory, we frequently encounter multiplicative group structures whose elements are $n \times n$ matrices. Such a group is called a *matrix group* if its identity element is the $n \times n$ identity matrix I. To investigate these groups, we have at our disposal, and shall freely apply, the machinery of linear algebra.

For example, if

$$A_k = \begin{pmatrix} \cos(2\pi k / m) & -\sin(2\pi k / m) \\ \sin(2\pi k / m) & \cos(2\pi k / m) \end{pmatrix}$$

then $(\{A_0, A_1, \ldots, A_{m-1}\}, \cdot)$ is a real matrix group of order m isomorphic to $\mathcal{C}_m$. The matrix A_k represents a counterclockwise rotation of the plane about the origin through an angle $(2\pi k / m)$.

The matrices

$$\begin{pmatrix} 1 & 0 \\ 0 & 1 \end{pmatrix}, \begin{pmatrix} -1 & 0 \\ 0 & -1 \end{pmatrix}, \begin{pmatrix} -i & 0 \\ 0 & i \end{pmatrix}, \begin{pmatrix} i & 0 \\ 0 & -i \end{pmatrix}, \begin{pmatrix} 0 & 1 \\ -1 & 0 \end{pmatrix}, \begin{pmatrix} 0 & -1 \\ 1 & 0 \end{pmatrix},$$

$$\begin{pmatrix} 0 & -i \\ -i & 0 \end{pmatrix}, \text{ and } \begin{pmatrix} 0 & i \\ i & 0 \end{pmatrix}$$

form a group under matrix multiplication. This is a complex matrix group of order 8 that is in fact isomorphic to the quaternion group $\mathcal{Q}$ of Exercise 47 in Chapter 3.

Since the identity of any matrix group is the identity matrix I and every element of a matrix group must have an inverse, every element must be a nonsingular matrix. All the nonsingular $n \times n$ matrices over a field **F** form a group $(\mathcal{GL}\,(n,\mathbf{F}),\,\cdot)$ called the *general linear group* of dimension n over $\mathbf{F}$. Any matrix group over the field $\mathbf{F}$ must be a subgroup of $\mathcal{GL}\,(n,\mathbf{F})$.

5.05 PROPOSITION. The determinant of any element of a finite matrix group must be an integral root of unity.

PROOF. Let A be an element of a matrix group of order m. Then, by Corollary 4.13, $A^m = I$. Hence $(\det A)^m = \det A^m = \det I = 1$. If G is a real matrix group, the determinant of any element of G is either $+1$ or -1. If G is a complex matrix group, the determinant of any element is of the form $e^{2\pi i k/m}$. $\square$

The orthogonal group $O(n)$ is a real matrix group and therefore any element must have determinant $+1$ or -1. The determinant function

$$\det : O(n) \longrightarrow \{1, -1\}$$

is a group morphism from $(O(n), \cdot)$ to $(\{1, -1\}, \cdot)$. The kernel, consisting of orthogonal matrices with determinant $+1$, is called the *special orthogonal group* of dimension n and is denoted by $SO(n) = \{A \in O(n) | \det A = +1\}$. This is a normal subgroup of $O(n)$ of index 2. The elements of $SO(n)$ are called *proper rotations*, whereas the elements in the other coset of $O(n)$ by $SO(n)$, consisting of orthogonal matrices with determinant -1, are called *improper rotations*.

The complex matrices that preserve inner products in $\mathbf{C}^n$ are called unitary matrices. The *unitary group* of dimension n, $U(n)$, consists of all $n \times n$ complex unitary matrices under multiplication. The *special unitary group*, $SU(n)$, is the subgroup of $U(n)$ consisting of those matrices with determinant $+1$. The group $SU(3)$ obtained some publicity in 1964 when the Brookhaven National Laboratory discovered the fundamental particle called the omega-minus baryon. The existence and properties of this particle had been predicted by a theory that used $SU(3)$ as a symmetry group of elementary particles. (See Dyson [27].)

5.06 PROPOSITION. If $\lambda \in \mathbf{C}$ is an eigenvalue of any orthogonal or unitary matrix, then $|\lambda| = 1$.

PROOF. Let λ be an eigenvalue and $\mathbf{x}$ a corresponding nonzero eigenvector of the orthogonal or unitary matrix A. Then $A\mathbf{x} = \lambda\mathbf{x}$, and since A preserves distances,

$$|\mathbf{x}| = |A\mathbf{x}| = |\lambda\mathbf{x}| = |\lambda|\,|\mathbf{x}|.$$

Since $\mathbf{x}$ is nonzero, it follows that $|\lambda| = 1$. $\square$

The group $\{A_k | k = 0, 1, \ldots, m-1\}$ of rotations of the plane is a subgroup of SO(2), and the eigenvalues that occur are $e^{\pm(2\pi i k/m)}$. The matrix group isomorphic to the quaternion group $\mathfrak{Q}$ is a subgroup of SU(2), and the eigenvalues that occur are ± 1 and $\pm i$.

Cayley's Theorem 3.44 showed that any group could be represented by a group of permutations. Another way to represent groups is by means of matrices. A *matrix representation* of a group G is a group morphism $\phi: G \to \mathcal{GL}(n, \mathbf{F})$. This is equivalent to an action of G on an n-dimensional vector space over the field $\mathbf{F}$, by means of linear transformations. The representation is called *faithful* if the kernel of ϕ is the identity. In this case, ϕ is injective and G is isomorphic to $\mathrm{Im}\,\phi$, a subgroup of the general linear group. Matrix representations provide powerful tools for studying groups because they lend themselves readily to calculation, and, as a result, most physical applications of group theory use representations.

It is possible to prove that any representation of a finite group over the real or complex field may be changed by a similarity transformation into a representation that uses only orthogonal or unitary matrices, respectively. Therefore, a real or complex faithful representation allows us to view a group as a subgroup of O(n) or U(n), respectively.

Finite Groups in Two Dimensions

We determine all the finite subgroups of rotations (proper and improper) of the plane $\mathbf{R}^2$. That is, we find all the finite matrix subgroups of SO(2) and O(2). This was essentially done by Leonardo da Vinci when he determined the possible symmetries of a central building with chapels attached. See Thompson [30] and Weyl [31] for interesting accounts of symmetries in nature and in art.

5.07 PROPOSITION.

(i)   The set of proper rotations in two dimensions is

$$\mathrm{SO}(2) = \left\{ \begin{pmatrix} \cos\theta & -\sin\theta \\ \sin\theta & \cos\theta \end{pmatrix} \middle| \theta \in \mathbf{R} \right\}.$$

(ii) The set of improper rotations in two dimensions is

$$\left\{ \begin{pmatrix} \cos\theta & \sin\theta \\ \sin\theta & -\cos\theta \end{pmatrix} \middle| \theta \in \mathbf{R} \right\}.$$

(iii)   The eigenvalues of the proper rotation $\begin{pmatrix} \cos\theta & -\sin\theta \\ \sin\theta & \cos\theta \end{pmatrix}$ are $e^{\pm i\theta}$ and of any improper rotation are ± 1.

PROOF.   (i) Let $A = \begin{pmatrix} p & q \\ r & s \end{pmatrix} \in SO(2)$, so that $A\begin{pmatrix} 1 \\ 0 \end{pmatrix} = \begin{pmatrix} p \\ r \end{pmatrix}$ and $A\begin{pmatrix} 0 \\ 1 \end{pmatrix}$ $= \begin{pmatrix} q \\ s \end{pmatrix}$. Since A preserves distances, $p^2 + r^2 = 1$, and $q^2 + s^2 = 1$; thus there exists angles θ and ϕ such that $p = \cos\theta$, $r = \sin\theta$, $q = \sin\phi$, and $s = \cos\phi$. Therefore

$$\det A = ps - qr = \cos\theta\cos\phi - \sin\theta\sin\phi = \cos(\theta + \phi).$$

If A is proper, $\det A = 1$, and so $\theta + \phi = 2n\pi$. Hence $\phi = 2n\pi - \theta$, and A is of the form $\begin{pmatrix} \cos\theta & -\sin\theta \\ \sin\theta & \cos\theta \end{pmatrix}$. Conversely, if A is of this form, then $AA^T = I$ and $A \in O(2)$. Since $\det A = +1$, A is a proper rotation, and $A \in SO(2)$.

(ii)   One improper rotation in $\mathbf{R}^2$ is $\begin{pmatrix} 1 & 0 \\ 0 & -1 \end{pmatrix}$ and so the coset of improper rotations is

$$SO(2)\begin{pmatrix} 1 & 0 \\ 0 & -1 \end{pmatrix} = \left\{ \begin{pmatrix} \cos\theta & \sin\theta \\ \sin\theta & -\cos\theta \end{pmatrix} \middle| \theta \in \mathbf{R} \right\}.$$

(iii)   If λ is an eigenvalue of

$$\begin{pmatrix} \cos\theta & -\sin\theta \\ \sin\theta & \cos\theta \end{pmatrix}, \text{ then } \det\begin{pmatrix} (\cos\theta) - \lambda & -\sin\theta \\ \sin\theta & (\cos\theta) - \lambda \end{pmatrix} = 0.$$

Therefore, $\lambda^2 - 2\lambda\cos\theta + 1 = 0$ and $\lambda = \cos\theta \pm i\sin\theta = e^{\pm i\theta}$.

If λ is an eigenvalue of the improper rotation $\begin{pmatrix} \cos\theta & \sin\theta \\ \sin\theta & -\cos\theta \end{pmatrix}$, then

$$\det\begin{pmatrix} (\cos\theta) - \lambda & \sin\theta \\ \sin\theta & -(\cos\theta) - \lambda \end{pmatrix} = 0. \text{ Hence } \lambda^2 - 1 = 0, \text{ and } \lambda = \pm 1. \quad \square$$

The improper rotation $B = \begin{pmatrix} \cos\theta & \sin\theta \\ \sin\theta & -\cos\theta \end{pmatrix}$ always has an eigenvalue 1 and hence leaves an axis through the origin invariant because, for any corresponding eigenvector $\mathbf{x}$, $B\mathbf{x} = \mathbf{x}$. It can be verified that this axis of eigenvectors, corresponding to the eigenvalue 1, is a line through the origin making an angle $\theta/2$ with the first coordinate axis. The matrix B corresponds to a reflection of the plane about this axis.

Hence we see that an improper rotation is a reflection about a line through the origin, and conversely, it is easy to see that a reflection about a line through the origin is an improper rotation.

5.08 THEOREM.   If G is a finite subgroup of $SO(2)$, then G is isomorphic to $\mathcal{C}_n$ for some $n \in \mathbf{P}$.

PROOF.   By Proposition 5.07, every element $A \in G \subset SO(2)$ is a counter-clockwise rotation through an angle $\theta(A)$, where $0 \leqslant \theta(A) < 2\pi$. Since G is finite, we can choose an element $B \in G$ so that $\theta(B)$ is the smallest positive angle. For any $A \in G$, there exists an integer $r \geqslant 0$ such that $r\theta(B) \leqslant \theta(A) < (r+1)\theta(B)$. Since $\theta(AB^{-r}) = \theta(A) - r\theta(B)$, it follows that $0 \leqslant \theta(AB^{-r}) < \theta(B)$. Therefore, $\theta(AB^{-r}) = 0$, $AB^{-r} = I$, and $A = B^r$.

Hence $G = \{I, B, B^2, \ldots, B^r, \ldots, B^{n-1}\}$, and G is a finite cyclic group that must be isomorphic to $\mathcal{C}_n$ for some integer n. $\square$

5.09 THEOREM.   If G is a finite subgroup of $O(2)$, then G is isomorphic to either $\mathcal{C}_n$ or $\mathcal{D}_n$ for some $n \in \mathbf{P}$.

PROOF.   The kernel of the morphism $\det : G \rightarrow \{1, -1\}$ is a normal sub-group, H, of index 1 or 2 consisting of the proper rotations in G. By the previous theorem, H is a cyclic group of order n, generated by B, for example.

If G contains no improper rotations, then $G = H \cong \mathcal{C}_n$. If G does contain an improper rotation A, then

$$G = H \cup HA = \{I, B, B^2, \ldots, B^{n-1}, A, BA, B^2A, \ldots, B^{n-1}A\}.$$

Since A and $B^k A$ are reflections, $A = A^{-1}$ and $B^k A = (B^k A)^{-1} = A^{-1}B^{-k} = AB^{n-k}$. These relations completely determine the multiplication in G, and it is now clear that G is isomorphic to the dihedral group $\mathcal{D}_n$ by an isomorphism that takes B to a rotation through $2\pi/n$ and A to a reflection. $\square$

The above theorem shows that the only possible types of finite symmetries, fixing one point, of any geometric figure in the plane are the cyclic and dihedral groups. Examples of such symmetries abound in nature; the symmetry group of a snowflake is usually $\mathcal{D}_6$, and many flowers have 5 petals with symmetry group $\mathcal{C}_5$.

We have found all the possible *finite* symmetries in the plane that fix one point. However, there are figures in the plane that have *infinite* symmetry groups that fix one point; one example is the circular disc. The group of proper symmetries of this disc is the group $SO(2)$, whereas the group of all symmetries is the whole of $O(2)$.

PROPER ROTATIONS OF REGULAR SOLIDS

One class of symmetries that we know occurs in three dimensions is the class of symmetry groups of the regular solids, namely, the tetrahedron, cube, octahedron, dodecahedron, and icosahedron. In this section, we determine the *proper* rotation groups of these solids. These will all be subgroups of SO(3). We restrict our consideration to proper rotations because these are the only ones that can be physically performed on models in three dimensions; to physically perform an improper symmetry on a solid, we would require four dimensions!

5.10 THEOREM. Every element $A \in SO(3)$ has a fixed axis, and A is a rotation about that axis.

PROOF. Let λ_1, λ_2, and λ_3 be the eigenvalues of A. These are the roots of a cubic characteristic polynomial with real coefficients. Hence, at least one eigenvalue is real , and, if a second one is complex, the third is its complex conjugate. By Proposition 5.06, $|\lambda_1| = |\lambda_2| = |\lambda_3| = 1$. Since $\det A = \lambda_1 \lambda_2 \lambda_3 = 1$, we can relabel the eigenvalues, if necessary, so that either

$$\text{(i)} \quad \lambda_1 = \lambda_2 = \lambda_3 = 1$$

or   (ii) $\lambda_1 = 1, \lambda_2 = \lambda_3 = -1$

or   (iii) $\lambda_1 = 1, \lambda_2 = \bar{\lambda}_3 = e^{i\theta}$ (where $\theta \neq n\pi$).

In all cases there is an eigenvalue equal to 1. If $\mathbf{x}$ is a corresponding eigenvector, then $A\mathbf{x} = \mathbf{x}$, and A fixes the axis along the vector $\mathbf{x}$. We can change the coordinate axes so that A can be written in one of the following three forms.

$$\text{(i)} \begin{bmatrix} 1 & 0 & 0 \\ 0 & 1 & 0 \\ 0 & 0 & 1 \end{bmatrix} \quad \text{(ii)} \begin{bmatrix} 1 & 0 & 0 \\ 0 & -1 & 0 \\ 0 & 0 & -1 \end{bmatrix} \quad \text{(iii)} \begin{bmatrix} 1 & 0 & 0 \\ 0 & \cos\theta & -\sin\theta \\ 0 & \sin\theta & \cos\theta \end{bmatrix}.$$

The first matrix is the identity. The second is a rotation through π, and the third is a rotation through θ about the fixed axis. $\square$

A *regular solid* is a polyhedron in which all faces are congruent regular polygons and all vertices are incident with the same number of faces. There are five such solids, and they are illustrated in Figure 5.02; their structure is given in Table 5.1.

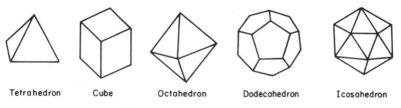

    Tetrahedron   Cube   Octahedron   Dodecahedron   Icosahedron

Figure 5.02. The regular solids.

Table 5.1. The regular solids

Polyhedron	# of Vertices	# of Edges	# of Faces	Faces	# of Faces at each vertex
Tetrahedron	4	6	4	Triangles	3
Cube	8	12	6	Squares	3
Octahedron	6	12	8	Triangles	4
Dodecahedron	20	30	12	Pentagons	3
Icosahedron	12	30	20	Triangles	5

The reader interested in making models of these polyhedra should consult Cundy and Rollett [26].

Given any polyhedron, we can construct its dual polyhedron in the following way. The vertices of the dual are the centers of the faces of the original polyhedron. Two centers are joined by an edge if the corresponding faces meet in an edge. The dual of a regular tetrahedron is another regular tetrahedron. The dual of a cube is an octahedron, and the dual of an octahedron is a cube. The dodecahedron and icosahedron are duals of each other. Any symmetry of a polyhedron will induce a symmetry on its dual and vice versa. Hence dual polyhedra will have the same rotation group.

5.11 THEOREM. The group of proper rotations of a regular tetrahedron is isomorphic to $\mathcal{C}_4$.

PROOF. Label the vertices of the tetrahedron 1, 2, 3, and 4. Then any rotation of the tetrahedron will permute these vertices. So, if G is the rotation group of the tetrahedron, we have a group morphism $f: G \to S_4$ whose kernel contains only the identity element. Hence, by the Morphism Theorem, G is isomorphic to $\operatorname{Im} f$.

We can use Theorem 4.47 to count the number of elements of G. The stabilizer of the vertex 1 is the set of elements fixing 1 and is $\{(1), (234), (243)\}$. The vertex 1 can be taken to any of the four vertices under G, and so the orbit of 1 is the set of four vertices. Hence $\#G = (\#\operatorname{Stab} 1)(\#\operatorname{Orb} 1) = 3 \cdot 4 = 12$.

There are two types of nontrivial elements in G that are illustrated in Figure 5.03 and Figure 5.04. There are rotations of order 3 about axes, each of which joins a vertex to the center of the opposite face. These rotations perform an even permutation of the vertices because each fixes one vertex and permutes the other three cyclically. There are also rotations of order 2 about axes, each of which joins the midpoints of a pair of opposite edges. (Two edges in a tetrahedron are said to be opposite if they

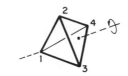

Figure 5.03.   The element (2 3 4).

Figure 5.04.   The element (1 2)∘(3 4).

do not meet.) The corresponding permutations interchange two pairs of vertices and, being products of two transpositions, are even.

Hence Imf consists of 12 permutations, all of which are even, and Im$f = \mathcal{Q}_4$.   □

The alternating group $\mathcal{Q}_4$ is sometimes called the *tetrahedral group*.

There are many different ways of counting the number of elements of G. One other way is as follows. Consider the tetrahedron sitting on a table, and shade in an equilateral triangle on the table where the bottom face rests, as in Figure 5.05. Any symmetry in G can be performed by picking up the tetrahedron, turning it, and replacing it on the table so that one face of the tetrahedron lies on top of the shaded equilateral triangle. Any of the *four* faces of the tetrahedron can be placed on the table, and each face can be placed on top of the shaded triangle in *three* different ways. Hence $\#G = 4 \cdot 3 = 12$. This really corresponds to applying Theorem 4.47 to the stabilizer and orbit of a *face* of the tetrahedron.

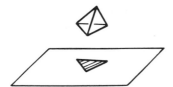

Figure 5.05

5.12 THEOREM.   The group of proper rotations of a regular octahedron and cube is isomorphic to $\mathcal{S}_4$.

PROOF.   There are four diagonals in a cube that join opposite vertices. Label these diagonals 1, 2, 3, and 4 as in Figure 5.06. Any rotation of the cube will permute these diagonals, and this defines a group morphism $f: G \to \mathcal{S}_4$, where G is the rotation group of the cube.

The stabilizer of any vertex of the cube is a cyclic group of order 3 that permutes the three adjacent vertices. The orbit of any vertex is the set of 8 vertices. Hence, by Theorem 4.47, $\#G = 3 \cdot 8 = 24$.

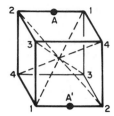

Figure 5.06. The diagonals of the cube.

Consider the rotation of order 2 about the line joining A to A' in Figure 5.06. The corresponding permutation is the transposition (12). Similarly, any other transposition is in Imf. Therefore, by Proposition 3.39, Im$f = \mathbb{S}_4$.

By the Morphism Theorem, $G/\mathrm{Ker}f \cong \mathbb{S}_4$ and $(\#G)/(\#\mathrm{Ker}f) = \#\mathbb{S}_4$. Since $\#G = \#\mathbb{S}_4 = 24$, it follows that $\#\mathrm{Ker}f = 1$, and f is an isomorphism.

The regular octohedron is dual to the cube, so it has the same rotation group. □

The symmetric group $\mathbb{S}_4$ is sometimes called the *octahedral group*.

5.13 THEOREM. The group of proper rotations of a regular dodecahedron and a regular icosahedron is isomorphic to $\mathcal{C}_5$.

PROOF. There are 30 edges of an icosahedron, and there are 15 lines through the center joining the midpoints of opposite edges. (The reflection of each edge in the center of the icosahedron is a parallel edge, called the opposite edge.) Given any one of these 15 lines, there are exactly two others that are perpendicular both to the first line and to each other. We call three such mutually perpendicular lines a triad. The 15 lines fall into five sets of triads. Label these triads 1, 2, 3, 4, and 5. Figure 5.07 shows the top half of an icosahedron where we have labeled the end points of each triad. (The existence of mutually perpendicular triads and the labeling of

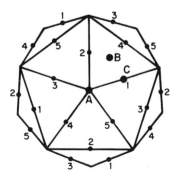

Figure 5.07. The ends of the triads of the icosahedron.

the diagram can best be seen by actually handling a model of the icosahedron.)

A rotation of the icosahedron permutes the five triads among themselves, and this defines a group morphism $f: G \to \mathbb{S}_5$, where G is the rotation group of the icosahedron.

The stabilizer of any vertex of the icosahedron is a group of order 5 that cyclically permutes the 5 adjacent vertices. The orbit of any vertex is the set of all 12 vertices. Hence, by Theorem 4.47, $\# G = 5 \cdot 12 = 60$.

There are three types of nontrivial elements in G. There are rotations of order 5 about axes through a vertex. The rotations about the vertex A in Figure 5.07 correspond to multiples of the cyclic permutation (12345), all of which are even. There are rotations of order 3 about axes through the center of a face. The rotations about an axis through the point B, in Figure 5.07, are multiples of (142) and are therefore even permutations. Finally, there are rotations of order 2 about the 15 lines joining midpoints of opposite edges. The permutation corresponding to a rotation about an axis through C, in Figure 5.07, is $(23) \circ (45)$, which is even.

Every 3-cycle occurs in the image of f so, by Proposition 3.43, $\mathrm{Im} f = \mathcal{C}_5$. Since G and $\mathcal{C}_5$ both have 60 elements, the Morphism Theorem implies that G is isomorphic to $\mathcal{C}_5$.

A regular dodecahedron is dual to the icosahedron, so it has the same rotation group. $\square$

The alternating group $\mathcal{C}_5$ is sometimes called the *icosahedral group*.

FINITE ROTATION GROUPS IN THREE DIMENSIONS

We now proceed to show that the only finite proper rotation groups in three dimensions are the three symmetry groups of the regular solids, $\mathcal{C}_4$, $\mathbb{S}_4$, and $\mathcal{C}_5$ together with the cyclic and dihedral groups, $\mathcal{C}_n$ and $\mathcal{D}_n$.

The unit sphere $S^2 = \{ \mathbf{x} \in \mathbf{R}^3 \mid |\mathbf{x}| = 1 \}$ is mapped to itself by every element of O(3). Every rotation group fixing the origin is determined by its action on the unit sphere S^2. By Theorem 5.10, every nonidentity element $A \in \mathrm{SO}(3)$ leaves precisely two antipodal points on S^2 fixed. That is, there exists $\mathbf{x} \in S^2$ such that $A(\mathbf{x}) = \mathbf{x}$ and $A(-\mathbf{x}) = -\mathbf{x}$. The points $\mathbf{x}$ and $-\mathbf{x}$ are called the *poles* of A. Let P be the set of poles of the nonidentity elements of a finite subgroup G of SO(3).

5.14 PROPOSITION. *G acts on the set, P, of poles of its nonidentity elements.*

PROOF. We show that G permutes the poles among themselves. Let A, B, be nonidentity elements of G, and let $\mathbf{x}$ be a pole of A. Then $(BAB^{-1})B(\mathbf{x}) = BA(\mathbf{x}) = B(\mathbf{x})$, so that $B(\mathbf{x})$ is a pole of BAB^{-1}. Therefore, the image of any pole is another pole, and G acts on the set of poles. $\square$

We classify the rotation groups by considering the number of elements in the stabilizers and orbits of the poles. Recall that the stabilizer of a pole $\mathbf{x}$, $\mathrm{Stab}\,\mathbf{x} = \{A \in G \mid A(\mathbf{x}) = \mathbf{x}\}$, is a subgroup of G, and that the orbit of $\mathbf{x}$, $\mathrm{Orb}\,\mathbf{x} = \{B(\mathbf{x}) \mid B \in G\}$, is a subset of P. In Table 5.2 we look at the stabilizers and orbits of the poles of the rotation groups we have already discussed.

Table 5.2. The poles of the finite rotation groups

Group G = #G = Symmetries of	$\mathcal{C}_n$ n n-agonal cone		$\mathcal{D}_n$ $2n$ n-agonal cylinder			$\mathcal{A}_4$ 12 tetrahedron		
Looking down on the pole, $\mathbf{x}$								
# Stab $\mathbf{x}$ =	n	n	2	2	n	2	3	3
# Orb $\mathbf{x}$ =	1	1	n	n	2	6	4	4

Group G = #G = Symmetries of	$\mathcal{S}_4$ 24 cube	or octahedron		$\mathcal{A}_5$ 60 dodecahedron	or icosahedron	
Looking down on the pole, $\mathbf{x}$	or	or	or	or	or	or
# Stab $\mathbf{x}$ =	2	3	4	2	3	5
# Orb $\mathbf{x}$ =	12	8	6	30	20	12

We take $\mathcal{C}_n$ to be the rotation group of a regular n-agonal cone whose base is a regular n-gon. (The sloping edges of the cone must not be equal to the base edges if $n = 3$.) $\mathcal{D}_n$ is the rotation group of a regular n-agonal cylinder whose base is a regular n-gon. (The vertical edges must not be equal to the base edges if $n = 4$.)

Each stabilizer group, $\mathrm{Stab}\,\mathbf{x}$, is a cyclic subgroup of rotations of the

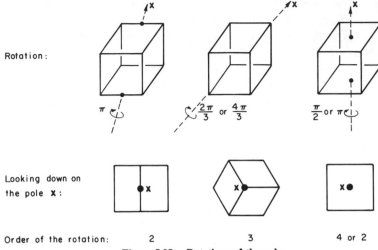

Rotation:

Looking down on
the pole **x**:

Order of the rotation: 2   3   4 or 2

Figure 5.08. Rotations of the cube.

solid about the axis through **x**. The orbit of **x**, Orb **x**, is the set of poles of
the same type as **x**. As a check on the number of elements in the stabilizers
and orbits, we have $\#G = (\#\text{Stab}\,\mathbf{x})(\#\text{Orb}\,\mathbf{x})$, for each pole **x**.

For example, the cube has three types of poles and four types of
nontrivial elements in its rotation group; these are illustrated in Figure
5.08.

5.15 THEOREM.   Any finite subgroup of SO(3) is isomorphic to one of

$$\mathcal{C}_n(n \geqslant 1), \mathcal{D}_n(n \geqslant 2), \mathcal{A}_4, \mathcal{S}_4 \text{ or } \mathcal{A}_5.$$

PROOF.   Let G be a finite subgroup of SO(3). Choose a set of poles
$\mathbf{x}_1, \ldots, \mathbf{x}_r$, one from each orbit. Let $p_i = \#\text{Stab}\,\mathbf{x}_i$ and $q_i = \#\text{Orb}\,\mathbf{x}_i$, so that
$p_i q_i = n = \#G$.

Each nonidentity element of G has two poles; thus the total number of
poles, counting repetitions, is $2(n-1)$. The pole $\mathbf{x}_i$ occurs as a pole of a
nonidentity element $p_i - 1$ times. There are q_i poles of the same type as $\mathbf{x}_i$.
Therefore, the total number of poles, counting repetitions, is

$$2(n-1) = \sum_{i=1}^{r} q_i(p_i - 1) = \sum_{i=1}^{r} (n - q_i)$$

and so

5.16.
$$2 - \frac{2}{n} = \sum_{i=1}^{r} \left(1 - \frac{1}{p_i}\right).$$

If G is not the trivial group, $n \geq 2$ and $1 \leq 2 - \dfrac{2}{n} < 2$. Since $\mathbf{x}_i$ is a pole of some nonidentity element, $\mathrm{Stab}\,\mathbf{x}_i$ contains a nonidentity element, and $p_i \geq 2$. Therefore $\dfrac{1}{2} \leq 1 - \dfrac{1}{p_i} < 1$. It follows from Equation 5.16 that the number of orbits, r, must be 2 or 3.

If there are just two orbits, it follows that

$$2 - \frac{2}{n} = 1 - \frac{1}{p_1} + 1 - \frac{1}{p_2}$$

and $2 = \dfrac{n}{p_1} + \dfrac{n}{p_2} = q_1 + q_2$. Hence $q_1 = q_2 = 1$, and $p_1 = p_2 = n$. This means that $\mathbf{x}_1 = \mathbf{x}_2$, and there is just one axis of rotation. Therefore, G is isomorphic to the cyclic group $\mathcal{C}_n$.

If there are three orbits, it follows that

$$2 - \frac{2}{n} = 1 - \frac{1}{p_1} + 1 - \frac{1}{p_2} + 1 - \frac{1}{p_3} \quad \text{and}$$

5.17.
$$1 + \frac{2}{n} = \frac{1}{p_1} + \frac{1}{p_2} + \frac{1}{p_3}.$$

Suppose $p_1 \leq p_2 \leq p_3$. If $p_1 \geq 3$, we would have

$$\frac{1}{p_1} + \frac{1}{p_2} + \frac{1}{p_3} \leq \frac{1}{3} + \frac{1}{3} + \frac{1}{3} = 1,$$

which is a contradiction, since $\dfrac{2}{n} > 0$. Hence $p_1 = 2$ and $q_1 = \dfrac{n}{2}$.

Now $1 + \dfrac{2}{n} = \dfrac{1}{2} + \dfrac{1}{p_2} + \dfrac{1}{p_3}$, so $\dfrac{1}{2} + \dfrac{2}{n} = \dfrac{1}{p_2} + \dfrac{1}{p_3}$. If $p_2 \geq 4$, we would have $\dfrac{1}{p_2} + \dfrac{1}{p_3} \leq \dfrac{1}{4} + \dfrac{1}{4} = \dfrac{1}{2}$, which is a contradiction, since $\dfrac{2}{n} > 0$. Hence p_2 is 2 or 3. The only possibilities are the following.

Case (i). $p_1 = 2, p_2 = 2, p_3 = \dfrac{n}{2}, n$ is even and $n \geq 4$, $q_1 = \dfrac{n}{2}, q_2 = \dfrac{n}{2}$ and $q_3 = 2$.

Case (ii). $p_1 = 2, p_2 = 3, p_3 = 3, n = 12, q_1 = 6, q_2 = 4$ and $q_3 = 4$.

Case (iii). $p_1 = 2, p_2 = 3, p_3 = 4, n = 24, q_1 = 12, q_2 = 8$ and $q_3 = 6$.

Case (iv). $p_1 = 2$, $p_2 = 3$, $p_3 = 5$, $n = 60$, $q_1 = 30$, $q_2 = 20$ and $q_3 = 12$.

If $p_2 = 2$ and $p_3 \geqslant 6$, $\dfrac{1}{p_2} + \dfrac{1}{p_3} \leqslant \dfrac{1}{3} + \dfrac{1}{6} = \dfrac{1}{2}$, which contradicts Equation 5.17, since $\dfrac{2}{n} > 0$.

Case (i). Let $H = \text{Stab}\, x_3$. This is a group of rotations about one axis, and it is a cyclic group of order $n/2$. Any other element A that is not in H is of order 2 and is a half turn. Therefore, $G = H \cup HA$, and G is isomorphic to $\mathcal{D}_{n/2}$ of order n.

Case (ii). Let y_1, y_2, y_3, and y_4 be the four poles in $\text{Orb}\, x_2$. Now $p_2 = \#\text{Stab}\, y_i = 3$; thus $\text{Stab}\, y_1$ permutes y_2, y_3, and y_4 as in Figure 5.09. Therefore, $|y_2 - y_1| = |y_3 - y_1| = |y_4 - y_1|$. We have similar results for $\text{Stab}\, y_2$ and $\text{Stab}\, y_3$. Hence y_1, y_2, y_3, and y_4 are the vertices of a regular tetrahedron, and G is a subgroup of the symmetries of this tetrahedron. Since $\# G = 12$, G must be the whole rotation group, $\mathcal{Q}_4$.

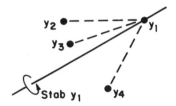

Figure 5.09

Case (iii). Let $y_1, y_2, \ldots, y_6$ be the six poles in $\text{Orb}\, x_3$. Since $p_3 = 4$, a rotation in $\text{Stab}\, y_i$ must fix two of the poles and rotate the other four cyclically. Hence $y_1, y_2, \ldots, y_6$ must lie at the vertices of a regular octahedron. Again, since $\# G = 24$, G must be the whole rotation group, $\mathcal{S}_4$, of this octahedron.

Case (iv). Let $y_1, y_2, \ldots, y_{12}$ be the twelve poles in $\text{Orb}\, x_3$. Any element of order 5 in G must permute these poles and hence must fix two poles and permute the others, as in Figure 5.10, in two disjoint 5-cycles, say (2 3 4 5 6) ∘ (7 8 9 10 11), where we denote the pole y_i by i. The points y_2, y_3, y_4,

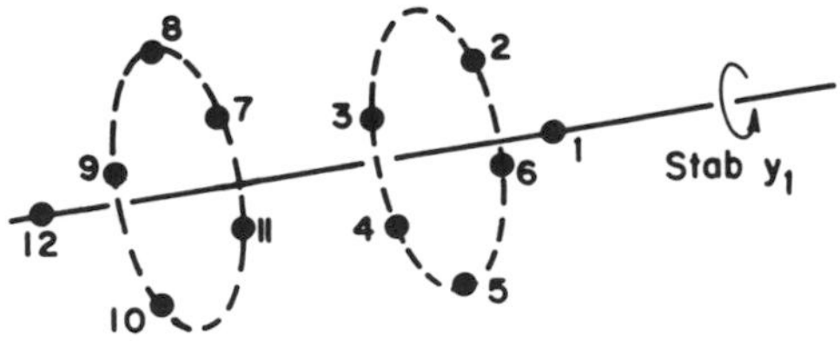

Figure 5.10

y_5, and y_6 form a regular pentagon and their distances from y_1 are all equal. Using similar results for rotations of order 5 about the other poles, we see that the poles are the vertices of an icosahedron, and the group G is the proper rotation group, $\mathcal{C}_5$, of this icosahedron. □

Throughout this section we have only considered *proper* rotations. However, if we allow *improper* rotations as well, it can be shown that a finite subgroup of O(3) is isomorphic to one of the groups in Theorem 5.15 or contains one of these groups as a normal subgroup of index two. See Coxeter [25; Sect. 15.5] for a more complete description of these improper rotation groups.

CRYSTALLOGRAPHIC GROUPS

This classification of finite symmetries in $\mathbf{R}^3$ has important applications in crystallography. Many chemical substances form crystals and their structures take the forms of crystalline lattices. A crystal lattice is always finite but, in order to study its symmetries, we create a mathematical model by extending this crystal lattice to infinity. We define an *ideal crystalline lattice* to be an infinite set of points in $\mathbf{R}^3$ of the form

$$n_1\mathbf{a}_1 + n_2\mathbf{a}_2 + n_3\mathbf{a}_3$$

where $\mathbf{a}_1$, $\mathbf{a}_2$, and $\mathbf{a}_3$ form a basis of $\mathbf{R}^3$ and $n_1, n_2, n_3 \in \mathbf{Z}$. Common salt forms a cubic crystalline lattice in which $\mathbf{a}_1$, $\mathbf{a}_2$, and $\mathbf{a}_3$ are orthogonal vectors of the same length. Figure 5.11 illustrates a crystalline lattice.

This use of the term "lattice" is not the same as that in Chapter 2, where a lattice referred to a special kind of partially ordered set. To avoid confusion, we always use the term "crystalline lattice" here.

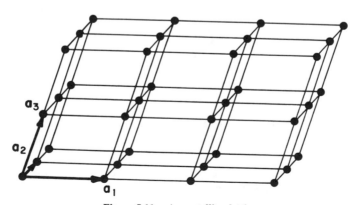

Figure 5.11. A crystalline lattice.

A subgroup of O(3) that leaves a crystalline lattice invariant is called a *crystallographic point group*. This is a finite subgroup of O(3) because there are only a finite number of crystalline lattice points that can be the images of $\mathbf{a}_1$, $\mathbf{a}_2$, and $\mathbf{a}_3$ when the origin is fixed.

However, not all finite subgroups of O(3) are crystallographic point groups. Suppose $A \in SO(3)$ leaves a crystalline lattice $\mathfrak{L}$ invariant. Then, by Theorem 5.10, A is a rotation through an angle θ, and the trace of A is $1 + 2\cos\theta$. If we choose a basis for $\mathbf{R}^3$ consisting of the vectors $\mathbf{a}_1, \mathbf{a}_2, \mathbf{a}_3$ of the crystalline lattice $\mathfrak{L}$, the matrix representing A will have integer entries. The trace is invariant under change of basis, so the trace of A must be an integer. Hence $2\cos\theta$ must be integral, and θ must be either $k\pi/2$ or $k\pi/3$, where $k \in \mathbf{Z}$. It follows that every element of a crystallographic point group in SO(3) can only contain elements of order 1, 2, 3, 4, or 6.

It can be shown that every crystallographic point group in SO(3) is isomorphic to one of $\mathcal{C}_1$, $\mathcal{C}_2$, $\mathcal{C}_3$, $\mathcal{C}_4$, $\mathcal{C}_6$, $\mathcal{D}_2$, $\mathcal{D}_3$, $\mathcal{D}_4$, $\mathcal{D}_6$, $\mathcal{A}_4$, or $\mathcal{S}_4$.

If we allow reflections, the only other such groups in O(3) must contain one of these groups as a normal subgroup of index two. Every one of these groups occurs in nature as the point group of at least one chemical crystal. See Coxeter [25; Sect. 15.6] or Lomont [28; Ch. 4, Sect. 4].

Exercises

1–7.   Find the group of proper rotations and the group of all rotations of the figures in Figure 5.12.

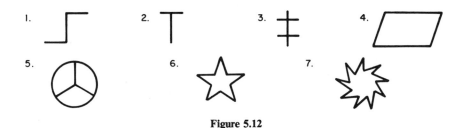

Figure 5.12

8.   Let G be the subgroup of O(2) isomorphic to $\mathcal{D}_n$. Find two matrices A and B so that any element of G can be written as a product of As and Bs.

9.   What is the group of proper rotations of a rectangular box of length 3 cm, depth 2 cm, and height 2 cm?

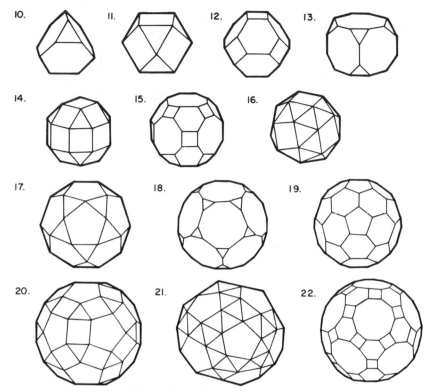

Figure 5.13.   The Archimedean solids.

10–22.   Find the proper rotation group of the 13 Archimedean solids in Figure 5.13. All the faces of these solids are regular polygons and all the vertices are similar. (See Cundy and Rollet [26] for methods on how to construct these solids.)

23.   It is possible to inscribe five cubes in a regular dodecahedron. One such cube is shown in Figure 5.14. Use these cubes to show that the rotation group of the dodecahedron is $\mathcal{C}_5$.

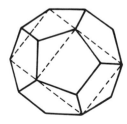

Figure 5.14.   A cube inside a dodecahedron.

24. What is the proper symmetry group of a cube in which three faces, coming together at one vertex, are painted green and the other faces are red?

25. Find the group of all rotations (both proper and improper) of a regular tetrahedron.

26. Let G be the full symmetry group of the cube. Define $f: G \to S_4$ as in Theorem 5.12. Find the kernel of f and the order of G.

27. Let the vertices of a tetrahedron be $(1,1,1)$, $(-1,-1,1)$, $(-1,1,-1)$, and $(1,-1,-1)$. Find matrices in SO(3) of orders 2 and 3 that leave the tetrahedron invariant.

28. Let the vertices of a cube be $(\pm 1, \pm 1, \pm 1)$. Find matrices in SO(3) of orders 2, 3, and 4 that leave the cube invariant.

29–31. Find the symmetry groups of the chemical molecules in Figure 5.15. (Assume that all of the C-C bonds are equivalent.)

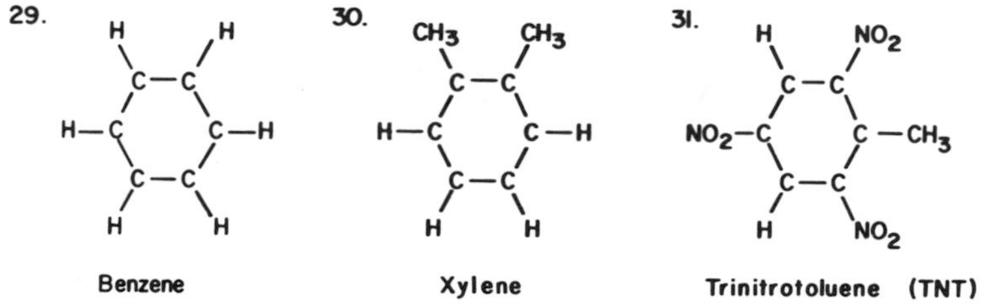

Benzene   **Xylene**   **Trinitrotoluene (TNT)**

Figure 5.15.

32–34. Find matrices in SO(3) that preserve the following crystalline lattices and find their crystallographic point groups. The points of the crystalline lattice are $n_1 a_1 + n_2 a_2 + n_3 a_3$, where $n_i \in \mathbf{Z}$ and the basis vectors a_i are given below.

32. $a_1 = (1,1,0)$, $a_2 = (-1,1,0)$, $a_3 = (0,1,1)$.

33. $a_1 = (1,0,0)$, $a_2 = (0,1,0)$, $a_3 = (0,0,2)$.

34. $a_1 = (1,0,0)$, $a_2 = (0, -3\sqrt{3}, 3)$, $a_3 = (0, 3\sqrt{3}, 3)$.

Pólya-Burnside Method of Enumeration

6

This chapter provides an introduction to the Pólya–Burnside method of counting the number of orbits of a set under the action of a symmetry group. If a group G acts on a set X and we know the number of elements of X, this method will enable us to count the number of different types of elements of X under G.

For example, how many different chemical compounds can be obtained by attaching a CH_3 or H radical to each carbon atom in the benzene ring of Figure 6.03? There are 2^6 different ways of attaching a CH_3 or H radical on paper, but these do not all give rise to different compounds because many are equivalent under a symmetry. There are six different ways of attaching one CH_3 radical and five H radicals, but they all give rise to the same compound. The dihedral group $\mathcal{D}_6$ acts on the 2^6 ways of attaching the radicals, and the number of different compounds is the number of orbits under the action of $\mathcal{D}_6$; that is, the number of formulae that cannot be obtained from each other by any rotation or reflection.

We have seen that the number of different switching circuits that can be obtained with n switches is 2^{2^n}. This number grows very quickly as n

becomes large. Table 2.11 gives the 16 switching functions of two variables; when $n=3$, there are 256 different circuits, and when $n=4$, there are 65,536 different circuits. However, many of these circuits are equivalent if we change the labels of the switches. That is, the symmetric group, $\mathbb{S}_n$, acts on the 2^{2^n} different circuits by permuting the labels of the switches. The number of nonequivalent circuits is the number of orbits under the action of $\mathbb{S}_n$.

BURNSIDE'S THEOREM

Let G be a finite group that acts on a finite set X. The following theorem describes the number of orbits in terms of the number of elements left fixed by each element of G. It was first proved by W. Burnside in 1911 and was called Burnside's Lemma; it was not until 1937 that its applicability to many combinatorial problems was discovered by G. Pólya.

6.01 BURNSIDE'S THEOREM. Let G be a finite group that acts on the elements of a finite set X. For each $g \in G$, let $\mathrm{Fix}\, g = \{x \in X \mid g(x) = x\}$, the set of elements of X left fixed by g. If N is the number of orbits of X under G, then

$$N = \frac{1}{\#G} \sum_{g \in G} \#\mathrm{Fix}\, g.$$

PROOF.   We count the set $S = \{(x,g) \in X \times G \mid g(x) = x\}$ in two different ways. Consider Table 6.1 whose columns are indexed by the elements of X

Table 6.1.   The elements of S correspond to the ones in this table

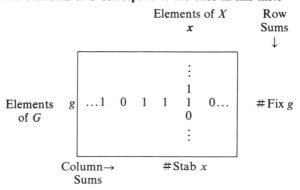

and whose rows are indexed by the elements of G. Put a value of 1 in the (x,g) position if $g(x)=x$; otherwise, let the entry be 0.

The sum of the entries in row g is the number of elements left fixed by g, $\#\mathrm{Fix}\,g$. The sum of the entries in column x is $\#\mathrm{Stab}\,x$, the number of elements of G that fix x.

We can count the number of elements of S either by totaling the row sums or by totaling the column sums. Hence

$$\#S=\sum_{g\in G}\#\mathrm{Fix}\,g=\sum_{x\in X}\#\mathrm{Stab}\,x.$$

Choose a set of representatives, $x_1,x_2,\ldots,x_N$, one from each orbit of X under G. If x is in the same orbit as x_i, then $\mathrm{Orb}\,x=\mathrm{Orb}\,x_i$ and, by Theorem 4.47, $\#\mathrm{Stab}\,x=\#\mathrm{Stab}\,x_i$. Hence

$$\sum_{g\in G}\#\mathrm{Fix}\,g=\sum_{i=1}^{N}\sum_{x\in\mathrm{Orb}\,x_i}\#\mathrm{Stab}\,x=\sum_{i=1}^{N}(\#\mathrm{Orb}\,x_i)(\#\mathrm{Stab}\,x_i)$$

$$=N\cdot(\#G),\qquad\text{by Theorem 4.47.}$$

The theorem now follows.   □

NECKLACE PROBLEMS

6.02 EXAMPLE. Three black and six white beads are strung onto a circular wire. This necklace can be rotated and turned over. How many different types of necklaces can be made assuming that beads of the same color are indistinguishable?

SOLUTION.   Position the three black and six white beads at the vertices of a regular 9-gon. If the 9-gon is fixed, there are $9\cdot8\cdot7/3!=84$ different ways of doing this. Two such arrangements are equivalent if there is an element of the symmetry group of the regular 9-gon, $\mathcal{D}_9$, which takes one arrangement into the other. $\mathcal{D}_9$ permutes the different arrangements, and the number of nonequivalent arrangements is the number, N, of orbits under $\mathcal{D}_9$. We can now use the Burnside Theorem to find N.

Table 6.2 lists all the different types of elements of $\mathcal{D}_9$ and the number of fixed points for each type. For example, consider the reflection $g\in\mathcal{D}_9$ about the line joining vertex 2 to the center of the circle, which is illustrated in Figure 6.01. Then the arrangements that are fixed under g occur when the black beads are at vertices 1 2 3 ,9 2 4 ,8 2 5 , or

7 2 6. Hence $\#\mathrm{Fix}\,g=4$. There are nine reflections about a line; one through each vertex. Therefore, the total number of fixed points contributed by these types of elements is $9\cdot4=36$. A rotation of order 3 in $\mathcal{D}_9$ will fix an arrangement if the black beads are at vertices 1 4 7, 2 5 8, or 3 6 9; hence there are three arrangements that are fixed. If an arrangement is fixed under a rotation of order 9, all the beads must be the same color; hence $\#\mathrm{Fix}\,g=0$ if g has order 9. Table 6.2 shows that the sum of all the numbers of fixed points is 126. By Theorem 6.01, $N=\dfrac{1}{\#\,\mathcal{D}_9}\Sigma_{g\in\mathcal{D}_9}\#\mathrm{Fix}\,g$

$=\dfrac{126}{18}=7$, and there are seven different types of necklaces. $\quad\square$

Table 6.2.   **The action of $\mathcal{D}_9$ on the necklaces**

Type of element, $g\in\mathcal{D}_9$	Order of g	Number, s, of such elements	$\#\mathrm{Fix}\,g$	$s\cdot\#\mathrm{Fix}\,g$
Identity	1	1	84	84
Reflection about a line	2	9	4	36
Rotation through $2\pi/3$ or $4\pi/3$	3	2	3	6
Other rotations	9	6	0	0

$$\Sigma\,\#\mathrm{Fix}\,g=126$$

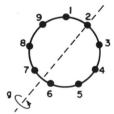

Figure 6.01. $\mathcal{D}_9$ **acting on the necklace.**

In this example, it is easy to determine all the seven types. They are illustrated in Figure 6.02.

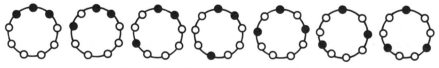

Figure 6.02.   **The seven types of necklaces.**

6.03 EXAMPLE.   Find the number of different chemical compounds that can be obtained by attaching CH_3 or H radicals to a benzene ring.

SOLUTION.   The carbon atoms are placed at the six vertices of a regular hexagon, and there are 2^6 ways of attaching CH_3 or H radicals. The dihedral group $\mathcal{D}_6$ acts on these 2^6 ways, and we wish to find the number of orbits.

Consider a reflection, g, about a line through opposite vertices. The order of g is 2, and there are three such reflections, through the three opposite pairs of vertices. #Fix$g$ can be determined by looking at Figure 6.03. If a configuration is fixed by g, the radical in place 2 must be the same as the radical in place 6, and also the radicals in places 3 and 5 must be equal to each other. Hence the radicals in places 1, 2, 3, and 4 can be chosen arbitrarily, and this can be done in 2^4 ways.

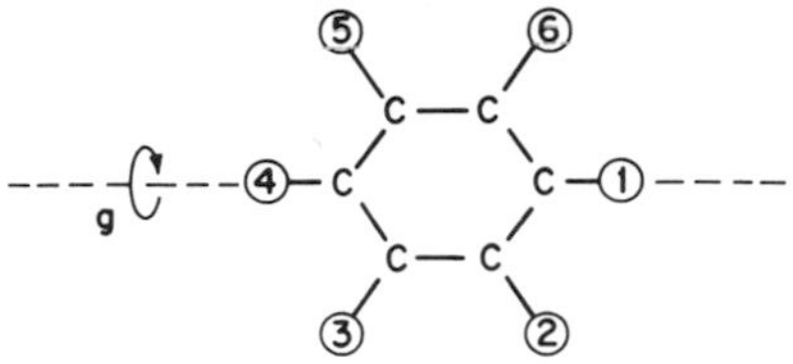

Figure 6.03.   A benzene ring.

The number of configurations left fixed by each element of $\mathcal{D}_6$ is given in Table 6.3. To check that we have not omitted any elements, we add the column containing the numbers of elements, and this should equal the order of the group $\mathcal{D}_6$. It follows from the Burnside Theorem that the number of orbits is $156 / \# \mathcal{D}_6 = 156/12 = 13$. Hence there are 13 different types of molecules obtainable.   □

Table 6.3.   The action of $\mathcal{D}_6$ on the compounds

Type of element, $g \in \mathcal{D}_6$	Order of g	Number, s, of such elements	#Fix$g$	$s \cdot \#$Fix$g$
Identity	1	1	2^6	64
Reflection in a line through opposite vertices, e.g., $(26) \circ (35) \circ (1) \circ (4)$	2	3	2^4	48
Reflection in a line through midpoints of opposite sides, e.g., $(56) \circ (14) \circ (23)$	2	3	2^3	24
Rotation through $\pm \pi/3$, e.g., (123456)	6	2	2	4
Rotation through $\pm 2\pi/3$, e.g., $(135) \circ (246)$	3	2	2^2	8
Rotation through π, $(14) \circ (25) \circ (36)$	2	1	2^3	8
		$\# \mathcal{D}_6 = 12$		$\Sigma = 156$

COLORING POLYHEDRA

6.04 EXAMPLE. How many ways is it possible to color the vertices of a cube, if n colors are available?

SOLUTION. If the cube is fixed, the eight vertices can each be colored in n ways, giving a total of n^8 colorings. The rotation group of the cube, $\mathbb{S}_4$, permutes these colorings among themselves, and the number of orbits is the number of distinct colorings taking the rotations into account. We can calculate the number of orbits using the Burnside Theorem.

There are five types of elements in the rotation group of the cube. We take an element g, in each type and determine the vertices that have to have the same color in order that the coloring be invariant under g.

Figure 6.04 illustrates the different types of rotations; vertices that have to have the same color are shaded in the same way. Table 6.4 gives the number of fixed colorings.

By the Burnside Theorem, the number of orbits and hence the number of colorings is

$$\frac{1}{\#\,\mathbb{S}_4} \sum_{g \in \mathbb{S}_4} \#\,\mathrm{Fix}\,g = \frac{1}{24}(n^8 + 17n^4 + 6n^2).$$

This shows, incidently, that $n^8 + 17n^4 + 6n^2$ is divisible by 24 for all $n \in \mathbf{P}$.

□

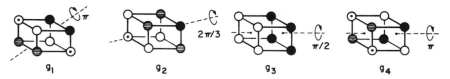

Figure 6.04.   Types of rotations of the cube.

Table 6.4.   Colorings of the vertices of the cube

Type of element, g_i, illustrated in Figure 6.04	Order of g_i	Number, s, of such elements	Number, r, of choices of colors	$\#\,\mathrm{Fix}\,g_i$ which is n^r	$s \cdot \#\,\mathrm{Fix}\,g_i$
Identity	1	1	8	n^8	n^8
g_1	2	$6 \cdot 1 = 6$	4	n^4	$6n^4$
g_2	3	$4 \cdot 2 = 8$	4	n^4	$8n^4$
g_3	4	$3 \cdot 2 = 6$	2	n^2	$6n^2$
g_4	2	$3 \cdot 1 = 3$	4	n^4	$3n^4$
		$\#\,\mathbb{S}_4 = 24$			$\sum = n^8 + 17n^4 + 6n^2$

6.05 EXAMPLE. How many ways is it possible to color a regular dodecahedron so that five of its faces are black and the other seven are white?

SOLUTION. The number of ways of choosing five faces of a fixed dodecahedron to be colored black is

$$\binom{12}{5} = \frac{12 \cdot 11 \cdot 10 \cdot 9 \cdot 8}{5!} = 792.$$

The different types of elements in the rotation group, $\mathcal{Q}_5$, of the dodecahedron are shown in Figure 6.05. The numbers of elements of a given type, in Table 6.5, are calculated as follows. An element of order 3 is a rotation about an axis through opposite vertices. Since there are 20 vertices, there are ten such axes. There are two nonidentity rotations of order 3 about each axis; thus the total number of elements of order 3 is $10 \cdot 2 = 20$. The elements of orders 2 and 5 can be counted in a similar way.

g_1 $\qquad\qquad$ g_2 $\qquad\qquad$ g_3

Figure 6.05.   Types of rotations of a dodecahedron.

Table 6.5.   Colorings of the dodecahedron

Type of element, g_i in Figure 6.05	Order of g_i	Number, s, of such elements	#Fixg_i	$s \cdot$ #Fixg_i
Identity	1	1	792	792
g_1	2	15	0	0
g_2	3	$10 \cdot 2 = 20$	0	0
g_3	5	$6 \cdot 4 = 24$	2	48
		#$\mathcal{Q}_5 = 60$		$\sum$ #Fix$g = 840$

If $g_2 \in \mathcal{Q}_5$ is of order 3, we can calculate #Fixg_2 as follows. The element g_2 does not fix any face and permutes the faces in disjoint 3-cycles. Now five black faces cannot be permuted by disjoint 3-cycles without fixing two faces, so #Fix$g_2 = 0$. Similarly, #Fix$g_1 = 0$ if g_1 is a 2-cycle. If g_3 is of order 5, then g_3 is a rotation about an axis through the centers of two opposite faces, and these two faces are fixed. The other ten faces are permuted in two disjoint 5-cycles; either of these 5-cycles can be black; thus $\#\operatorname{Fix} g_3 = 2$.

It follows from Table 6.5 and from the Burnside Theorem that the number of different colorings is $840/60 = 14$. □

Any face coloring of the dodecahedron corresponds to a vertex coloring of its dual, the icosahedron.

COUNTING SWITCHING CIRCUITS

The Burnside Theorem can still be applied when the sets to be enumerated do not have any geometric symmetry. In this case, the symmetry group is usually the full permutation group S_n.

Consider the different switching circuits obtained by using three switches. We can think of these as black boxes with three binary inputs x_1, x_2, and x_3 and one binary output $f(x_1, x_2, x_3)$, as in Figure 6.06. Two circuits, f and g, are called equivalent if there is a permutation π of the variables so that $f(x_1, x_2, x_3) = g(x_{\pi 1}, x_{\pi 2}, x_{\pi 3})$. Equivalent circuits can be obtained from each other by just permuting the wires outside the black boxes, as in Figure 6.07.

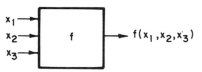

Figure 6.06. A switching circuit.

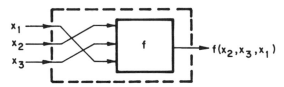

Figure 6.07. A permutation of the inputs.

6.06 EXAMPLE. Find the number of switching circuits using three switches that are not equivalent under permutations of the inputs.

SOLUTION.   There are eight possible inputs using three binary variables and hence there are $2^8 = 256$ circuits to consider. The symmetric group S_3 acts on these 256 circuits, and we wish to find the number of different equivalence classes; that is, the number of orbits.

Table 6.6 lists the number of circuits left fixed by the different types of elements in S_3. For example, if the switching function $f(x_1, x_2, x_3)$ is fixed by the transposition (12) of the input variables, then $f(0, 1, 0) = f(1, 0, 0)$, and $f(0, 1, 1) = f(1, 0, 1)$. The values of f for the inputs $(0, 0, 0)$, $(0, 0, 1)$, $(0, 1, 0)$, $(0, 1, 1)$, $(1, 1, 0)$, and $(1, 1, 1)$ can be chosen arbitrarily in 2^6 ways.

Table 6.6. The action of S_3 on the inputs of the switches

Type of element $g \in S_3$	Number, s, of such elements	# Fix g	$s \cdot$ # Fix g
Identity	1	2^8	$2^8 = 256$
Transposition	3	2^6	$3 \cdot 2^6 = 192$
3-cycle	2	2^4	$2 \cdot 2^4 = 32$
	$\# S_3 = 6$		$\sum = 480$

By Burnside's Theorem and Table 6.6, the number of nonequivalent circuits is $480 / \# S_3 = 480/6 = 80$. $\square$

However, this number can be further reduced if we allow permutations and complementation of the three variables. In a circuit consisting of two-state switches, the variable x_i can be complemented by simply reversing each of the switches controlled by x_i. The resulting circuit is just as simple and the cost is the same as the original one. In transistor networks, we can just permute the input wires and add NOT gates as in Figure 6.08.

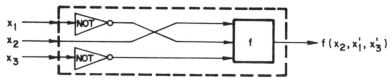

Figure 6.08. Permutation and complementation of inputs.

The eight input values of a three-variable switching circuit can be considered as the vertices of a three-dimensional cube, as shown in Figure 6.09. The six faces of this cube are defined by the equations $x_1 = 0, x_1 = 1, x_2 = 0, x_2 = 1, x_3 = 0, x_3 = 1$. The group that permutes and complements the variables takes each face to another face and takes opposite faces to opposite faces. Hence the group is the complete symmetry group, G, of the cube. There is a morphism $\psi : G \to \{1, -1\}$, which sends proper rotations to

1 and improper rotations to -1; the kernel of ψ is the group of proper rotations of the cube which, by the Morphism Theorem, must be a normal subgroup of index 2. Therefore, the order of G is $2 \cdot 24 = 48$.

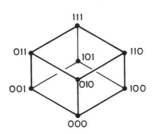

Figure 6.09.   The cube of input values.

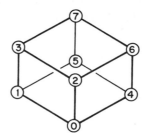

Figure 6.10.   Labeling of the cube.

6.07 EXAMPLE.   Find the number of switching circuits involving three switches that are nonequivalent under permutations and complementation of the variables.

SOLUTION.   Each Boolean function in three variables defines a coloring of the vertices of the cube of input values. A vertex is colored black if the function is 1 for the corresponding input value. It is colored white if the function takes the value 0 at that input value.

We can represent the complete symmetry group, G, of the cube by means of permutations of the vertices labeled $0, 1, 2, 3, 4, 5, 6, 7$ in Figure 6.10. Since the group of proper rotations of the cube is a normal subgroup of index 2 in G, every element of G can be written as a proper rotation π or as $\pi \circ \rho$, where ρ is the reflection of the cube in its center.

There are 2^8 different switching functions of three variables, and Table 6.7 describes the number of circuits that are fixed by the action of each element of the group G on the eight inputs. For example, consider the element $g = (01) \circ (67) \circ (34) \circ (25)$. If a switching function f is fixed under the action of g, then the images of the input values corresponding to the vertices 0 and 1 must be the same; that is, $f(0,0,0) = f(0,0,1)$. Similarly, the images of the input values corresponding to the vertices 6 and 7 are the same, and $f(1,1,0) = f(1,1,1)$. Also $f(0,1,1) = f(1,0,0)$ and $f(0,1,0) = f(1,0,1)$. Hence the values of $f(0,0,0)$, $f(1,1,0)$, $f(0,1,1)$, and $f(0,1,0)$ can be chosen arbitrarily in 2^4 ways, and $\#\mathrm{Fix}\,g = 2^4$. In general, if the function f is fixed under g, the images of the input values, corresponding to the vertices in any one cycle of g, must be the same. Hence $\#\mathrm{Fix}\,g$ is 2^r, where r is the number of disjoint cycles in the permutation representation of g.

Table 6.7. The symmetries of a cube acting on the three-variable switching functions

Type of element, g, in the symmetry group of the cube	Order of g	Number, s, of such elements	# Fix g	$s \cdot$ # Fix g
Proper rotations				
Identity	1	1	2^8	256
Rotation about a line joining midpoints of opposite edges, e.g., $(01) \circ (67) \circ (34) \circ (25)$	2	6	2^4	96
Rotation about a line joining opposite vertices, e.g., $(124) \circ (365) \circ (0) \circ (7)$	3	8	2^4	128
Rotation about a line joining centers of opposite faces, e.g., $(0264) \circ (1375)$	4	6	2^2	24
Rotation about a line joining centers of opposite faces, e.g., $(06) \circ (24) \circ (17) \circ (35)$	2	3	2^4	48
Improper rotations				
Reflection in the center, $\rho = (07) \circ (16) \circ (25) \circ (34)$	2	1	2^4	16
Reflection in a diagonal plane, e.g., $(01) \circ (67) \circ (34) \circ (25) \circ \rho$ $= (06) \circ (17) \circ (2) \circ (3) \circ (4) \circ (5)$	2	6	2^6	384
Reflection and rotation, e.g., $(124) \circ (365) \circ \rho$ $= (07) \circ (154623)$	6	8	2^2	32
Reflection and rotation, e.g., $(0264) \circ (1375) \circ \rho$ $= (0563) \circ (1472)$	4	6	2^2	24
Reflection in a central plane, e.g., $(06) \circ (24) \circ (17) \circ (35) \circ \rho$ $= (01) \circ (23) \circ (45) \circ (67)$	2	3	2^4	48
		$\# G = 48$		$\sum = 1056$

It follows from Table 6.7 and the Burnside Theorem that the number of nonequivalent circuits is $1056/\#G = 1056/48 = 22$.   □

We can reduce this number slightly more by complementing the function, as well as the variables; this corresponds to adding a NOT gate to the output. The group acting is now a combination of a cyclic group of order 2 with the complete symmetry groups of the cube.

The numbers of nonequivalent circuits for five or fewer switches given in Table 6.8 were obtained from Harrison [14].

Table 6.8.   The number of different types of switching functions

Number of Switches, n	1	2	3	4	5
Number of Boolean functions, 2^{2^n}	4	16	256	65,536	4,294,967,296
Nonequivalent functions under permutations of inputs	4	12	80	3,984	37,333,248
Nonequivalent functions under permutation and complementation of inputs	3	6	22	402	1,228,158
Nonequivalent functions under permutation and complementation of inputs and outputs	2	4	14	222	616,126

In 1951, the Harvard Computing Laboratory laboriously calculated all the nonequivalent circuits using four switches and the best way to design each of them. It was not until later that it was realized that the Pólya theory could be applied to this problem.

In many examples, it is quite difficult to calculate $\#\mathrm{Fix}\,g$ for every element g of the group G. Pólya's most important contribution to this theory of enumeration was to show how $\#\mathrm{Fix}\,g$ can be calculated, using what are called cycle index polynomials. This saves much individual calculation, and the results on nonequivalent Boolean functions in Table 6.8 can easily be calculated. However, it is still a valuable exercise to tackle a few enumeration problems without using cycle index polynomials, since this gives a better understanding of the Pólya Theory. For example, we see in Tables 6.3, 6.4, and 6.7 that $\#\mathrm{Fix}\,g$ is always of the form n^r, where r is the number of disjoint cycles in g.

Further information on the Pólya Theory can be obtained from Liu [33], Harrison [14], or Stone [12].

Exercises

1–4. Find the number of different types of circular necklaces that could be made from the following beads, assuming that all the beads are used on one necklace.

1. Three black and three white beads.

2. Four black, three white, and one red bead.

3. Seven black and five white beads.

4. Five black, six white, and three red beads.

5. How many different circular necklaces containing ten beads can be made using beads of at most two colors?

6. Five neutral members and two members from each of two warring factions are to be seated around a circular armistice table. In how many nonequivalent ways, under the action of $\mathcal{D}_9$, can they be seated if no two members of opposing factions sit next to each other?

7. How many different chemical compounds can be made by attaching H, CH_3, C_2H_5, or Cl radicals to the four bonds of a carbon atom? The radicals lie at the vertices of a regular tetrahedron, and the group is the tetrahedral group $\mathcal{A}_4$.

8. How many different chemical compounds can be made by attaching H, CH_3, or OH radicals to each of the carbon atoms in the benzene ring of Figure 6.03? (Assume that all of the C-C bonds in the ring are equivalent.)

9. How many ways can the vertices of a cube be colored using, at most, three colors?

10. How many ways can the vertices of a regular tetrahedron be colored using, at most, n colors?

11. How many different tetrahedra can be made from n types of resistors when each edge contains one resistor?

12. How many ways can the faces of a regular dodecahedron be colored using, at most, n colors?

13–16. Find the number of different colorings of the faces of the following solids.

13. A regular tetrahedron with two white faces and two black faces.

14. A cube with two white, one black, and three red faces.

15. A regular icosahedron with four black faces and 16 white faces.

16. A regular dodecahedron with five black faces, two white faces, and five green faces.

17. How many ways can the faces of a cube be colored with six different colors, if all the faces are to be a different color?

18. (i) Find the number of binary relations, on a set with four elements, that are not equivalent under permutations of the four elements.

(ii) Find the number of equivalence relations, on a set with four elements, that are not equivalent under permutations of the four elements.

19. How many different patchwork quilts, four patches long and three patches wide, can be made from five red and seven blue squares, assuming that the quilts cannot be turned over?

20. If the quilts in Exercise 19 could be turned over, how many different patterns are possible?

21. Find the number of ways of distributing three blue balls, two red balls, and four green balls into three piles.

22. If the cyclic group $\mathcal{C}_n$, generated by g, operates on a set S, show that the number of orbits is

$$\frac{1}{n} \sum_{d|n} \# \mathrm{Fix}\, g^{n/d} \cdot \phi(d)$$

where the Euler ϕ-function, $\phi(d)$, is the number of integers from 1 to d that are relatively prime to d.

23. Some transistor switching devices are sealed in a can with three input sockets at the vertices of an equilateral triangle. The three input wires are connected to a plug that will fit into the input sockets as shown in Figure 6.11. How many different cans are needed to produce any Boolean function of three input variables?

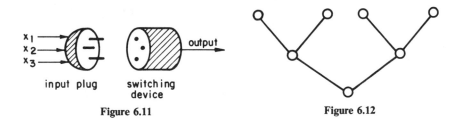

Figure 6.11	Figure 6.12

24. How many different ways can the elements of the poset in Figure 6.12 be colored using, at most, n colors?

25. Verify that the number of nonequivalent switching functions of four variables, under permutation of the inputs, is 3984.

Monoids and Machines | 7

For many purposes, a group is too restrictive an algebraic concept, and we need a more general object. In the theory of machines, or automata theory, and in the mathematical study of languages and programming, algebraic objects arise naturally that have a single binary operation that is associative and has an identity. These are called monoids. The instructions to a digital machine consist of a sequence of input symbols that is fed into the machine. Two such sequences can be combined by following one by the other, and, since this operation is associative, these input sequences form a monoid; the identity is the empty sequence that leaves the machine alone. Even though inverses do not necessarily exist in monoids, many of the general notions from group theory can be applied to these objects; for example, we can define subobjects, morphisms, and quotient objects.

MONOIDS AND SEMIGROUPS

7.01 DEFINITION.   A *monoid* $(M, \star)$ consists of a set M together with a binary operation $\star$ on M such that

(i) $a \star (b \star c) = (a \star b) \star c$ for all $a, b, c \in M$.   (Associativity)

(ii) There exists an *identity* $e \in M$ such that $a \star e = e \star a = a$ for all $a \in M$.

All groups are monoids. However, more general objects such as $(\mathbf{N}, +)$ and $(\mathbf{N}, \cdot)$, which do not have inverses, are also monoids.

A monoid $(M, \star)$ is called *commutative* if the operation $\star$ is commutative. The algebraic objects $(\mathbf{N}, +)$, $(\mathbf{N}, \cdot)$, $(\mathbf{Z}, +)$, $(\mathbf{Z}, \cdot)$, $(\mathbf{Q}, +)$, $(\mathbf{Q}, \cdot)$, $(\mathbf{R}, +)$, $(\mathbf{R}, \cdot)$, $(\mathbf{C}, +)$, $(\mathbf{C}, \cdot)$, $(\mathbf{Z}_n, +)$, and $(\mathbf{Z}_n, \cdot)$ are all commutative monoids.

However, $(\mathbf{Z}, -)$ is *not* a monoid because subtraction is not associative. In general, $(a - b) - c \neq a - (b - c)$.

Sometimes an algebraic object would be a monoid but for the fact that it lacks an identity element; such an object is called a semigroup. Hence a *semigroup* $(S, \star)$ is just a set S together with an associative binary operation, $\star$. For example, $(\mathbf{P}, +)$ is a semigroup, but not a monoid, because the set of positive integers, $\mathbf{P}$, does not contain the zero element.

Just as one of the basic examples of a group consists of the permutations of any set, a basic example of a monoid is the set of transformations of any set. A transformation is just a function (not necessarily a bijection) from a set to itself. In fact, the analogue of Cayley's Theorem holds for monoids, and it can be shown that every monoid can be represented as a transformation monoid.

7.02 PROPOSITION.   Let X be any set and let $X^X = \{f : X \to X\}$ be the set of all functions from X to itself. Then $(X^X, \circ)$ is a monoid, called the *monoid of transformations* of X.

PROOF. If $f, g \in X^X$, then the composition $f \circ g \in X^X$. Composition of functions is always associative, because, if $f, g, h \in X^X$, then

$$f \circ (g \circ h)(x) = f(g(h(x))) \quad \text{and} \quad (f \circ g) \circ h(x) = f(g(h(x)))$$

for all $x \in X$. The identity function $1_X : X \to X$ defined by $1_X(x) = x$ is the identity for composition. Hence $(X^X, \circ)$ is a monoid. $\square$

7.03 EXAMPLE. If $X = \{0, 1\}$, write out the table for the transformation monoid $(X^X, \circ)$.

SOLUTION. X^X has four elements, e, f, g, h, defined as follows.

$$e(0) = 0 \qquad f(0) = 0 \qquad g(0) = 1 \qquad h(0) = 1$$

$$e(1) = 1 \qquad f(1) = 0 \qquad g(1) = 0 \qquad h(1) = 1$$

The table for $(X^X, \circ)$ is shown in Table 7.1. For example, $g \circ f(0) = g(f(0))$

Table 7.1.   The transformation monoid of $\{0,1\}$

$\circ$	e	f	g	h
e	e	f	g	h
f	f	f	f	f
g	g	h	e	f
h	h	h	h	h

$= g(0) = 1$, and $g \circ f(1) = g(f(1)) = g(0) = 1$. Therefore, $g \circ f = h$. The other compositions can be calculated in a similar manner. $\square$

7.04 EXAMPLE.   Prove that $(\mathbf{Z}, \star)$ is a commutative monoid, where $x \star y = 6 - 2x - 2y + xy$ for $x, y \in \mathbf{Z}$.

SOLUTION.   For any $x, y \in \mathbf{Z}$, $x \star y \in \mathbf{Z}$, and $x \star y = y \star x$, so that $\star$ is a commutative binary operation on $\mathbf{Z}$. Now

$$x \star (y \star z) = x \star (6 - 2y - 2z + yz) = 6 - 2x + (-2 + x)(6 - 2y - 2z + yz)$$

$$= -6 + 4x + 4y + 4z - 2xy - 2xz - 2yz + xyz.$$

Also

$$(x \star y) \star z = (6 - 2x - 2y + xy) \star z = 6 + (-2 + z)(6 - 2x - 2y + xy) - 2z$$

$$= -6 + 4x + 4y + 4z - 2xy - 2xz - 2yz + xyz$$

$$= x \star (y \star z).$$

Hence $\star$ is associative.

Suppose that $e \star x = x$. Then $6 - 2e - 2x + ex = x$, and $6 - 2e - 3x + ex = 0$. This implies $(x - 2)(e - 3) = 0$. Hence $e \star x = x$ for all $x \in \mathbf{Z}$ if and only if $e = 3$. Therefore, $(\mathbf{Z}, \star)$ is a commutative monoid with 3 as the identity. $\square$

Since the operation in a monoid, $(M, \star)$, is associative, we can omit the parentheses when writing down a string of symbols combined by $\star$. We write the element $x_1 \star (x_2 \star x_3) = (x_1 \star x_2) \star x_3$ simply as $x_1 \star x_2 \star x_3$.

7.05 DEFINITION.   In any monoid $(M, \star)$ with identity e, the powers of any element $a \in M$ are defined by

$$a^0 = e, a^1 = a, a^2 = a \star a, \ldots, a^n = a \star a^{n-1} \quad \text{for} \quad n \in \mathbf{N}.$$

The monoid $(M, \star)$ is said to be *generated* by the subset A if every element of M can be written as a finite combination of the powers of elements of A. That is, each element $m \in M$ can be written as

$$m = a_1^{r_1} \star a_2^{r_2} \star \cdots \star a_n^{r_n} \qquad \text{for some } a_1, a_2, \ldots, a_n \in A.$$

For example, the monoid $(\mathbf{P}, \cdot)$ is generated by all the prime numbers. The monoid $(\mathbf{N}, +)$ is generated by the element 1, since each element can be written as the sum of n copies of 1, where $n \in \mathbf{N}$. A monoid generated by one element is called a *cyclic monoid*.

A finite cyclic *group* is also a cyclic monoid. However, the infinite cyclic group $(\mathbf{Z}, +)$ is not a cyclic monoid; it needs at least two elements to generate it, for example, 1 and -1. Not all finite cyclic monoids are groups. For example, the points in Figure 7.01 correspond to the elements of a cyclic monoid, and the arrows correspond to multiplication by the element c.

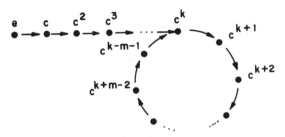

Figure 7.01.   A finite cyclic monoid.

A computer receives its information from an input terminal that feeds in a sequence of symbols, usually binary digits consisting of 0s and 1s. If one sequence is fed in after another, the computer receives one long sequence that is the *concatenation* (or juxtaposition) of the two sequences. These input sequences together with the binary operation of concatenation form a monoid that is called the free monoid generated by the input symbols.

Let A be any set (sometimes called the alphabet), and let A^n be the set of n-tuples of elements in A. In this chapter, we write an n-tuple as a string of elements of A without any symbols between them. The elements of A^n are called *words of length n* in A. A word of length 0 is an empty string; this empty word is denoted by Λ. For example, if $A = \{a, b\}$, then *baabbaba* $\in A^8$, $A^0 = \{\Lambda\}$ and,

$$A^3 = \{aaa, aab, aba, abb, baa, bab, bba, bbb\}.$$

7.06 DEFINITION.   Let

$$\mathrm{FM}(A) = A^0 \cup A \cup A^2 \cup A^3 \cup \ldots = \bigcup_{n=0}^{\infty} A^n.$$

Then $(FM(A), \star)$ is called the *free monoid generated by* A, where the operation $\star$ is concatenation, and the identity is the empty word Λ.

Another common notation for $FM(A)$ is A^*.

If we do not include the empty word, Λ, we obtain the *free semigroup* generated by A; this is often denoted by A^+.

If α and β are words of length m and n, then $\alpha \star \beta$ is the word of length $m + n$ obtained by placing α to the left of β.

If A consists of a single element, a, then the monoid $FM(A) = \{\Lambda, a, aa, aaa, aaaa, \dots\}$ and, for example, $aaa \star aa = aaaaa$. This free monoid, generated by one element, is commutative.

If $A = \{0, 1\}$, then $FM(A)$ consists of all the finite sequences of 0s and 1s,

$$FM(A) = \{\Lambda, 0, 1, 00, 01, 10, 11, 000, 001, \dots\}.$$

We have $010 \star 1110 = 0101110$ and $1110 \star 010 = 1110010$, so $FM(A)$ is not commutative.

If $A = \{a, b, c, d, \dots, y, z, \square, \cdot\}$, the letters of the alphabet together with a space, $\square$, and a period, then

$$\text{the}\square\text{sky} \in FM(A) \quad \text{and} \quad \text{the}\square\text{sky} \star \text{is}\square\text{b} \star \text{lue.} = \text{the}\square\text{sky}\square\text{is}\square\text{blue.}$$

Of course, any nonsense string of letters is also in $FM(A)$; for example, $pqb.a\square..\square xxu \in FM(A)$.

There is an important theorem that characterizes free monoids in terms of monoid morphisms.

7.07 DEFINITION. If $(M, \star)$ and $(N, \cdot)$ are two monoids, with identities e_M and e_N, respectively, then the function $f: M \to N$ is a *monoid morphism* from $(M, \star)$ to $(N, \cdot)$ if

$$\text{(i)} \quad f(x \star y) = f(x) \cdot f(y) \qquad \text{for all } x, y \in M$$
and $\qquad$ (ii) $\quad f(e_M) = e_N$.

A *monoid isomorphism* is simply a bijective monoid morphism.

For example, $f: (\mathbf{N}, +) \to (\mathbf{P}, \cdot)$ defined by $f(n) = 2^n$ is a monoid morphism because

$$f(n + m) = 2^{n+m} = 2^n \cdot 2^m = f(n) \cdot f(m) \qquad \text{for all } m, n \in \mathbf{N}.$$

However, $f: \mathbf{N} \to \mathbf{N}$ defined by $f(x) = x^2$ is *not* a monoid morphism from $(\mathbf{N}, +)$ to $(\mathbf{N}, +)$. We have $f(x + y) = (x + y)^2$, whereas $f(x) + f(y) = x^2 + y^2$. Hence $f(1 + 1) = 4$, whereas $f(1) + f(1) = 2$.

7.08 THEOREM. Let $(FM(A), \star)$ be the free monoid generated by A and let $i: A \to FM(A)$ be the function that maps each element of A into the corresponding word of length 1, so that $i(a) = a$.

Then if $l:A \rightarrow M$ is any *function* into the underlying set of any monoid $(M, \cdot)$, there is a unique monoid morphism $h:(\mathrm{FM}(A), \star) \rightarrow (M, \cdot)$ such that $h \circ i = l$, which is illustrated in Figure 7.02.

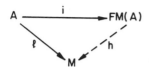

Figure 7.02.   The function l factors through the free monoid FM(A).

PROOF.   If h satisfies $h \circ i = l$, then h must be defined on words of length 1 by $h(a) = l(a)$. Once a morphism has been defined on its generators, it is completely determined as follows. Let α be a word of length $n \geqslant 2$ in FM(A). Write α as $\beta \star c$ where β is of length $n-1$ and c is of length 1. Then $h(\alpha) = h(\beta \star c) = h(\beta) \circ h(c) = h(\beta) \circ l(c)$. Hence h can be determined by using induction on the word length. In fact, if $\alpha = a_1 a_2 \ldots a_n$, where $a_i \in A$, then $h(\alpha) = l(a_1) \cdot l(a_2) \cdots l(a_n)$. Finally, let $h(\Lambda)$ be the identity of M. $\square$

FINITE-STATE MACHINES

We now look at mathematical models of sequential machines. These are machines that accept a finite set of inputs in sequential order. At any one time, the machine can be in one of a finite set of internal configurations or states. There may be a finite set of outputs. These outputs and internal states depend not only on the previous input but also on the stored information in the machine, that is, on the previous state of the machine. A pushbutton elevator is an example of such a machine. A digital computer is a very complex finite state machine. It can be broken down into its component parts, each of which is also a machine. The RS and JK flip-flops, discussed in Exercise 65 and 66 of Chapter 2, are examples of two widely used components.

For simplicity, we only consider machines with a finite set of inputs and a finite set of states. We do not mention any outputs explicitly, because the state set can be enlarged, if necessary, to include any outputs. The states can be arranged so that a particular state always gives rise to a certain output.

7.09 DEFINITION.   A *finite-state machine*, (S, I, m) consists of a set of *states* $S = \{s_1, s_2, \ldots, s_n\}$, a set of *input values* $I = \{i_1, i_2, \ldots, i_t\}$, and a *transi-*

tion function

$$m : I \times S \to S,$$

which describes how each input value changes the states. If the machine is in state s_p and an input i_q is applied, the machine will change to state $m(i_q, s_p)$.

For example, consider a pushbutton elevator that travels between two levels, 1 and 2, and stops at the lower level 1 when not in use. We take the time for the elevator to travel from one level to the other to be the basic time interval, and the controlling machine can change states at the end of each interval. We allow the machine three inputs, so that $I = \{0, 1, 2\}$.

$$\text{Input} = \begin{cases} 0 \text{ if no button is pressed in the previous time interval} \\ 1 \text{ if button 1 is pressed in the previous time interval} \\ 2 \text{ if button 2 or both buttons are pressed} \\ \quad \text{in the previous time interval.} \end{cases}$$

Since the elevator is to stop at the bottom when not in use, we only consider states that end with the elevator going down. Let the set of states be

$$S = \{\text{Stop, Down, Up-down, Down-up-down}\}.$$

For example, in the "Up-down" state, the elevator is traveling up, but must remember to come down. If no button is pressed or just button 1 is pressed while it is going up, the machine will revert to the "Down" state when the elevator reaches level 2. On the other hand, if someone arrives at level 1 and presses button 2, the machine will change to the "Down-up-down" state when the elevator reaches level 2.

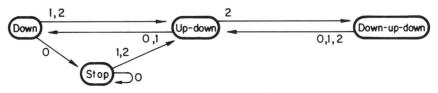

Figure 7.03.   The state diagram of the elevator.

The machine can be pictured by the *state diagram* in Figure 7.03. If the input i causes the machine to change from state s_p to state s_q, we draw an arrow labelled i from s_p to s_q in the diagram.

As another example, consider the following machine that checks the parity of the number of ones fed into it. The set of states is $S = \{\text{Start, Even, Odd}\}$, and the set of input values is $I = \{0, 1\}$. The function

$m : I \times S \rightarrow S$ is described by Table 7.2, and the state diagram is given in Figure 7.04.

Table 7.2.   The transition function of the parity checker

	Next state	
Initial	Input	
state	0	1
Start	Even	Odd
Even	Even	Odd
Odd	Odd	Even

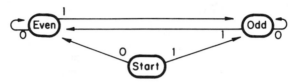

Figure 7.04.   The state diagram of the parity checker.

If any sequence of 0s and 1s is fed into this machine, it will be in the even state if there is an even number of 1s in the sequence, and in an odd state otherwise.

Let I be the set of input values for any finite state machine with state set S and function $m : I \times S \rightarrow S$. Each input value defines a function from the set of states to itself, the image of any state being the subsequent state produced by the given input. Hence we have a function

$$\tilde{m} : I \rightarrow S^S$$

where S^S is the set of functions from S to itself, and $\tilde{m}(i) : S \rightarrow S$ is defined by $[\tilde{m}(i)](s) = m(i, s)$.

Any set of input values can be fed into the machine in sequence. The set of all such input sequences is the underlying set of the free monoid of input values, FM(I). By Theorem 7.08, the function $\tilde{m} : I \rightarrow S^S$ can be extended to a monoid morphism

$$h : (\mathrm{FM}(I), \star) \rightarrow (S^S, \circ)$$

where $h(i_1 i_2 \ldots i_r) = \tilde{m}(i_1) \circ \tilde{m}(i_2) \circ \cdots \circ \tilde{m}(i_r)$. Note that the input value i_r is fed into the machine first, and we can visualize this feeding of the input sequence in Figure 7.05. (The reader should be aware that many authors

Figure 7.05. An input sequence being fed into a machine.

use the opposite convention in which the left input is fed into the machine first.)

For example, in the machine that checks the parity of the number of 1s in a sequence, the state set is $S = \{Start, Even, Odd\}$ with functions

$$\tilde{m} : \{0, 1\} \to S^S \quad \text{and} \quad h : FM(\{0, 1\}) \to S^S.$$

The morphism h is defined by

$$h(\text{sequence}) = \begin{cases} \tilde{m}(0), \text{ if the sequence contains an even number of 1s} \\ \tilde{m}(1), \text{ if the sequence contains an odd number of 1s} \\ \text{identity function on } S, \text{ if the sequence is empty.} \end{cases}$$

QUOTIENT MONOIDS AND THE MONOID OF A MACHINE

We have seen that different input sequences may have the same effect on a machine. For example, in the machine that checks the parity of the number of 1s in a sequence,

$$h(0101101) = h(0000) = h(11) = h(0);$$

thus the sequences 0101101, 0000, 11, and 0 cannot be distinguished by the machine.

In any machine with n states, the input sequences can have at most $\#(S^S) = n^n$ different effects. Since there are an infinite number of sequences in FM(I), there must always be many different input sequences that have the same effect.

The effect that an input has on a finite-state machine defines an equivalence relation on the input monoid FM(I). The monoid of a machine will be the quotient monoid of FM(I) by this relation. It will always be a finite monoid with, at most, n^n elements.

7.10 DEFINITION.   The equivalence relation R on the set M is called a *congruence relation* on the monoid $(M, \star)$ if aRb implies that

$(a \star c)R(b \star c)$ and $(c \star a)R(c \star b)$ for all $c \in M$. The *congruence class* containing the element $a \in M$ is the set

$$[a] = \{ x \in M \,|\, xRa \}.$$

7.11 PROPOSITION. If R is a congruence relation on the monoid $(M, \star)$, the quotient set $M/R = \{ [a] \| a \in M \}$ is a monoid under the operation defined by

$$[a] \star [b] = [a \star b].$$

This monoid is called the *quotient monoid* of M by R.

PROOF. We first have to verify that the operation is well defined on congruence classes. Suppose $[a] = [a']$ and $[b] = [b']$ so that aRa' and bRb'. Then $(a \star b)R(a \star b')$ and $(a \star b')R(a' \star b')$. Since R is transitive, $(a \star b)R(a' \star b')$ and $[a \star b] = [a' \star b']$. This shows that $\star$ is well defined on M/R. The associativity of $\star$ in M/R follows from the associativity of $\star$ in M. If e is the identity of M, then $[e]$ is the identity of M/R. Hence $(M/R, \star)$ is a monoid. $\square$

Let (S, I, m) be a finite-state machine and let the effect of an input sequence be given by

$$h : \mathrm{FM}(I) \to S^S.$$

Define the relation R on $\mathrm{FM}(I)$ by

$$\alpha R\beta \quad \text{if and only if} \quad h(\alpha) = h(\beta).$$

This is easily verified to be an equivalence relation. Furthermore, it is a congruence relation on the free monoid $(\mathrm{FM}(I), \star)$, because, if $\alpha R\beta$, then $h(\alpha) = h(\beta)$, and $h(\alpha \star \gamma) = h(\alpha) \circ h(\gamma) = h(\beta) \circ h(\gamma) = h(\beta \star \gamma)$; thus $(\alpha \star \gamma)R(\beta \star \gamma)$ and similarly, $(\gamma \star \alpha)R(\gamma \star \beta)$.

7.12 DEFINITION. The quotient monoid $(\mathrm{FM}(I)/R, \star)$ is called the *monoid of the machine* (S, I, m).

We can apply the same construction to the free semigroup of input sequences to obtain the *semigroup of the machine*.

The monoid of a machine reflects the capability of the machine to respond to the input sequences. There are an infinite number of sequences in $\mathrm{FM}(I)$, whereas the number of elements in the quotient monoid is less than or equal to n^n. Two sequences are in the same congruence class if and only if they have the same effect on the machine.

A Morphism Theorem for Monoids can be proven in a similar way to the Morphism Theorem for Groups 4.29 (see Exercise 24). Applying this to the monoid morphism $h: FM(I) \to S^S$, it follows that the quotient monoid $FM(I)/R$ is isomorphic to $\text{Im}\, h$. This isomorphism assigns to each congruence class a unique transition between states.

7.13 EXAMPLE.   Draw the state diagram and find the monoid of the following machine (S, I, m). The machine has two states, s_0 and s_1, and two input symbols, 0 and 1. The effects of the input symbols are given by the functions $h(0), h(1): S \to S$, defined in Table 7.3.

Table 7.3

Initial state	Next state	
	$h(0)$	$h(1)$
s_0	s_0	s_1
s_1	s_0	s_0

Table 7.4

Initial state	End state			
	$h(00)$	$h(01)$	$h(10)$	$h(11)$
s_0	s_0	s_0	s_1	s_0
s_1	s_0	s_0	s_1	s_1

SOLUTION.   Let us calculate the effect of inputs of length 2. We have $h(ij) = h(i) \circ h(j)$, where j is fed into the machine first. It follows from Tables 7.3 and 7.4 that $h(00) = h(01) = h(0)$ and $[00] = [01] = [0]$ in the monoid of the machine. There are only four functions from $\{s_0, s_1\}$ to $\{s_0, s_1\}$, and these are $h(0)$, $h(1)$, $h(10)$, and $h(11)$. Hence the monoid of the machine consists of the four congruence classes $[0]$, $[1]$, $[10]$, and $[11]$. The table of this quotient monoid is given in Table 7.5, and the state diagram is given in Figure 7.06.

Table 7.5. The monoid of the machine

$\star$	$[0]$	$[1]$	$[10]$	$[11]$
$[0]$	$[0]$	$[0]$	$[0]$	$[0]$
$[1]$	$[10]$	$[11]$	$[0]$	$[1]$
$[10]$	$[10]$	$[10]$	$[10]$	$[10]$
$[11]$	$[0]$	$[1]$	$[10]$	$[11]$

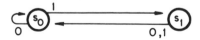

Figure 7.06.   The state diagram.

For example, $[1] \star [10] = [110]$. Since $h(110)(s_0) = s_0$ and $h(110)(s_1) = s_0$, it follows that $[110] = [0]$.

Notice that [11] is the identity; thus, in the monoid of the machine, $[\Lambda] = [11]$. ☐

7.14 EXAMPLE. Describe the monoid of the machine ({Start, Even, Odd}, {0, 1}, m) that determines the parity of the number of 1s in the input.

SOLUTION. We have already seen that any input sequence with an even number of 1s has the same effect as 0 and that any sequence with an odd number of 1s has the same effect as 1. It follows from Table 7.6 that the monoid of the machine contains the three elements [Λ], [0], and [1]. The table for this monoid is given in Table 7.7. ☐

Table 7.6

Initial state	Next state		
	$h(\Lambda)$	$h(0)$	$h(1)$
Start	Start	Even	Odd
Even	Even	Even	Odd
Odd	Odd	Odd	Even

Table 7.7. The monoid of the parity checker machine

★	[Λ]	[0]	[1]
[Λ]	[Λ]	[0]	[1]
[0]	[0]	[0]	[1]
[1]	[1]	[1]	[0]

Finite-state machines can easily be designed to recognize certain types of input sequences. For example, most numbers inside a computer are in binary form and have a check digit attached to them so that there is always an even number of 1s in each sequence. This is used to detect any machine errors (see Chapter 14). A finite-state machine like Example 7.14 can be used to perform a parity check on all the sequences of numbers in the computer. The machine can be designed to signal a parity check error whenever it ends in the "Odd" state.

Let us now look at a machine that will recognize the pattern 010 in any binary input sequence that is fed into the machine. Figure 7.07 is the state diagram of such a machine. If the machine is initialized in state s_1, it will be in state s_4 if and only if the preceeding inputs were 010, and, in this case, the machine sends an output signal.

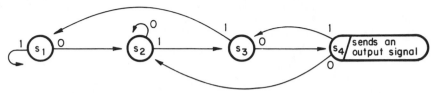

Figure 7.07.   The state diagram of a machine that recognizes the sequence 010.

This machine has four states; thus the total possible number of different functions between states is $4^4 = 256$. Table 7.8 shows that the input sequences of length 0, 1, and 2 all have different effects on the various states. However, seven of the eight sequences of length 3 have the same effect as sequences of length 2. The only input sequence with a different effect is 010, the sequence that the machine is designed to recognize. Therefore, the only sequences of length 4 that we check are those whose initial inputs are 010, namely, 0010 and 1010.

Table 7.8.   The effects of the input sequences on the states of the machine

Initial state	Λ	0	1	00	01	10	11	000	001	010	011	100	101	110
						End state for various input sequences								
s_1	s_1	s_2	s_1	s_2	s_2	s_3	s_1	s_2	s_2	s_4	s_2	s_3	s_3	s_1
s_2	s_2	s_2	s_3	s_2	s_4	s_3	s_1	s_2	s_2	s_4	s_2	s_3	s_3	s_1
s_3	s_3	s_4	s_1	s_2	s_2	s_3	s_1	s_2	s_2	s_4	s_2	s_3	s_3	s_1
s_4	s_4	s_2	s_3	s_2	s_4	s_3	s_1	s_2	s_2	s_4	s_2	s_3	s_3	s_1

Initial state	111	0010	1010
		End state	
s_1	s_1	s_2	s_3
s_2	s_1	s_2	s_3
s_3	s_1	s_2	s_3
s_4	s_1	s_2	s_3

We can use the *tree diagram* in Figure 7.08 to check that we have covered all the possible transition functions obtainable by any input sequence. We label the nodes of the tree by input sequences. At any node α, there will be two upward branches ending in the nodes $0 \star \alpha$ and $1 \star \alpha$, corresponding to the two input symbols. We prune the tree at node α, if α gives rise to the same transition function as another node β in the tree. The tree must eventually stop growing because there are only a finite number of transition functions. Every input sequence has the same effect as one of

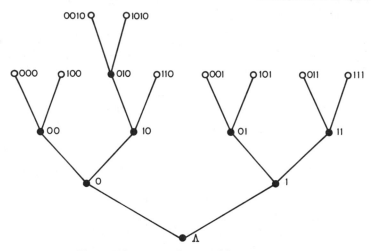

Figure 7.08.   A tree diagram of input sequences.

the solid black nodes in Figure 7.08. These nodes provide a complete set of representatives for the monoid of the machine.

Therefore, the monoid of the machine that recognizes the sequence 010 contains only eight elements, namely, [Λ], [0], [1], [00], [01], [10], [11], and [010], out of a possible 256 transition functions between states. Its table is given in Table 7.9.

Table 7.9.   The monoid of the machine that recognizes 010

★	[Λ]	[0]	[1]	[00]	[01]	[10]	[11]	[010]
[Λ]	[Λ]	[0]	[1]	[00]	[01]	[10]	[11]	[010]
[0]	[0]	[00]	[01]	[00]	[00]	[010]	[00]	[00]
[1]	[1]	[10]	[11]	[10]	[10]	[11]	[11]	[10]
[00]	[00]	[00]	[00]	[00]	[00]	[00]	[00]	[00]
[01]	[01]	[010]	[00]	[010]	[010]	[00]	[00]	[010]
[10]	[10]	[10]	[10]	[10]	[10]	[10]	[10]	[10]
[11]	[11]	[11]	[11]	[11]	[11]	[11]	[11]	[11]
[010]	[010]	[010]	[010]	[010]	[010]	[010]	[010]	[010]

For further reading on the mathematical structure of finite-state machines see Stone [12], Hartmanis and Stearns [34], or Kohavi [35].

Exercises

1–13.   Are the following structures semigroups or monoids or neither? Give the identity of each monoid.

1. $(\mathbf{N}, \text{GCD})$.
2. $(\mathbf{Z}, \lceil)$ where $a \lceil b = a$.
3. $(\mathbf{R}, \star)$ where $x \star y = \sqrt{x^2 + y^2}$.
4. $(\mathbf{R}, \star)$ where $x \star y = \sqrt[3]{x^3 + y^3}$.
5. $(\mathbf{Z}_3, -)$.
6. $(\mathbf{R}, |\ |)$ where $|\ |$ is the absolute value.
7. $(\mathbf{Z}, \max)$ where $\max(m, n)$ is the larger of m and n.
8. $(\mathbf{Z}, \star)$ where $x \star y = x + y + xy$.
9. (S, GCD) where $S = \{1, 2, 3, 4, 5, 6\}$.
10. $(X, \max)$ where X is the set of real valued functions on the unit interval $[0, 1]$ and if $f, g \in X$, then $\max(f, g)$ is the function on X defined by

$$\max(f, g)(x) = \max(f(x), g(x)).$$

11. (T, LCM) where $T = \{1, 2, 4, 5, 10, 20\}$.
12. The set of all relations on a set X, where the composition of two relations R and S is the relation RS defined by $xRSz$ if and only if, for some $y \in X$, xRy and ySz.
13. $(\{a, b, c\}, \star)$ where the table for $\star$ is given in Table 7.10.

Table 7.10

$\star$	a	b	c
a	a	b	c
b	b	a	a
c	c	a	a

14–17. Write out the tables for the following monoids and semigroups.

14. (S, GCD) where $S = \{1, 2, 3, 4, 6, 8, 12, 24\}$.
15. (T, GCD) where $T = \{1, 2, 3, 4\}$.
16. $(X^X, \circ)$ where $X = \{1, 2, 3\}$.
17. $(\{e, c, c^2, c^3, c^4\}, \cdot)$ where multiplication by c is indicated by an arrow in Figure 7.09.

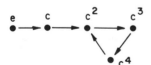

Figure 7.09

18. Find all the commutative monoids on the set $S = \{e, a, b\}$ with identity e.

19. Are all the elements of the free semigroup generated by $\{0,1,2,3,4,5,6,7,8,9\}$ simply the nonnegative integers written in the base 10?

20. A *submonoid* of a monoid $(M, \cdot)$ is a subset N of M containing the identity and such that $x \cdot y \in N$, for all $x, y \in N$. Find all the submonoids of the monoid given in Exercise 17.

21. Prove that there is a monoid isomorphism between $(\mathrm{FM}(\{a\}), \star)$ and $(\mathbf{N}, +)$.

22. *(Representation Theorem for Monoids)* Prove that any monoid $(M, \star)$ is isomorphic to a submonoid of $(M^M, \circ)$. This gives a representation of any monoid as a monoid of transformations.

23. Prove that any cyclic monoid is either isomorphic to $(\mathbf{N}, +)$ or is isomorphic to a monoid of the form shown in Figure 7.01, for some values of k and m.

24. *(Morphism Theorem for Monoids)* Let $f: (M, \star) \rightarrow (N, \cdot)$ be a morphism of monoids. Let R be the relation on M defined by $m_1 R m_2$ if and only if $f(m_1) = f(m_2)$. Prove that the quotient monoid $(M/R, \star)$ is isomorphic to the submonoid $(\mathrm{Im} f, \cdot)$ of $(N, \cdot)$.

25. An *automorphism* of a monoid M is an isomorphism from M to itself. Prove that the set of all automorphisms of a monoid M forms a group under composition.

26. A machine has three states, s_1, s_2, and s_3 and two input symbols, α and β. The effect of the input symbols on the states is given by Table 7.11.

Table 7.11

Initial state	Next state	
	$h(\alpha)$	$h(\beta)$
s_1	s_1	s_1
s_2	s_3	s_1
s_3	s_2	s_1

Draw the state diagram and find the monoid of this machine.

27. Prove that every finite monoid is the monoid of some finite state machine.

28–30.   Draw state diagrams of machines with the given input set, I, that will recognize the given sequence.

28.   1101 where $I = \{0, 1\}$.   **29.**   0101 where $I = \{0, 1\}$.

30.   2131 where $I = \{1, 2, 3\}$.

31–34. Which of the following relations are congruence relations on the monoid (**N**, +)? *Find the quotient monoid when the relation is a congruence relation.*

31. aRb if $a - b$ is even.   **32.** aRb if $a > b$.

33. aRb if $a = 2^r b$ for some $r \in \mathbf{Z}$.   **34.** aRb if $10 | a - b$.

35–37. The machines in Tables 7.12, 7.13, and 7.14 have state set $S = \{s_1, s_2, s_3\}$ and input set $I = \{0, 1\}$.

Table 7.12

Initial state	Next state	
	$h(0)$	$h(1)$
s_1	s_2	s_1
s_2	s_1	s_2
s_3	s_3	s_2

Table 7.13

Initial state	Next state	
	$h(0)$	$h(1)$
s_1	s_2	s_1
s_2	s_3	s_1
s_3	s_3	s_2

Table 7.14

Initial state	Next state	
	$h(0)$	$h(1)$
s_1	s_2	s_1
s_2	s_3	s_3
s_3	s_1	s_1

35. Draw the table of the monoid of the machine defined by Table 7.12.
36. Draw the table of the monoid of the machine defined by Table 7.13.
37. Find the number of elements in the monoid of the machine defined by Table 7.14.

38. Find the number of elements in the semigroup of the machine, given by Figure 7.03, that controls the elevator.

39. Find the monoid of the machine in Figure 7.10.

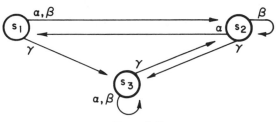

Figure 7.10

40. A *serial adder*, illustrated in Figure 7.11, is a machine that adds two numbers in binary form. The two numbers are fed in together, one digit at a time, starting from the right end. Their sum appears as the output. The machine has input symbols 00, 01, 10, and 11, corresponding to the right-most digits of the numbers. Figure 7.12 gives the state diagram of such a machine, where the symbol "s_{ij}/j" indicates that the machine is in state s_{ij} and emits an output j. The carry digit is the number i of the state s_{ij}. Find the monoid of this machine.

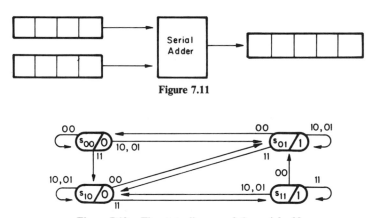

Figure 7.11

Figure 7.12. The state diagram of the serial adder.

41–44. The circuits in Figure 7.13 represent the internal structures of some finite-state machines constructed from transistor circuits. These circuits are controlled by a clock, and the rectangular boxes denote delays of one time unit.

The input symbols are 0 and 1 and are fed in at unit time intervals. The internal states of the machines are described by the contents of the delays. Draw the state diagram and find the elements in the semigroup of each machine.

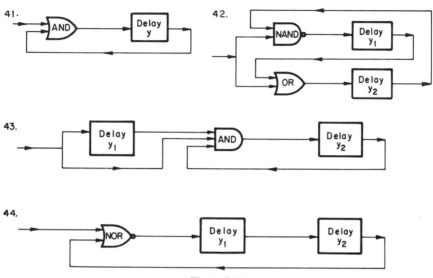

Figure 7.13

45. In the spring, a plant bud has to have the right conditions in order to develop. One particular bud has to have a rainy day followed by two warm days, without being interrupted by cool or freezing days, in order to develop. Furthermore, if a freezing day occurs after the bud has developed, the bud dies. Draw a state diagram for such a bud using the input symbols *R*, *W*, *C*, *F* to stand for rainy, warm, cool, and freezing days, respectively. What is the number of elements in the resulting monoid of this bud?

46. A dog can either be passive, angry, frightened, or angry and frightened, in which case he bites. If you give him a bone, he becomes passive. If you remove one of his bones, he becomes angry, and, if he is already frightened, he will bite you. If you threaten him, he becomes frightened, but, if he is already angry, he will bite. Write out the table of the monoid of the dog.

Rings and Fields $\boxed{8}$

The familiar number systems of the real or complex numbers contain two basic binary operations, namely, addition and multiplication. Group theory is not sufficient to capture all of the algebraic structure of these number systems, because a group only deals with one binary operation. It is possible to consider the integers as a group $(\mathbf{Z}, +)$ and the nonzero integers as a monoid $(\mathbf{Z}^*, \cdot)$, but this still neglects the relation between addition and multiplication, namely, the fact that multiplication is distributive over addition. We therefore consider algebraic structures with two binary operations modeled after these number systems. A ring is a structure that has the minimal properties we would expect of addition and multiplication. A field is a more specialized ring in which division by nonzero elements is always possible.

In this chapter, we look at the basic properties of rings and fields and consider many examples. In later chapters we construct new number systems with properties similar to the familiar systems.

RINGS

8.01 DEFINITION. A *ring* $(R, +, \cdot)$ is a set R, together with two binary operations $+$ and $\cdot$ on R satisfying the following axioms. For any elements $a, b, c \in R$,

(i)	$(a + b) + c = a + (b + c)$.	(Associativity of addition)
(ii)	$a + b = b + a$.	(Commutativity of addition)
(iii)	there exists $0 \in R$, called the *zero*, such that $a + 0 = a$.	(Existence of an additive identity)
(iv)	there exists $(-a) \in R$ such that $a + (-a) = 0$.	(Existence of an additive inverse)
(v)	$(a \cdot b) \cdot c = a \cdot (b \cdot c)$.	(Associativity of multiplication)
(vi)	there exists $1 \in R$ such that $1 \cdot a = a \cdot 1 = a$.	(Existence of multiplicative identity)
(vii)	$a \cdot (b + c) = a \cdot b + a \cdot c$ and $(b + c) \cdot a = b \cdot a + c \cdot a$.	(Distributivity)

Axioms (i)–(iv) are equivalent to saying that $(R, +)$ is an Abelian group, and axioms (v) and (vi) are equivalent to saying that $(R, \cdot)$ is a monoid.

The ring $(R, +, \cdot)$ is called a *commutative ring* if

(viii) $a \cdot b = b \cdot a$ for all $a, b \in R$.   (Commutativity of multiplication)

The integers under addition and multiplication satisfy all of the above axioms so that $(\mathbf{Z}, +, \cdot)$ is a commutative ring. Also, $(\mathbf{Q}, +, \cdot)$, $(\mathbf{R}, +, \cdot)$, and $(\mathbf{C}, +, \cdot)$ are all commutative rings. If there is no confusion about the operations, we write only R for the ring $(R, +, \cdot)$. Therefore, the above rings would be referred to as $\mathbf{Z}$, $\mathbf{Q}$, $\mathbf{R}$, or $\mathbf{C}$. Moreover, if we refer to a ring R without explicitly defining its operations, it can be assumed that they are addition and multiplication.

Many authors do not require a ring to have a multiplicative identity, and most of the results we prove can be verified to hold for these objects as well. Exercise 49 at the end of this chapter shows that such an object can always be embedded in a ring that does have a multiplicative identity.

The set of all $n \times n$ square matrices with real coefficients forms a ring $(\mathfrak{M}(n \times n; \mathbf{R}), +, \cdot)$, which is *not commutative* if $n > 1$, because matrix multiplication is not commutative.

The elements "even" and "odd" form a commutative ring ($\{$even, odd$\}, +, \cdot$) where the operations are given by Table 8.1. "Even" is the zero of this ring, and "odd" is the multiplicative identity. This is really a special case of the following example when $n = 2$.

Table 8.1. The ring of odd and even integers

+	even	odd		·	even	odd
even	even	odd		even	even	even
odd	odd	even		odd	even	odd

8.02 EXAMPLE. Show that $(\mathbf{Z}_n, +, \cdot)$ is a commutative ring, where addition and multiplication on congruence classes, modulo n, are defined by $[x]+[y]=[x+y]$ and $[x]\cdot[y]=[xy]$.

SOLUTION. It follows, from Example 4.22, that $(\mathbf{Z}_n, +)$ is an Abelian group.

Since multiplication on congruence classes is defined in terms of representatives, it must be verified that it is well defined. Suppose that $[x]=[x']$ and $[y]=[y']$, so that $x \equiv x'$ and $y \equiv y' \bmod n$. This implies that $x = x' + kn$ and $y = y' + ln$ for some $k, l \in \mathbf{Z}$. Now $x \cdot y = (x' + kn) \cdot (y' + ln) = x' \cdot y' + (ky' + lx' + kln)n$, so $x \cdot y \equiv x' \cdot y' \bmod n$ and $[x \cdot y] = [x' \cdot y']$. This shows that multiplication is well defined.

The remaining axioms now follow from the definitions of addition and multiplication and from the properties of the integers. The zero is $[0]$, and the unit is $[1]$. The left distributive law is true, for example, because

$$[x]\cdot([y]+[z]) = [x]\cdot[y+z] = [x\cdot(y+z)]$$

$$= [x\cdot y + x\cdot z], \quad \text{by distributivity in } \mathbf{Z}$$

$$= [x\cdot y] + [x\cdot z] = [x]\cdot[y] + [x]\cdot[z]. \quad \square$$

8.03 EXAMPLE. Construct the addition and multiplication tables for the ring $(\mathbf{Z}_5, +, \cdot)$.

SOLUTION. We denote the congruence class $[x]$ just by x. The tables are given in Table 8.2. $\square$

Table 8.2. The ring $(\mathbf{Z}_5, +, \cdot)$

+	0	1	2	3	4
0	0	1	2	3	4
1	1	2	3	4	0
2	2	3	4	0	1
3	3	4	0	1	2
4	4	0	1	2	3

$\cdot$	0	1	2	3	4
0	0	0	0	0	0
1	0	1	2	3	4
2	0	2	4	1	3
3	0	3	1	4	2
4	0	4	3	2	1

8.04 EXAMPLE. Show that $(\mathbf{Q}(\sqrt{2}), +, \cdot)$ is a commutative ring where $\mathbf{Q}(\sqrt{2}) = \{a + b\sqrt{2} \in \mathbf{R} | a, b \in \mathbf{Q}\}$.

SOLUTION. The set $\mathbf{Q}(\sqrt{2})$ is a subset of $\mathbf{R}$, and the addition and multiplication is the same as that of real numbers. First of all, we check that $+$ and $\cdot$ are binary operations on $\mathbf{Q}(\sqrt{2}\,)$. If $a,b,c,d \in \mathbf{Q}$, we have

$$(a+b\sqrt{2}\,)+(c+d\sqrt{2}\,)=(a+c)+(b+d)\sqrt{2} \in \mathbf{Q}(\sqrt{2}\,)$$

since $(a+c)$ and $(b+d) \in \mathbf{Q}$. Also

$$(a+b\sqrt{2}\,)\cdot(c+d\sqrt{2}\,)=(ac+2bd)+(ad+bc)\sqrt{2} \in \mathbf{Q}(\sqrt{2}\,)$$

since $(ac+2bd)$ and $(ad+bc) \in \mathbf{Q}$.

We now check that axioms (i)–(viii) of a commutative ring are valid in $\mathbf{Q}(\sqrt{2}\,)$.

(i) and (ii)   Addition of real numbers is associative and commutative.

(iii)   The zero is $0=0+0\sqrt{2} \in \mathbf{Q}(\sqrt{2}\,)$.

(iv)   The additive inverse of $a+b\sqrt{2}$ is $(-a)+(-b)\sqrt{2} \in \mathbf{Q}(\sqrt{2}\,)$, since $(-a)$ and $(-b) \in \mathbf{Q}$.

(v)   Multiplication of real numbers is associative.

(vi)   The multiplicative identity is $1=1+0\sqrt{2} \in \mathbf{Q}(\sqrt{2}\,)$.

(vii)   The distributive axioms hold for real numbers and hence hold for elements of $\mathbf{Q}(\sqrt{2}\,)$.

(viii)   Multiplication of real numbers is commutative.   □

We have already investigated one algebraic system with two binary operations, namely, a Boolean algebra. The Boolean algebra of subsets of a set is not a ring under the operations of union and intersection, because neither of these operations has inverses. However, the symmetric difference does have an inverse, and we can make a Boolean algebra into a ring using this operation and the operation of intersection.

8.05 EXAMPLE. $(\mathscr{P}(X),\Delta,\cap)$ is a commutative ring for any set X.

SOLUTION.   The axioms (i)–(viii) of a commutative ring follow from Propositions 2.02 and 2.04. The zero is $\varnothing$, and the identity is X.   □

In the above ring, $A \cap A = A$ for every element A in the ring. Such rings are called *Boolean rings*, since they are all derivable from Boolean algebras. (See Exercise 13.)

8.06 EXAMPLE.   Construct the tables for the ring $(\mathscr{P}(X),\Delta,\cap)$ where $X=\{a,b,c\}$.

SOLUTION.   Let $A=\{a\}$, $B=\{b\}$, and $C=\{c\}$, so that $\overline{A}=\{b,c\}$, $\overline{B}=\{a,c\}$, and $\overline{C}=\{a,b\}$. Therefore, $\mathscr{P}(X)=\{\varnothing,A,B,C,\overline{A},\overline{B},\overline{C},X\}$. The tables for the symmetric difference and intersection are given in Table 8.3.   □

Table 8.3.   The ring $\mathscr{P}(\{a,b,c\})$

$\triangle$	$\varnothing$	A	B	C	$\bar{A}$	$\bar{B}$	$\bar{C}$	X
$\varnothing$	$\varnothing$	A	B	C	$\bar{A}$	$\bar{B}$	$\bar{C}$	X
A	A	$\varnothing$	$\bar{C}$	$\bar{B}$	X	C	B	$\bar{A}$
B	B	$\bar{C}$	$\varnothing$	$\bar{A}$	C	X	A	$\bar{B}$
C	C	$\bar{B}$	$\bar{A}$	$\varnothing$	B	A	X	$\bar{C}$
$\bar{A}$	$\bar{A}$	X	C	B	$\varnothing$	$\bar{C}$	$\bar{B}$	A
$\bar{B}$	$\bar{B}$	C	X	A	$\bar{C}$	$\varnothing$	$\bar{A}$	B
$\bar{C}$	$\bar{C}$	B	A	X	$\bar{B}$	$\bar{A}$	$\varnothing$	C
X	X	$\bar{A}$	$\bar{B}$	$\bar{C}$	A	B	C	$\varnothing$

$\cap$	$\varnothing$	A	B	C	$\bar{A}$	$\bar{B}$	$\bar{C}$	X
$\varnothing$	$\varnothing$	$\varnothing$	$\varnothing$	$\varnothing$	$\varnothing$	$\varnothing$	$\varnothing$	$\varnothing$
A	$\varnothing$	A	$\varnothing$	$\varnothing$	$\varnothing$	A	A	A
B	$\varnothing$	$\varnothing$	B	$\varnothing$	B	$\varnothing$	B	B
C	$\varnothing$	$\varnothing$	$\varnothing$	C	C	C	$\varnothing$	C
$\bar{A}$	$\varnothing$	$\varnothing$	B	C	$\bar{A}$	C	B	$\bar{A}$
$\bar{B}$	$\varnothing$	A	$\varnothing$	C	C	$\bar{B}$	A	$\bar{B}$
$\bar{C}$	$\varnothing$	A	B	$\varnothing$	B	A	$\bar{C}$	$\bar{C}$
X	$\varnothing$	A	B	C	$\bar{A}$	$\bar{B}$	$\bar{C}$	X

The following properties are useful in manipulating elements of any ring.

8.07 PROPOSITION.   If $(R, +, \cdot)$ is a ring, then for all $a, b \in R$

(i) $a \cdot 0 = 0 \cdot a = 0$.
(ii) $a \cdot (-b) = (-a) \cdot b = -(a \cdot b)$.
(iii) $(-a) \cdot (-b) = a \cdot b$.
(iv) $(-1) \cdot a = -a$.
(v) $(-1) \cdot (-1) = 1$.

PROOF.   (i) $a \cdot 0 = a \cdot (0 + 0) = a \cdot 0 + a \cdot 0$. Adding $-(a \cdot 0)$ to each side, we obtain $0 = a \cdot 0$. Similarly, $0 \cdot a = 0$.

(ii) $a \cdot (-b) + a \cdot b = a \cdot (-b + b) = a \cdot 0 = 0$, using (i).   Therefore $a \cdot (-b) = -(a \cdot b)$. Similarly, $(-a) \cdot b = -(a \cdot b)$.

(iii) $(-a) \cdot (-b) = -(a \cdot (-b)) = -(-(a \cdot b)) = a \cdot b$ by (ii) and Proposition 3.09.
(iv)   By (ii), $(-1) \cdot a = -(1 \cdot a) = -a$.
(v)   By (iii), $(-1) \cdot (-1) = 1 \cdot 1 = 1$. $\square$

8.08 PROPOSITION. If $0 = 1$, the ring contains only one element and is called the *trivial ring*. All other rings are called *nontrivial*.

PROOF. For any element, a, in a ring in which $0 = 1$, we have $a = a \cdot 1 = a \cdot 0 = 0$. Therefore, the ring only contains the element 0. It can be verified that this forms a ring with the operations defined by $0 + 0 = 0$ and $0 \cdot 0 = 0$. □

INTEGRAL DOMAINS AND FIELDS

One very useful property of the familiar number systems is the fact that if $ab = 0$, then either $a = 0$ or $b = 0$. This property allows us to cancel nonzero elements, because, if $ab = ac$ and $a \neq 0$, then $a(b - c) = 0$, and so $b = c$. However, this property does not hold for all rings. For example, in $\mathbf{Z}_4$, we have $[2] \cdot [2] = [0]$, and we cannot always cancel since $[2] \cdot [1] = [2] \cdot [3]$, but $[1] \neq [3]$.

8.09 DEFINITION. If $(R, +, \cdot)$ is a commutative ring, a nonzero element $a \in R$ is called a *zero divisor* if there exists a nonzero element $b \in R$ such that $a \cdot b = 0$.

8.10 DEFINITION. A nontrivial commutative ring is called an *integral domain* if it has no zero divisors. Hence a nontrivial commutative ring is an integral domain if $a \cdot b = 0$ always implies that $a = 0$ or $b = 0$.

As the name implies, the integers form an integral domain. Also, $\mathbf{Q}$, $\mathbf{R}$, and $\mathbf{C}$ are integral domains. However, $\mathbf{Z}_4$ is *not*, because $[2]$ is a zero divisor. Neither is $(\mathcal{P}(X), \Delta, \cap)$, because every nonempty proper subset of X is a zero divisor. $\mathfrak{M}(n \times n; \mathbf{R})$ is *not* an integral domain because, for one thing, it fails to be commutative.

8.11 PROPOSITION. If a is a nonzero element of an integral domain R and $a \cdot b = a \cdot c$, then $b = c$.

PROOF. If $a \cdot b = a \cdot c$, then $a \cdot (b - c) = a \cdot b - a \cdot c = 0$. Since R is an integral domain, it has no zero divisors, and, since $a \neq 0$, it follows that $(b - c) = 0$. Hence $b = c$. □

Generally speaking, it is possible to add, subtract, and multiply elements in a ring, but it is not always possible to divide. Even in an integral domain, where elements can be canceled, it is not always possible to divide by nonzero elements. For example, if $x, y \in \mathbf{Z}$, then $2x = 2y$ implies that

$x = y$, but not all elements in **Z** can be divided by 2.

The most useful number systems are those in which we can divide by nonzero elements. Such systems are called fields.

8.12 DEFINITION. A *field* is a ring in which the nonzero elements form an Abelian group under multiplication. In other words, a field is a nontrivial commutative ring R satisfying the following extra axiom.

(ix) For each nonzero element $a \in R$ there exists $a^{-1} \in R$ such that $a \cdot a^{-1} = 1$.

The rings **Q**, **R**, and **C** are all fields, but the integers do not form a field.

8.13 PROPOSITION. A field has no zero divisors and is an integral domain.

PROOF. Let $a \cdot b = 0$ in a field **F**. If $a \neq 0$, there exists an inverse $a^{-1} \in F$ and $b = (a^{-1} \cdot a) \cdot b = a^{-1}(a \cdot b) = a^{-1} \cdot 0 = 0$. Hence either $a = 0$ or $b = 0$, and **F** is an integral domain. $\square$

8.14 THEOREM. A finite integral domain is a field.

PROOF. Let $D = \{x_0, x_1, x_2, \ldots, x_n\}$ be a finite integral domain with x_0 as zero and x_1 as one. We have to show that every nonzero element of D has a multiplicative inverse.

If x_i is nonzero, we show that the set $x_i D = \{x_i x_0, x_i x_1, x_i x_2, \ldots, x_i x_n\}$ is the same as the set D. If $x_i x_j = x_i x_k$, then, by the cancellation property, $x_j = x_k$. Hence all the elements $x_i x_0, x_i x_1, x_i x_2, \ldots, x_i x_n$ are distinct, and $x_i D$ is a subset of D with the same number of elements. Therefore, $x_i D = D$.

There is some element, x_j, such that $x_i x_j = x_1 = 1$. Hence $x_j = x_i^{-1}$, and D is a field. $\square$

We show in the next chapter that $\mathbf{Z}_n$ is a finite integral domain, and hence a field, if and only if n is prime. Of course, as the integers demonstrate, an *infinite* integral domain is not necessarily a field.

8.15 EXAMPLE. Is $(\mathbf{Q}(\sqrt{2}), +, \cdot)$ an integral domain or a field?

SOLUTION. From Example 8.04, we know that $\mathbf{Q}(\sqrt{2})$ is a commutative ring. If $a + b\sqrt{2}$ is a nonzero element, so that at least one of a and b is not zero, its inverse in the real numbers is

$$\frac{1}{a + b\sqrt{2}} = \frac{a - b\sqrt{2}}{(a + b\sqrt{2})(a - b\sqrt{2})} = \frac{a}{a^2 - 2b^2} - \frac{b\sqrt{2}}{a^2 - 2b^2}.$$

However, this inverse is an element of $\mathbf{Q}(\sqrt{2})$, and so $\mathbf{Q}(\sqrt{2})$ is a field (and an integral domain). □

SUBRINGS AND MORPHISMS OF RINGS

8.16 DEFINITION. If $(R, +, \cdot)$ is a ring, a nonempty subset S of R is called a *subring* of R if, for all $a, b \in S$,

(i) $a + b \in S$,
(ii) $-a \in S$,
(iii) $a \cdot b \in S$,
(iv) $1 \in S$.

Conditions (i) and (ii) imply that $(S, +)$ is a subgroup of $(R, +)$, and can be replaced by the condition $a - b \in S$.

8.17 PROPOSITION. If S is a subring of $(R, +, \cdot)$, then $(S, +, \cdot)$ is a ring.

PROOF. Conditions (i) and (iii) of the above definition guarantee that S is closed under addition and multiplication. Condition (iv) shows that $1 \in S$. It follows from Proposition 3.11 that $(S, +)$ is a group. $(S, +, \cdot)$ satisfies the remaining axioms for a ring because they hold in $(R, +, \cdot)$. □

For example, $\mathbf{Z}$, $\mathbf{Q}$, and $\mathbf{R}$ are all subrings of $\mathbf{C}$. Let D be the set of $n \times n$ real diagonal matrices. Then D is a subring of the ring of all $n \times n$ real matrices, $\mathfrak{M}(n \times n; \mathbf{R})$, because the sum, difference, and product of two diagonal matrices is another diagonal matrix. Note that D is commutative even though $\mathfrak{M}(n \times n; \mathbf{R})$ is not.

8.18 EXAMPLE. Show that $\mathbf{Q}(\sqrt{2}) = \{a + b\sqrt{2} \mid a, b \in \mathbf{Q}\}$ is a subring of $\mathbf{R}$.

SOLUTION. Let $a + b\sqrt{2}, c + d\sqrt{2} \in \mathbf{Q}(\sqrt{2})$.
Then

(i) $(a + b\sqrt{2}) + (c + d\sqrt{2}) = (a + c) + (b + d)\sqrt{2} \in \mathbf{Q}(\sqrt{2})$.
(ii) $-(a + b\sqrt{2}) = (-a) + (-b)\sqrt{2} \in \mathbf{Q}(\sqrt{2})$.
(iii) $(a + b\sqrt{2}) \cdot (c + d\sqrt{2}) = (ac + 2bd) + (ad + bc)\sqrt{2} \in \mathbf{Q}(\sqrt{2})$.
(iv) $1 = 1 + 0\sqrt{2} \in \mathbf{Q}(\sqrt{2})$. □

A morphism between two rings is a function between their underlying sets that preserves the two operations of addition and multiplication and also the element 1. Many authors use the term 'homomorphism' instead of 'morphism'.

8.19 DEFINITION. Let $(R, +, \cdot)$ and $(S, +, \cdot)$ be two rings. The function $f: R \to S$ is called a *ring morphism* if, for all $a, b \in R$,

$$\text{(i)} \quad f(a+b) = f(a) + f(b),$$
$$\text{(ii)} \quad f(a \cdot b) = f(a) \cdot f(b),$$

and $\quad$ (iii) $\quad f(1) = 1$.

If the operations in the two rings are denoted by different symbols, for example, if the rings are $(R, +, \cdot)$ and $(S, \oplus, \bigcirc)$, then the conditions for $f: R \to S$ to be a ring morphism are

$$\text{(i)} \quad f(a+b) = f(a) \oplus f(b)$$
$$\text{(ii)} \quad f(a \cdot b) = f(a) \bigcirc f(b)$$

and $\quad$ (iii) $\quad f(1_R) = 1_S$ where 1_R and 1_S are the respective identities.

A *ring isomorphism* is a bijective ring morphism. If there is an isomorphism between the rings R and S, we say R and S are *isomorphic rings* and write $R \cong S$.

A ring morphism, f, from $(R, +, \cdot)$ to $(S, +, \cdot)$ is, in particular, a group morphism from $(R, +)$ to $(S, +)$. Therefore, by Proposition 3.24, $f(0) = 0$ and $f(-a) = -f(a)$ for all $a \in R$.

The inclusion function, $i: S \to R$, of any subring S into a ring R is always a ring morphism. The function $f: \mathbf{Z} \to \mathbf{Z}_n$, defined by $f(x) = [x]$, which maps an integer to its equivalence class modulo n, is a ring morphism from $(\mathbf{Z}, +, \cdot)$ to $(\mathbf{Z}_n, +, \cdot)$.

8.20 EXAMPLE. If X is a one element set, show that $f: \mathcal{P}(X) \to \mathbf{Z}_2$ is a ring isomorphism between $(\mathcal{P}(X), \triangle, \cap)$ and $(\mathbf{Z}_2, +, \cdot)$, where $f(\emptyset) = [0]$ and $f(X) = [1]$.

SOLUTION. We can check that f is a morphism by testing all the possibilities for $f(A \triangle B)$ and $f(A \cap B)$. Since the rings are commutative, they are

$$f(\emptyset \triangle \emptyset) = f(\emptyset) = [0] = f(\emptyset) + f(\emptyset)$$

$$f(\emptyset \triangle X) = f(X) = [1] = f(\emptyset) + f(X)$$

$$f(X \triangle X) = f(\emptyset) = [0] = f(X) + f(X)$$

$$f(\varnothing \cap \varnothing) = f(\varnothing) = [0] = f(\varnothing) \cdot f(\varnothing)$$

$$f(\varnothing \cap X) = f(\varnothing) = [0] = f(\varnothing) \cdot f(X)$$

$$f(X \cap X) = f(X) = [1] = f(X) \cdot f(X).$$

Both rings contain only two elements, and f is a bijection; therefore, f is an isomorphism. $\square$

If $f: R \to S$ is an isomorphism between two finite rings, the addition and multiplication tables of S will be the same as those of R if we replace each $a \in R$ by $f(a) \in S$. For example, Tables 8.4 and 8.5 illustrate the isomorphism of Example 8.20.

Table 8.4. The ring $\mathscr{P}(X)$ when X is a point

$\triangle$	$\varnothing$	X		$\cap$	$\varnothing$	X
$\varnothing$	$\varnothing$	X		$\varnothing$	$\varnothing$	$\varnothing$
X	X	$\varnothing$		X	$\varnothing$	X

Table 8.5. The ring $\mathbf{Z}_2$

$+$	$[0]$	$[1]$		$\cdot$	$[0]$	$[1]$
$[0]$	$[0]$	$[1]$		$[0]$	$[0]$	$[0]$
$[1]$	$[1]$	$[0]$		$[1]$	$[0]$	$[1]$

The following ring isomorphism between linear transformations and matrices is the crux of much of linear algebra.

8.21 EXAMPLE. The linear transformations from $\mathbf{R}^n$ to itself form a ring, $(\mathscr{L}(\mathbf{R}^n, \mathbf{R}^n), +, \circ)$, under addition and composition. Show that the function

$$f: \mathscr{L}(\mathbf{R}^n, \mathbf{R}^n) \to \mathscr{M}(n \times n; \mathbf{R})$$

is a ring morphism, where f assigns to each linear transformation its $n \times n$ coefficient matrix, with respect to the standard basis of $\mathbf{R}^n$.

SOLUTION. If α is a linear transformation from $\mathbf{R}^n$ to itself, then

$$\alpha \begin{pmatrix} x_1 \\ \vdots \\ x_n \end{pmatrix} = \begin{pmatrix} a_{11}x_1 + \cdots + a_{1n}x_n \\ \vdots \qquad \vdots \\ a_{n1}x_1 + \cdots + a_{nn}x_n \end{pmatrix} \quad \text{and} \quad f(\alpha) = \begin{pmatrix} a_{11} \dots a_{1n} \\ \vdots \qquad \vdots \\ a_{n1} \dots a_{nn} \end{pmatrix}.$$

Matrix addition and multiplication is defined so that $f(\alpha + \beta) = f(\alpha) + f(\beta)$ and $f(\alpha \circ \beta) = f(\alpha) \cdot f(\beta)$. Also, if ι is the identity linear transformation then $f(\iota)$ is the identity matrix.

Any matrix defines a linear transformation, so that f is surjective. Furthermore, f is injective, because any matrix can arise from only one linear transformation, since the jth column of the matrix must be the image of the jth basis vector. Hence f is an isomorphism. $\quad \square$

8.22 EXAMPLE.   Show that $f: \mathbf{Z}_{24} \rightarrow \mathbf{Z}_4$, defined by $f([x]_{24}) = [x]_4$ is a ring morphism.

PROOF.   Since the function is defined in terms of representatives of equivalence classes, we first check that it is well defined. If $[x]_{24} = [y]_{24}$, then $x \equiv y \bmod 24$ and $24 | (x - y)$. Hence $4 | (x - y)$ and $[x]_4 = [y]_4$, which shows that f is well defined.

We now check the conditions for f to be a ring morphism.

(i) $f([x]_{24} + [y]_{24}) = f([x + y]_{24}) = [x + y]_4 = [x]_4 + [y]_4$.

(ii) $f([x]_{24} \cdot [y]_{24}) = f([xy]_{24}) = [xy]_4 = [x]_4 \cdot [y]_4$.

(iii) $f([1]_{24}) = [1]_4$. $\square$

NEW RINGS FROM OLD

This section introduces various methods for constructing new rings from given rings. These include the direct product of rings, matrix rings, polynomial rings, rings of sequences, and rings of formal power series. Perhaps the most important class of rings constructible from given rings is the class of quotient rings. Their construction is analogous to that of quotient groups and is discussed in Chapter 10.

8.23 DEFINITION.   If $(R, +, \cdot)$ and $(S, +, \cdot)$ are two rings, their *product* is the ring $(R \times S, +, \cdot)$ whose underlying set is the Cartesian product of R and S and whose operations are defined by

$$(r_1, s_1) + (r_2, s_2) = (r_1 + r_2, s_1 + s_2) \quad \text{and} \quad (r_1, s_1) \cdot (r_2, s_2) = (r_1 \cdot r_2, s_1 \cdot s_2).$$

It is readily verified that these operations do indeed define a ring structure on $R \times S$ whose zero is $(0_R, 0_S)$, where 0_R and 0_S are the zeros of R and S, and whose multiplicative identity is $(1_R, 1_S)$, where 1_R and 1_S are the identities in R and S.

The product construction can be iterated any number of times. For example, $(\mathbf{R}^n, +, \cdot)$ is a commutative ring, where $\mathbf{R}^n$ is the n-fold product of $\mathbf{R}$ with itself.

8.24 EXAMPLE. Write down the addition and multiplication tables for $\mathbf{Z}_2 \times \mathbf{Z}_3$.

SOLUTION. Let $\mathbf{Z}_2 = \{0, 1\}$ and $\mathbf{Z}_3 = \{0, 1, 2\}$. Then $\mathbf{Z}_2 \times \mathbf{Z}_3 = \{(0,0), (0,1), (0,2), (1,0), (1,1), (1,2)\}$. The addition and multiplication tables are given in Table 8.6. In calculating these, it must be remembered that addition and multiplication are performed modulo 2 in the first coordinate and modulo 3 in the second coordinate. $\square$

Table 8.6. The ring $\mathbf{Z}_2 \times \mathbf{Z}_3$

+	(0,0)	(0,1)	(0,2)	(1,0)	(1,1)	(1,2)
(0,0)	(0,0)	(0,1)	(0,2)	(1,0)	(1,1)	(1,2)
(0,1)	(0,1)	(0,2)	(0,0)	(1,1)	(1,2)	(1,0)
(0,2)	(0,2)	(0,0)	(0,1)	(1,2)	(1,0)	(1,1)
(1,0)	(1,0)	(1,1)	(1,2)	(0,0)	(0,1)	(0,2)
(1,1)	(1,1)	(1,2)	(1,0)	(0,1)	(0,2)	(0,0)
(1,2)	(1,2)	(1,0)	(1,1)	(0,2)	(0,0)	(0,1)

·	(0,0)	(0,1)	(0,2)	(1,0)	(1,1)	(1,2)
(0,0)	(0,0)	(0,0)	(0,0)	(0,0)	(0,0)	(0,0)
(0,1)	(0,0)	(0,1)	(0,2)	(0,0)	(0,1)	(0,2)
(0,2)	(0,0)	(0,2)	(0,1)	(0,0)	(0,2)	(0,1)
(1,0)	(0,0)	(0,0)	(0,0)	(1,0)	(1,0)	(1,0)
(1,1)	(0,0)	(0,1)	(0,2)	(1,0)	(1,1)	(1,2)
(1,2)	(0,0)	(0,2)	(0,1)	(1,0)	(1,2)	(1,1)

We know that $\mathbf{Z}_2 \times \mathbf{Z}_3$ and $\mathbf{Z}_6$ are isomorphic as *groups*; we now show that they are isomorphic as *rings*.

8.25 THEOREM. $\mathbf{Z}_m \times \mathbf{Z}_n$ is isomorphic as a ring to $\mathbf{Z}_{mn}$ if and only if $\text{GCD}(m, n) = 1$.

PROOF. If the $\text{GCD}(m, n) = 1$, it follows from Theorems 4.36 and 3.25 that the function

$$f : \mathbf{Z}_{mn} \to \mathbf{Z}_m \times \mathbf{Z}_n$$

defined by $f([x]_{mn}) = ([x]_m, [x]_n)$ is a group isomorphism. However, this function preserves multiplication because

$$f([x]_{mn} \cdot [y]_{mn}) = f([xy]_{mn}) = ([xy]_m, [xy]_n)$$
$$= ([x]_m, [x]_n) \cdot ([y]_m, [y]_n) = f([x]_{mn}) \cdot f([y]_{mn}).$$

Also $f([1]_{mn}) = ([1]_m, [1]_n)$; thus f is a ring isomorphism.

It was shown in the discussion following Corollary 4.37 that, if $\text{GCD}(m,n) \neq 1$, $\mathbf{Z}_m \times \mathbf{Z}_n$ and $\mathbf{Z}_{mn}$ are not isomorphic as groups, and hence they cannot be isomorphic as rings. $\square$

We can extend this result by induction to show the following.

8.26 THEOREM. Let $m = m_1 \cdot m_2 \cdots m_r$ where $\text{GCD}(m_i, m_j) = 1$ if $i \neq j$. Then $\mathbf{Z}_{m_1} \times \mathbf{Z}_{m_2} \times \cdots \times \mathbf{Z}_{m_r}$ is a ring isomorphic to $\mathbf{Z}_m$. $\square$

8.27 COROLLARY. Let $n = p_1^{\alpha_1} p_2^{\alpha_2} \cdots p_r^{\alpha_r}$ be a decomposition of the integer n into powers of distinct primes. Then $\mathbf{Z}_n \cong \mathbf{Z}_{p_1^{\alpha_1}} \times \mathbf{Z}_{p_2^{\alpha_2}} \times \cdots \times \mathbf{Z}_{p_r^{\alpha_r}}$ as rings. $\square$

8.28 DEFINITION.   If R is a commutative ring, we can construct the *ring of $n \times n$ matrices with entries from R*, $(\mathfrak{M}(n \times n; R), +, \cdot)$. Addition and multiplication are performed as in real matrices.

For example, $(\mathfrak{M}(n \times n; \mathbf{Z}_2), +, \cdot)$ is the ring of $n \times n$ matrices with 0 and 1 entries. Addition and multiplication is performed modulo 2. This is a noncommutative ring with $2^{(n^2)}$ elements.

If R is a commutative ring, a *polynomial*, $p(x)$, in the indeterminate x over the ring R is an expression of the form

$$p(x) = a_0 + a_1 x + a_2 x^2 + \cdots + a_n x^n$$

where $a_0, a_1, a_2, \ldots, a_n \in R$ and $n \in \mathbf{N}$. The element a_i is called the *coefficient* of x^i in $p(x)$. If the coefficient of x^i is zero, the term $0x^i$ may be omitted, and if the coefficient of x^i is one, $1x^i$ may be written simply as x^i. If n is the largest integer for which $a_n \neq 0$, we say that $p(x)$ has *degree n* and write $\deg p(x) = n$. If all the coefficients of $p(x)$ are zero, then $p(x)$ is called the *zero polynomial*, and its degree is not defined.

For example, $4x^2 - \sqrt{3}$ is a polynomial over $\mathbf{R}$ of degree 2, $ix^4 - (2+i)x^3 + 3x$ is a polynomial over $\mathbf{C}$ of degree 4, and $x^7 + x^5 + x^4 + 1$ is a polynomial over $\mathbf{Z}_2$ of degree 7. The number 5 is a polynomial over $\mathbf{Z}$ of degree 0; the zero polynomial and the polynomials of degree 0 are called *constant polynomials* because they contain no x terms.

8.29 DEFINITION.   The set of all polynomials in x with coefficients from the commutative ring R is denoted by $R[x]$. That is,

$$R[x] = \{a_0 + a_1 x + a_2 x^2 + \cdots + a_n x^n | a_i \in R, n \in \mathbf{N}\}.$$

This forms a ring $(R[x], +, \cdot)$ called the *polynomial ring with coefficients from R* when addition and multiplication of the polynomials

$$p(x) = \sum_{i=0}^{n} a_i x^i \quad \text{and} \quad q(x) = \sum_{i=0}^{m} b_i x^i$$

are defined by

$$p(x) + q(x) = \sum_{i=0}^{\max(m,n)} (a_i + b_i) x^i$$

and

$$p(x) \cdot q(x) = \sum_{k=0}^{m+n} c_k x^k \quad \text{where} \quad c_k = \sum_{i+j=k} a_i b_j.$$

With a little effort, it can be verified that $(R[x], +, \cdot)$ satisfies all the axioms for a commutative ring. The zero is the zero polynomial, and the multiplicative identity is the constant polynomial 1.

For example, in $Z_5[x]$, the polynomial ring with coefficients in the integers modulo 5, we have

$$(2x^3 + 2x^2 + 1) + (3x^2 + 4x + 1) = 2x^3 + 4x + 2$$

and

$$(2x^3 + 2x^2 + 1) \cdot (3x^2 + 4x + 1) = x^5 + 4x^4 + 4x + 1.$$

When working in $Z_n[x]$, the coefficients, but *not* the exponents, are reduced modulo n.

8.30 PROPOSITION.   If R is an integral domain and $p(x)$ and $q(x)$ are nonzero polynomials in $R[x]$, then

$$\deg(p(x) \cdot q(x)) = \deg p(x) + \deg q(x).$$

PROOF.   Let $\deg p(x) = n$, $\deg q(x) = m$ and let $p(x) = a_0 + \cdots + a_n x^n$, $q(x) = b_0 + \cdots + b_m x^m$ where $a_n \neq 0$, $b_m \neq 0$. Then the coefficient of the highest power of x in $p(x) \cdot q(x)$ is $a_n b_m$, which is nonzero, since R has no zero divisors. Hence $\deg(p(x) \cdot q(x)) = m + n$. $\square$

If the coefficient ring is not an integral domain, the degree of a product may be less than the sum of the degrees. For example, $(2x^3 + x) \cdot (3x) = 3x^2$ in $\mathbf{Z}_6[x]$.

8.31 COROLLARY. If R is an integral domain, then so is $R[x]$.

PROOF. If $p(x)$ and $q(x)$ are nonzero elements of $R[x]$, then $p(x) \cdot q(x)$ is also nonzero by Proposition 8.30. Hence $R[x]$ has no zero divisors. $\square$

The construction of a polynomial ring can be iterated to obtain the ring of polynomials in n variables $x_1, \ldots, x_n$, with coefficients from R. We define inductively $R[x_1, \ldots, x_n] = R[x_1, \ldots, x_{n-1}][x_n]$. For example,

$$R[x,y] = R[x][y] = \{f_0 + f_1 y + f_2 y^2 + \cdots + f_n y^n \mid f_i \in R[x]\}$$
$$= \{(a_{00} + a_{10}x + a_{20}x^2 + a_{30}x^3 + \cdots + a_{r0}x^r) + (a_{01} + a_{11}x + \cdots + a_{s1}x^s)y$$
$$+ (a_{02} + \cdots + a_{t2}x^t)y^2 + \cdots + (a_{0n} + \cdots + a_{mn}x^m)y^n \mid a_{ij} \in R\}$$
$$= \{a_{00} + a_{10}x + a_{01}y + a_{20}x^2 + a_{11}xy + a_{02}y^2 + \cdots + a_{mn}x^m y^n \mid a_{ij} \in R\}.$$

Clearly, we can prove by induction from Corollary 8.31 that $R[x_1, \ldots, x_n]$ is an integral domain if R is an integral domain.

8.32 PROPOSITION. Let R be a commutative ring and denote the infinite sequence of elements of R, $\langle a_0, a_1, a_2, \ldots \rangle$, by $\langle a_i \rangle$. Define addition, $+$, and convolution, $*$, of two such sequences by

$$\langle a_i \rangle + \langle b_i \rangle = \langle a_i + b_i \rangle$$

and

$$\langle a_i \rangle * \langle b_i \rangle = \left\langle \sum_{j+k=i} a_j b_k \right\rangle = \langle a_0 b_i + a_1 b_{i-1} + \cdots + a_i b_0 \rangle.$$

The set of all such sequences forms a commutative ring $(R^{\mathbf{N}}, +, *)$ called the *ring of sequences* in R. If R is an integral domain, so is $R^{\mathbf{N}}$.

PROOF. Addition is clearly associative and commutative. The zero element is the zero sequence $\langle 0 \rangle = \langle 0, 0, 0, \ldots \rangle$, and the negative of $\langle a_i \rangle$ is $\langle -a_i \rangle$.

Now

$$(\langle a_i\rangle * \langle b_i\rangle) * \langle c_i\rangle = \left\langle \sum_{j+k=i} a_j b_k \right\rangle * \langle c_i\rangle$$

$$= \left\langle \sum_{l+m=i} \left(\sum_{j+k=m} a_j b_k \right) c_l \right\rangle = \left\langle \sum_{j+k+l=i} a_j b_k c_l \right\rangle.$$

Similarly, bracketing the sequences in the other way, we obtain the same result, which shows that convolution is associative.

Convolution is clearly commutative and the distributive laws hold because

$$\langle a_i\rangle * (\langle b_i\rangle + \langle c_i\rangle) = \left\langle \sum_{j+k=i} a_j (b_k + c_k) \right\rangle$$

$$= \left\langle \sum_{j+k=i} a_j b_k \right\rangle + \left\langle \sum_{j+k=i} a_j c_k \right\rangle = \langle a_i\rangle * \langle b_i\rangle + \langle a_i\rangle * \langle c_i\rangle.$$

The identity in the ring of sequences is $\langle 1,0,0,\dots\rangle$ because

$$\langle 1,0,0,\dots\rangle * \langle a_0,a_1,a_2,\dots\rangle = \langle 1a_0, 1a_1+0a_0, 1a_2+0a_1+0a_0,\dots\rangle$$

$$= \langle a_0,a_1,a_2,\dots\rangle.$$

Therefore, $(R^{\mathbf{N}}, +, *)$ is a commutative ring.

Suppose a_q and b_r are the first nonzero elements in the nonzero sequences $\langle a_i\rangle$ and $\langle b_i\rangle$, respectively. Then the element in the $(q+r)$th position of their convolution is

$$\sum_{j+k=q+r} a_j b_k = a_0 b_{q+r} + a_1 b_{q+r-1} + \cdots + a_q b_r + a_{q+1} b_{r-1} + \cdots + a_{q+r} b$$

$$= 0 \quad + \quad 0 \quad + \cdots + a_q b_r + \quad 0 \quad + \cdots + \quad 0 \quad = a_q b_r.$$

Hence, if R is an integral domain, this element is not zero and the ring of sequences has no zero divisors. $\square$

The ring of sequences cannot be a field because $\langle 0,1,0,0,\dots\rangle$ has no inverse. For any sequence $\langle b_i\rangle$, $\langle 0,1,0,0,\dots\rangle * \langle b_0, b_1, b_2,\dots\rangle = \langle 0, b_0, b_1,\dots\rangle$, which can never be the identity in the ring.

8.33 DEFINITION. A *formal power series* in x with coefficients from a commutative ring R is an expression of the form.

$$a_0 + a_1 x + a_2 x^2 + \cdots = \sum_{i=0}^{\infty} a_i x^i \quad \text{where} \quad a_i \in R.$$

In contrast to a polynomial, these power series can have an infinite number of nonzero terms.

We denote the set of all such formal power series by $R[[x]]$. The term "formal" is used to indicate that questions of convergence of these series are not considered. Indeed, over many rings, such as $\mathbf{Z}_n$, convergence would not be meaningful.

Addition and multiplication are defined in $R[[x]]$ by

$$\left(\sum_{i=0}^{\infty} a_i x^i \right) + \left(\sum_{i=0}^{\infty} b_i x^i \right) = \sum_{i=0}^{\infty} (a_i + b_i) x^i$$

and

$$\left(\sum_{i=0}^{\infty} a_i x^i \right) \cdot \left(\sum_{i=0}^{\infty} b_i x^i \right) = \sum_{i=0}^{\infty} \left(\sum_{j+k=i} a_j b_k \right) x^i.$$

It can be verified that these formal power series do form a ring, $(R[[x]], +, \cdot)$, and that the polynomial ring, $R[x]$, is the subring consisting of those power series with only a finite number of nonzero terms.

The ring of sequences $(R^{\mathbf{N}}, +, *)$ is isomorphic to the ring of formal power series $(R[[x]], +, \cdot)$. The function $f: R^{\mathbf{N}} \to R[[x]]$ that is defined by $f(\langle a_0, a_1, a_2, \ldots \rangle) = a_0 + a_1 x + a_2 x^2 + \cdots$ is clearly a bijection. It follows from the definitions of addition, multiplication, and convolution in these rings, that f is a ring morphism.

FIELD OF FRACTIONS

We can always add, subtract, and multiply elements in any ring, but we cannot always divide. However, if the ring is an integral domain, it is possible to enlarge it so that division by nonzero elements is possible. In other words, we can construct a field containing the given ring as a subring. This is precisely what we did on page 85 when constructing the rational numbers from the integers.

If the original ring did have zero divisors or was noncommutative, it could not possibly be a subring of any field, because fields cannot contain zero divisors or pairs of noncommutative elements.

8.34 THEOREM. If R is an integral domain, it is possible to construct a field **F** so that

> (i) R is isomorphic to a subring, R', of **F**

and (ii) every element of **F** can be written as $p \cdot q^{-1}$ for suitable $p, q \in R'$.

F is called the *field of fractions* of R (or sometimes the *field of quotients* of R).

PROOF. Consider the set $R \times R^* = \{(a,b) | a, b \in R, b \neq 0\}$, consisting of pairs of elements of R, the second being nonzero. Define a relation $\sim$ on $R \times R^*$ by

$$(a,b) \sim (c,d) \quad \text{if and only if} \quad ad = bc \text{ in } R.$$

We verify that this is an equivalence relation.

(i) $(a,b) \sim (a,b)$, since $ab = ba$.
(ii) If $(a,b) \sim (c,d)$, then $ad = bc$. This implies $cb = da$ and $(c,d) \sim (a,b)$.
(iii) If $(a,b) \sim (c,d)$ and $(c,d) \sim (e,f)$, then $ad = bc$ and $cf = de$. This implies that

$$(af - be)d = (ad)f - b(ed) = bcf - bcf = 0.$$

Since R has no zero divisors and $d \neq 0$, it follows that $af = be$ and $(a,b) \sim (e,f)$. Hence the relation $\sim$ is reflexive, symmetric, and transitive.

Denote the equivalence class containing (a,b) by a/b and the set of equivalence classes by **F**. Define addition and multiplication in **F** by

$$\frac{a}{b} + \frac{c}{d} = \frac{ad + bc}{bd} \qquad \text{and} \qquad \frac{a}{b} \cdot \frac{c}{d} = \frac{ac}{bd}.$$

These operations on equivalence classes are defined in terms of particular representatives, so it should be checked that they are well defined. If $a/b = a'/b'$ and $c/d = c'/d'$, then $ab' = a'b$ and $cd' = c'd$. Hence

$$(ad + bc)(b'd') = (ab')dd' + bb'(cd') = (a'b)dd' + bb'(c'd)$$

$$= (a'd' + b'c')(bd)$$

and therefore, $(ad + bc)/bd = (a'd' + b'c')/b'd'$, which shows that addition is well defined. Also $acb'd' = a'c'bd$; thus $ac/bd = a'c'/b'd'$, which shows that multiplication is well defined.

It can now be verified that $(\mathbf{F}, +, \cdot)$ is a field. The zero is $0/1$, and the identity is $1/1$. For example, the distributive laws hold because

$$\frac{a}{b} \cdot \left(\frac{c}{d} + \frac{e}{f} \right) = \frac{a}{b} \cdot \frac{cf + de}{df} = \frac{a(cf + de)}{bdf} = \frac{a(cf + de)}{bdf} \cdot \frac{b}{b} = \frac{ac}{bd} + \frac{ae}{bf}$$

$$= \frac{a}{b} \cdot \frac{c}{d} + \frac{a}{b} \cdot \frac{e}{f}.$$

The inverse of any nonzero element a/b is b/a. The remaining axioms for a field are straightforward to check.

The ring R is isomorphic to the subring $R' = \{r/1 | r \in R\}$ of $\mathbf{F}$ by an isomorphism that maps r to $r/1$.

Any element a/b in the field $\mathbf{F}$ can be written as $a/b = (a/1) \cdot (1/b) = (a/1) \cdot (b/1)^{-1}$ where $a/1, \ b/1 \in R'$. $\square$

If we take R to be integers in the above construction, we obtain the rational numbers as the field of fractions.

The field of fractions of a polynomial ring $R[x]$ is called the *field of rational functions with coefficients in R*. Its elements can be considered as fractions of one polynomial over a nonzero polynomial.

Convolution Fractions

We now present an application of the field of fractions that has important implications in analysis. This example is of a different type than most of the applications in this book. It can be omitted, without loss of continuity, by those readers not interested in analysis or applied mathematics.

We construct the field of fractions of a set of continuous functions that will be used to explain two mathematical techniques that have been used successfully by engineers and physicists for many years, but that have been mistrusted by mathematicians because they did not have a firm mathematical basis. One such technique was introduced by O. Heaviside in 1893 in dealing with electrical circuits; this is called the *operational calculus*, and it enabled him to solve partial differential equations by manipulating differential operators as if they were algebraic quantities. The second such technique is the use of *impulse functions* in applied mathematics and mathematical physics. In 1926, when solving problems in relativistic quantum mechanics, P. Dirac introduced his so-called "*delta function*," $\delta(x)$, which has the property that

$$\delta(x) = 0, \text{ if } x \neq 0, \quad \text{and} \quad \int_{-\infty}^{\infty} \delta(x) \, dx = 1.$$

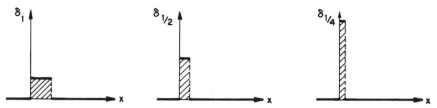

Figure 8.01. The Dirac delta "function" is the limit of δ_k as k tends to zero.

Using the usual definition of functions, no such object exists. However, it can be pictured in Figure 8.01 as the limit, as k tends to zero, of the functions $\delta_k(x)$ where

$$\delta_k(x) = \begin{cases} 1/k & \text{if } 0 \leqslant x \leqslant k. \\ 0 & \text{otherwise.} \end{cases}$$

Each function $\delta_k(x)$ vanishes outside the interval $0 \leqslant x \leqslant k$ and has the property that

$$\int_{-\infty}^{\infty} \delta_k(x)\,dx = 1.$$

Consider the set, $C[0, \infty)$, of real valued functions that are continuous in the interval $0 \leqslant x < \infty$. We define the operations of addition and convolution on this set so that the algebraic structure $(C[0, \infty), +, *)$ is *nearly* an integral domain; convolution does not have an identity; therefore, the structure fails to satisfy Axiom (vi) of a ring. However, it is still possible to embed this structure in its field of fractions. The Polish mathematician Jan Mikusinski constructed this field of fractions and called such elements "operators" or "generalized functions." The Dirac delta function is a generalized function and is in fact the identity for convolution in the field of fractions.

Define addition and convolution of two functions f and g in $C[0, \infty)$ by

$$(f+g)(x) = f(x) + g(x) \qquad \text{and} \qquad (f*g)(x) = \int_0^x f(t)g(x-t)\,dt.$$

This convolution of functions is the continuous analogue of convolution of sequences. This can be seen by writing the ith term of the sequence

$$\langle a_i \rangle * \langle b_i \rangle \text{ as } \sum_{t=0}^{i} a_t b_{i-t}.$$

It is clear that addition is associative and commutative, and the zero function is the additive identity. Also the negative of $f(x)$ is $-f(x)$.

Convolution is commutative because

$$(f*g)(x) = \int_0^x f(t)\,g(x-t)\,dt$$

$$= -\int_x^0 f(x-u)\,g(u)\,du, \qquad \text{substituting } u = x-t$$

$$= \int_0^x g(u)\,f(x-u)\,du = (g*f)(x).$$

Convolution is associative because

$$(f*(g*h))(x) = \int_0^x f(t)(g*h)(x-t)\,dt$$

$$= \int_0^x f(t)\left[\int_0^{x-t} g(u)\,h(x-t-u)\,du\right]dt$$

$$= \int_0^x f(t)\left[\int_t^x g(w-t)\,h(x-w)\,dw\right]dt, \qquad \text{putting } u = w-t$$

$$= \int\!\!\int_T f(t)\,g(w-t)\,h(x-w)\,dw\,dt, \quad \text{where } T \text{ is the triangle in Figure 8.02,}$$

$$= \int_0^x \int_0^w f(t)\,g(w-t)\,h(x-w)\,dt\,dw, \qquad \text{changing the order of integration}$$

$$= \int_0^x (f*g)(w)\,h(x-w)\,dw = ((f*g)*h)(x).$$

The distributive laws follow because

$$((f+g)*h)(x) = \int_0^x (f(t)+g(t))h(x-t)\,dt$$

$$= \int_0^x f(t)h(x-t)\,dt + \int_0^x g(t)h(x-t)\,dt$$

$$= (f*h)(x) + (g*h)(x).$$

If f is a function that is the identity under convolution, then $f*h = h$, for all functions h. If we take h to be the function defined by $h(x) = 1$, for all

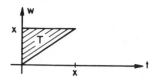

Figure 8.02

$0 \leqslant x < \infty$, then

$$(f*h)(x) = \int_0^x f(t)\,dt = 1, \qquad \text{for all } x \geqslant 0.$$

There is no function f in $C[0, \infty)$ with this property, although the Dirac delta "function" does have this property. Hence $(C[0, \infty), +, *)$ satisfies all the axioms for a commutative ring except for the existence of an identity under convolution.

Furthermore, there are no zero divisors under convolution; that is, $f*g = 0$ implies $f = 0$ or $g = 0$. This is a hard result in analysis, which is known as Titchmarsh's Theorem. Proofs can be found in Erdelyi [39], Marchand [40], or Mikusinski [41].

However, we can still construct the field of fractions of this algebraic object in exactly the same way as we did in Theorem 8.34. For example, the even integers under addition and multiplication, $(2\mathbf{Z}, +, \cdot)$ is also an algebraic object that satisfies all the axioms for an integral domain except for the fact that multiplication has no identity. The field of fractions of $2\mathbf{Z}$ is the set of rational numbers; every rational number can be written in the form $2r/2s$, where $2r, 2s \in 2\mathbf{Z}$.

The field of fractions of $(C[0, \infty), +, *)$ is called the *field of convolution fractions*, and its elements are sometimes called *generalized functions*, *distributions*, or *operators*. Elements of this field are the abstract entities f/g, where f and g are functions. There is a bijection between the set of elements of the form $f*g/g$ and the set $C[0, \infty)$. It is possible to interpret other convolution fractions as impulse functions, discontinuous functions, and even differential or integral operators. The Dirac delta function can be defined to be the identity of this field under convolution; therefore $\delta = f/f$, for any nonzero function f.

The Heaviside step function illustrated in Figure 8.03 is defined by $h(x) = 1$, if $x \geqslant 0$, and $h(x) = 0$, if $x < 0$. The function is continuous when restricted to the nonnegative numbers and, in some sense, is the integral of the Dirac delta function. Convolution by h acts as an integral operator on any continuous function because

$$(h*f)(x) = (f*h)(x) = \int_0^x f(t)h(x-t)\,dt = \int_0^x f(t)\,dt.$$

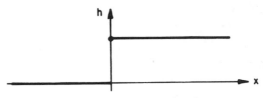

Figure 8.03.   The Heaviside step function.

Hence $h*f$ is the integral of f. We can use this to define integration of any generalized function. Take the integral of the convolution fraction f/g to be the fraction $(h*f)/g$.

Denote the inverse of the Heaviside step function by s, so that $s = h/h*h$. This element s is not a genuine function, but only a convolution fraction. Convolution by s acts, in some sense, as a differential operator in the field of convolution fractions. It is not exactly the usual differential operator, because convolution by s and by h must commute, and $s*h = h*s$ must be the identity. If $f(x)$ is a continuous function, we know from the calculus, that the derivative of $\int_0^x f(t)\,dt$ is just $f(x)$; however, $\int_0^x f'(t)\,dt$ is not just $f(x)$ but is $f(x) - f(0)$. In fact, if the function f has a derivative,

$$(s*f)(x) = f'(x) + f(0)\delta(x)$$

where $\delta(x)$ is the identity in the field of convolution fractions. Now, when we calculate $h*s*f$, which is equivalent to integrating $s*f$ from 0 to x, we obtain the function f back again.

By repeated convolution with s or h, a generalized function can be differentiated or integrated any number of times, the result being another generalized function. We can even differentiate or integrate a fractional number of times. These operations s and h can be used to explain Heaviside's operational calculus, in which differential and integral operators are manipulated like algebraic symbols.

For further information on the algebraic aspects of generalized functions and distributions see Erdelyi [39], Marchand [40], or Mikusinski [41].

Exercises

1. Write out the tables for the ring $\mathbf{Z}_4$.
2. Write out the tables for the ring $\mathbf{Z}_2 \times \mathbf{Z}_2$.

3–12.   Which of the following systems are rings under addition and multiplication? Give reasons.

3. $\{a+b\sqrt{5}\,|\,a,b\in\mathbf{Z}\}$.   4.   **N**.

5. $\{a+b\sqrt{2}+c\sqrt{3}\,|\,a,b,c\in\mathbf{Z}\}$.   6. $\{a+\sqrt[3]{2}\,b\,|\,a,b\in\mathbf{Q}\}$.

7.   All 2×2 real matrices with zero determinant.

8.   All rational numbers that can be written with denominator 2.

9.   All rational numbers that can be written with an odd denominator.

10. $(\mathbf{Z},+,\times)$ where $+$ is the usual addition and $a\times b=0$ for all $a,b\in\mathbf{Z}$.

11.   The set $A=\{a,b,c\}$ with tables given in Table 8.7.

12.   The set $A=\{a,b,c\}$ with tables given in Table 8.8.

Table 8.7

+	a	b	c
a	a	b	c
b	b	c	a
c	c	a	b

·	a	b	c
a	a	a	a
b	a	b	c
c	a	c	c

Table 8.8

+	a	b	c
a	a	b	c
b	b	c	a
c	c	a	b

·	a	b	c
a	a	a	a
b	a	c	b
c	a	b	c

13.   A ring R is called a *Boolean ring* if $a^2=a$ for all $a\in R$.

(i)   Show that $(\mathscr{P}(X),\Delta,\cap)$ is a Boolean ring for any set X.

(ii)   Show that $\mathbf{Z}_2$ and $\mathbf{Z}_2\times\mathbf{Z}_2$ are Boolean rings.

(iii)   Prove that if R is Boolean, then $2a=0$ for all $a\in R$.

(iv)   Prove that any Boolean ring is commutative.

(v)   If $(R,\wedge,\vee,')$ is any Boolean algebra, show that $(R,\Delta,\wedge)$ is a Boolean ring where $a\,\Delta\,b=(a\wedge b')\vee(a'\wedge b)$.

(vi)   If $(R,+,\cdot)$ is a Boolean ring, show that $(R,\wedge,\vee,')$ is a Boolean algebra where $a\wedge b=a\cdot b$, $a\vee b=a+b+a\cdot b$ and $a'=1+a$.

This shows that there is a one-to-one correspondence between Boolean algebras and Boolean rings.

14.   If A and B are subrings of a ring R, prove that $A\cap B$ is also a subring of R.

15.   Prove that the only subring of $\mathbf{Z}_n$ is itself.

16–20.   Which of the following sets are subrings of $\mathbf{C}$? Give reasons.

16. $\{0+ib|b\in\mathbf{R}\}$.

17. $\{a+ib|a,b\in\mathbf{Q}\}$.

18. $\{a+b\sqrt{-7}\,|a,b\in\mathbf{Z}\}$.

19. $\{z\in\mathbf{C}||z|\leqslant 1\}$.

20. $\{a+ib|a,b\in\mathbf{Z}\}$.

21–26.   Which of the following rings are integral domains and which are fields?

21. $\mathbf{Z}_2\times\mathbf{Z}_2$.

22. $(\mathcal{P}(\{a\}),\triangle,\cap)$.

23. $\{a+bi|a,b\in\mathbf{Q}\}$.

24. $\mathbf{Z}\times\mathbf{R}$.

25. $\{a+b\sqrt{2}\,|a,b\in\mathbf{Z}\}$.

26. $\mathbf{R}[x]$.

27.   Prove that the set $C(\mathbf{R})$ of continuous real valued functions defined on the real line forms a ring $(C(\mathbf{R}),+,\cdot)$, where addition and multiplication of two functions $f,g\in C(\mathbf{R})$ is given by

$$(f+g)(x)=f(x)+g(x)\qquad\text{and}\qquad(f\cdot g)(x)=f(x)\cdot g(x).$$

28–33.   Find all the zero divisors in the following rings.

28. $\mathbf{Z}_4$.

29. $\mathbf{Z}_{10}$.

30. $\mathbf{Z}_4\times\mathbf{Z}_2$.

31. $(\mathcal{P}(X),\triangle,\cap)$.

32. $\mathfrak{M}(2\times 2;\mathbf{Z}_2)$.

33. $\mathfrak{M}(n\times n;\mathbf{R})$.

34.   Let $(R,+,\cdot)$ be a ring in which $(R,+)$ is a cyclic group. Prove that $(R,+,\cdot)$ is commutative ring.

35.   Show that $S=\left\{\begin{pmatrix} a & b \\ -b & a \end{pmatrix}\Big|a,b\in\mathbf{R}\right\}$ is a subring of $\mathfrak{M}(2\times 2;\mathbf{R})$ isomorphic to $\mathbf{C}$.

36.   Show that $\mathbf{H}=\left\{\begin{pmatrix} \alpha & \beta \\ -\bar{\beta} & \bar{\alpha} \end{pmatrix}\Big|\alpha,\beta\in\mathbf{C}\right\}$ is a subring of $\mathfrak{M}(2\times 2;\mathbf{C})$, where $\bar{\alpha}$ is the conjugate of α. This is called the *ring of quaternions* and in some sense generalizes the complex numbers. However, it is not a field because it fails to be commutative.

37.   Find all the ring morphisms from $\mathbf{Z}$ to $\mathbf{Z}_6$.

38.   Find all the ring morphisms from $\mathbf{Z}_{15}$ to $\mathbf{Z}_3$.

39.   Find all the ring morphisms from $\mathbf{Z}\times\mathbf{Z}$ to $\mathbf{Z}\times\mathbf{Z}$.

40.   Find all the ring morphisms from $\mathbf{Z}_7$ to $\mathbf{Z}_4$.

41.   If $(A,+)$ an Abelian group, the set of endomorphisms of A, End(A), consists of all the group morphisms from A to itself. Show that $(\text{End}(A), +, \circ)$ is a ring under addition and composition, where $(f+g)(a)$ $= f(a) + g(a)$, for $f,g \in \text{End}(A)$. This is called the *endomorphism ring* of A.

42. Describe the endomorphism ring $\text{End}(\mathbf{Z}_2 \times \mathbf{Z}_2)$. Is it commutative?

43. Prove that $10^n \equiv 1 \bmod 9$ for all $n \in \mathbf{N}$. Then prove that an integer is divisible by 9 if and only if the sum of its digits is divisible by 9.

44. Find the number of nonisomorphic rings with three elements.

45. Prove that $R[x] \cong R[y]$.

46. Prove that $R[x,y] \cong R[y,x]$.

47. Let $(R, +, \cdot)$ be a ring. Define the operations $\oplus$ and $\bigcirc$ on R by

$$r \oplus s = r + s + 1 \qquad \text{and} \qquad r \bigcirc s = r \cdot s + r + s.$$

(i)   Prove that $(R, \oplus, \bigcirc)$ is a ring.

(ii) What are the additive and multiplicative identities of $(R, \oplus, \bigcirc)$?

(iii) Prove that $(R, \oplus, \bigcirc)$ is isomorphic to $(R, +, \cdot)$.

48. Let a and b be elements of a commutative ring. For each positive integer n, prove the Binomial Theorem:

$$(a+b)^n = a^n + \binom{n}{1} a^{n-1} b + \cdots + \binom{n}{k} a^{n-k} b^k + \cdots + b^n.$$

49. Let $(R, +, \cdot)$ be an algebraic object that satisfies all the axioms for a ring except for the multiplicative identity. Define addition and multiplication in $R \times \mathbf{Z}$ by

$$(a,n) + (b,m) = (a+b, n+m) \qquad \text{and} \qquad (a,n) \cdot (b,m) = (ab + ma + nb, nm).$$

Show that $(R \times \mathbf{Z}, +, \cdot)$ is a ring that contains a subset in one-to-one correspondence with R that has all the properties of the algebraic object $(R, +, \cdot)$.

50. If R and S are commutative rings, prove that the ring of sequences $(R \times S)^{\mathbf{N}}$ is isomorphic to $R^{\mathbf{N}} \times S^{\mathbf{N}}$.

51. If $\mathbf{F}$ is a field, show that the field of fractions of $\mathbf{F}$ is isomorphic to $\mathbf{F}$.

52. Describe the field of fractions of the ring $(\{a + ib \mid a, b \in \mathbf{Z}\}, +, \cdot)$.

53. Let $(S, \star)$ be a commutative semigroup that satisfies the cancellation law; that is, $a \star b = a \star c$ implies $b = c$. Show that $(S, \star)$ can be embedded in a group.

54. Let $T = \{f : \mathbf{R} \to \mathbf{R} \mid f(x) = a \cos x + b \sin x, \ a, b \in \mathbf{R}\}$. Define addition of two such trigonometric functions in the usual way and define convolution by

$$(f*g)(x)=\int_{0}^{2\pi} f(t)\,g(x-t)\,dt.$$

Show that $(T,+,*)$ is a field.

55. Let $T_n=\left\{ f:\mathbf{R}\to\mathbf{R}\,\middle|\,f(x)=\dfrac{a_0}{2}+\sum_{r=1}^{n}(a_r\cos rx+b_r\sin rx),\ a_r,b_r\in\mathbf{R}\right\}.$

Show that $(T_n,+,*)$ is a commutative ring where addition and convolution are defined as in the previous exercise. What is the multiplicative identity? Is the ring an integral domain?

Polynomial and Euclidean Rings

9

Polynomial functions and the solution of polynomial equations are an extremely basic part of mathematics. One of the important uses of ring and field theory is to extend a field to a larger field so that a given polynomial has a root. For example, the complex number field can be obtained by enlarging the real field so that all quadratic equations will have solutions.

Before we are able to extend fields, we need to investigate the ring of polynomials, $F[x]$, with coefficients in a field F. This polynomial ring has many properties in common with the ring of integers; both $F[x]$ and Z are integral domains, but not fields. Moreover, both rings have division and Euclidean algorithms. These algorithms are extremely useful, and rings with such algorithms are called Euclidean rings.

DIVISION ALGORITHM

Long division of integers gives a method for dividing one integer by another to obtain a quotient and a remainder. The fact that this is always possible is stated formally in the following division algorithm.

9.01 DIVISION ALGORITHM FOR INTEGERS. If a and b are integers and b is nonzero, then there exist unique integers q and r such that

$$a = qb + r \quad \text{and} \quad 0 \leqslant r < |b|.$$

The integer r is called the *remainder* in the division of a by b, and q is called the *quotient*.

PROOF.   We prove the existence of the integers q and r by considering first the case $a \geqslant 0$ and then the case $a < 0$.

If $a \geqslant 0$, we prove the result by induction on a. The basis for the induction is provided by the trivial case in which $0 \leqslant a < |b|$. We can then take $q = 0$ and $r = a$.

Now suppose that $k \geqslant |b|$ and that q and r exist for all positive integers, a, less than k. Then $0 \leqslant k - |b| < k$, and we can write

$$k - |b| = q_1 b + r \quad \text{where} \quad 0 \leqslant r < |b|.$$

Hence $k = qb + r$ where $q = q_1 + 1$, if $b > 0$, and $q = q_1 - 1$, if $b < 0$. This completes the induction step, and the result is true for all $a \geqslant 0$.

If $a < 0$, then $-a > 0$, and we can write

$$-a = q_2 b + r_2 \quad \text{where} \quad 0 \leqslant r_2 < |b|.$$

If $r_2 = 0$, then $a = (-q_2)b$. If $r_2 \neq 0$, then $a = qb + r$ where $0 < r = |b| - r_2 < |b|$, $q = -q_2 - 1$, if $b > 0$, and $q = -q_2 + 1$, if $b < 0$.

To prove the uniqueness of q and r, suppose that $a = q_1 b + r_1 = q_2 b + r_2$, where $0 \leqslant r_2 \leqslant r_1 < |b|$. Then $(q_2 - q_1)b = r_1 - r_2$. The right-hand side of this equation is nonnegative and less than $|b|$, whereas the left-hand side is a multiple of b. Hence each side must be zero, from which it follows that $r_1 = r_2$ and $q_1 = q_2$. $\square$

What other rings, besides integers, have a division algorithm? In a field, we can always divide any element exactly by a nonzero element. If a ring contains zero divisors, the cancellation property does not hold, and we cannot expect to obtain a unique quotient. This leaves integral domains, and the following kinds contain a useful generalization of the division algorithm.

9.02 DEFINITION. An integral domain R is called a *Euclidean ring* if, for each nonzero element $a \in R$, there exists a nonnegative integer $d(a)$ such that

(i)   if a and b are nonzero elements of R, then $d(a) \leqslant d(ab)$.

(ii)   for every pair of elements $a, b \in R$ with $b \neq 0$, there exist elements $q, r \in R$ such that

$$a = qb + r \quad \text{where} \quad r = 0 \quad \text{or} \quad d(r) < d(b). \quad \text{(Division Algorithm)}$$

We know that the integers form a Euclidean ring, when we take $d(a)$ to be the absolute value of a. A field is trivially a Euclidean ring when $d(a) = 1$ for all nonzero elements a of the field. We now show that the ring of polynomials, with coefficients in a field, is a Euclidean ring when we take $d(g(x))$ to be the degree of the polynomial $g(x)$.

9.03 DIVISION ALGORITHM FOR POLYNOMIALS. Let $f(x), g(x)$ be elements of the polynomial ring $\mathbf{F}[x]$, with coefficients in the *field* $\mathbf{F}$. If $g(x)$ is not the zero polynomial, there exist unique polynomials $q(x), r(x) \in \mathbf{F}[x]$ such that

$$f(x) = q(x) \cdot g(x) + r(x)$$

where either $r(x)$ is the zero polynomial or $\deg r(x) < \deg g(x)$.

PROOF.   If $f(x)$ is the zero polynomial or $\deg f(x) < \deg g(x)$, then writing $f(x) = 0 \cdot g(x) + f(x)$, we see that the requirements of the algorithm are fulfilled.

If $\deg f(x) = \deg g(x) = 0$, then $f(x)$ and $g(x)$ are nonzero constant polynomials a_0 and b_0, respectively. Now $f(x) = (a_0 b_0^{-1}) g(x)$, and the algorithm holds.

We prove the other cases by induction on the degree of $f(x)$. Suppose that, when we divide by a fixed polynomial $g(x)$, the division algorithm holds for polynomials of degree less than n. Let $f(x) = a_0 + \cdots + a_n x^n$ and $g(x) = b_0 + \cdots + b_m x^m$ where $a_n \neq 0$, $b_m \neq 0$. If $n < m$, we have already shown that the algorithm holds.

Suppose $n \geq m$ and put

$$f_1(x) = f(x) - a_n b_m^{-1} x^{n-m} g(x)$$

so that $\deg f_1(x) < n$. By the induction hypothesis

$$f_1(x) = q_1(x) \cdot g(x) + r(x) \quad \text{where either } r(x) = 0 \text{ or } \deg r(x) < \deg g(x).$$

Hence $f(x) = a_n b_m^{-1} x^{n-m} g(x) + f_1(x) = \{a_n b_m^{-1} x^{n-m} + q_1(x)\} \cdot g(x) + r(x)$, which is a representation of the required form. The algorithm now follows by induction, starting with $n = m - 1$, if $m \neq 0$, or with $n = 0$, if $m = 0$.

The uniqueness of the quotient, $g(x)$, and of the remainder, $r(x)$, follows in a similar way to the uniqueness of the quotient and remainder in the division algorithm for integers. $\square$

The quotient and remainder polynomials can be calculated by long division of polynomials.

9.04 EXAMPLE.   Divide $x^3 + 2x^2 + x + 2$ by $x^2 + 2$ in $\mathbf{Z}_3[x]$.

SOLUTION.

$$
\begin{array}{r}
x + 2 \\
x^2 + 2 \overline{\smash{\big)}\, x^3 + 2x^2 + x + 2} \\
\underline{x^3 + + 2x} \\
2x^2 + 2x + 2 \\
\underline{2x^2 + 1} \\
2x + 1
\end{array}
$$

Hence $x^3 + 2x^2 + x + 2 = (x + 2)(x^2 + 2) + 2x + 1$. $\quad\square$

If we divide by a polynomial of degree 1, the remainder must be a constant. This constant can be found as follows.

9.05 REMAINDER THEOREM.   The remainder when the polynomial $f(x)$ is divided by $(x - \alpha)$ in $\mathbf{F}[x]$ is $f(\alpha)$.

PROOF.   By the division algorithm, there exist $q(x), r(x) \in \mathbf{F}[x]$ with $f(x) = q(x)(x - \alpha) + r(x)$, where $r(x) = 0$ or $\deg r(x) < 1$. The remainder is therefore a constant $r_0 \in \mathbf{F}$ and $f(x) = q(x)(x - \alpha) + r_0$. Substituting α for x, we obtain the result $f(\alpha) = r_0$. $\square$

9.06 FACTOR THEOREM.   The polynomial $(x - \alpha)$ is a factor of $f(x)$ in $\mathbf{F}[x]$ if and only if $f(\alpha) = 0$.

PROOF.   We can write $f(x) = q(x)(x - \alpha)$ for some $q(x) \in \mathbf{F}[x]$ if and only if $f(x)$ has remainder 0 when divided by $(x - \alpha)$. By the Remainder Theorem, this happens if and only if $f(\alpha) = 0$. $\square$

An element α is called a *root* of a polynomial $f(x)$ if $f(\alpha) = 0$. The Factor Theorem shows that $(x - \alpha)$ is a factor of $f(x)$ if and only if α is a root of $f(x)$.

9.07 THEOREM.   A polynomial of degree n over a field $\mathbf{F}$ has at most n roots in $\mathbf{F}$.

PROOF.   We prove the theorem by induction on the degree n. A polynomial of degree 0 consists of only a nonzero constant and therefore has no roots.

Assume that the theorem is true for polynomials of degree $n-1$ and let $f(x) \in \mathbf{F}[x]$ be a polynomial of degree n. If $f(x)$ has no roots, the theorem holds. If $f(x)$ does have roots, let α be one such root. By the Factor Theorem, we can write

$$f(x) = (x-\alpha)g(x),$$

and by Proposition 8.30, $\deg g(x) = n-1$.

Since the field $\mathbf{F}$ has no zero divisors, $f(\beta)=0$ if and only if $(\beta-\alpha)=0$ or $g(\beta)=0$. Therefore, any root of $f(x)$ is either equal to α or is a root of $g(x)$. By the induction hypothesis, $g(x)$ has, at most, $n-1$ roots and so $f(x)$ has, at most, n roots. $\square$

9.08 EXAMPLE.   Show that the Gaussian integers, $\mathbf{Z}[i] = \{a+ib \mid a,b \in \mathbf{Z}\}$, is a Euclidean ring with $d(a+ib) = a^2 + b^2$.

SOLUTION. $\mathbf{Z}[i]$ is a subring of the complex numbers, $\mathbf{C}$, and therefore is an integral domain.

If $z \in \mathbf{Z}[i]$, then $d(z) = z\bar{z}$, where $\bar{z}$ is the conjugate of z in the complex numbers. For any nonzero complex number z, $d(z) > 0$, and for two nonzero Gaussian integers z and w, $d(z \cdot w) = d(z) \cdot d(w)$.

To prove the division algorithm in $\mathbf{Z}[i]$, let z and w be Gaussian integers where $w \neq 0$. Then z/w is a complex number, $c+id$, where $c,d \in \mathbf{Q}$. Choose integers, a,b, as in Figure 9.01 so that $|c-a| \leqslant 1/2$ and $|d-b| \leqslant 1/2$. Then $z/w = a+ib+[(c-a)+i(d-b)]$ and $z = (a+ib)w+[(c-a)+i(d-b)]w$. Now $d([(c-a)+i(d-b)]w) = d((c-a)+i(d-b))d(w) = [(c-a)^2 + (d-b)^2]d(w) \leqslant (1/4+1/4)d(w) < d(w)$. Hence the division algorithm holds, and $\mathbf{Z}[i]$ is a Euclidean ring. $\square$

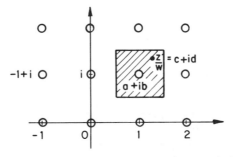

Figure 9.01.   The complex numbers with the elements of $\mathbf{Z}[i]$ circled.

EUCLIDEAN ALGORITHM

The division algorithm allows us to generalize the concepts of divisors and greatest common divisors to any Euclidean ring. Furthermore, we can produce a Euclidean algorithm that will enable us to calculate greatest common divisors.

If a, b, q are three elements in an integral domain such that $a = qb$, we say that b *divides* a or that b is a *factor* of a and write $b|a$. For example, $(2 + i)|(7 + i)$ in the Gaussian integers, $\mathbf{Z}[i]$, because $7 + i = (3 - i)(2 + i)$.

9.09 PROPOSITION.   Let a, b, c be elements in an integral domain R.

(i)   If $a|b$ and $a|c$, then $a|(b + c)$.
(ii)   If $a|b$, then $a|br$ for any $r \in R$.
(iii)   If $a|b$ and $b|c$, then $a|c$.

PROOF.   These results follow immediately from the definition of divisibility. $\square$

9.10 DEFINITION.   If a and b are elements in an integral domain R, then the element $g \in R$ is called *a greatest common divisor* of a and b, and is written $g = \mathrm{GCD}(a, b)$, if

        (i) $g|a$ and $g|b$
and   (ii)   if $c|a$ and $c|b$, then $c|g$.

The element $l \in R$ is called a *least common multiple* of a and b, and is written $l = \mathrm{LCM}(a, b)$, if

        (i) $a|l$ and $b|l$
and   (ii)   if $a|k$ and $b|k$, then $l|k$.

For example, 4 and -4 are greatest common divisors and 60 and -60 are least common multiples of 12 and 20 in $\mathbf{Z}$.

9.11 THEOREM.   Let R be a Euclidean ring. Any two elements a and b in R have a greatest common divisor g. Moreover, there exist $s, t \in R$ such that

$$g = sa + tb.$$

PROOF.   If a and b are both zero, their greatest common divisor is zero, because $r|0$ for any $r \in R$.

Suppose at least one of a and b is nonzero. Let g be a nonzero element for which $d(g)$ is minimal in the set $I = \{xa + yb \,|\, x, y \in R\}$. We can write $g = sa + tb$ for some $s, t \in R$.

Since R is a Euclidean ring, $a = hg + r$ where $r = 0$ or $d(r) < d(g)$. Therefore, $r = a - hg = a - h(sa + tb) = (1 - hs)a - htb \in I$. Since g was an element for which $d(g)$ was minimal in I, it follows that r must be zero, and $g|a$. Similarly, $g|b$.

If $c|a$ and $c|b$, so that $a = kc$ and $b = lc$, then $g = sa + tb = skc + tlc = (sk + tl)c$ and $c|g$. Therefore, $g = \text{GCD}(a,b)$. $\square$

The above theorem shows that greatest common divisors exist in a Euclidean ring, but does not give a method for finding them. Greatest common divisors in any Euclidean ring can be computed using the following Euclidean algorithm.

9.12 EUCLIDEAN ALGORITHM. Let a,b be elements of a Euclidean ring R and let b be nonzero. By repeated use of the division algorithm, we can write

$$
\begin{aligned}
a &= bq_1 + r_1 && \text{where } d(r_1) < d(b) \\
b &= r_1 q_2 + r_2 && \text{where } d(r_2) < d(r_1) \\
r_1 &= r_2 q_3 + r_3 && \text{where } d(r_3) < d(r_2) \\
&\ \ \vdots && \qquad \vdots \\
r_{k-2} &= r_{k-1} q_k + r_k && \text{where } d(r_k) < d(r_{k-1}) \\
r_{k-1} &= r_k q_{k+1} + 0.
\end{aligned}
$$

If $r_1 = 0$, then $b = \text{GCD}(a,b)$; otherwise $r_k = \text{GCD}(a,b)$.
Furthermore, elements $s, t \in R$ such that

$$\text{GCD}(a,b) = sa + tb$$

can be found by starting with the equation $r_k = r_{k-2} - r_{k-1} q_k$ and successively working up the above sequence of equations, each time replacing r_i in terms of r_{i-1} and r_{i-2}.

PROOF. This algorithm must terminate, because $d(b), d(r_1), d(r_2), \ldots$ is a decreasing sequence of nonnegative integers; thus, $r_{k+1} = 0$ for some $k + 1$. The proof of the algorithm follows by repeated application of the following lemma. $\square$

9.13 LEMMA. If $r_{i-1} = r_i q_{i+1} + r_{i+1}$, then $\text{GCD}(r_{i-1}, r_i) = \text{GCD}(r_i, r_{i+1})$.

PROOF. Let $g = \text{GCD}(r_{i-1}, r_i)$ so that $g|r_{i-1}$ and $g|r_i$. Since $r_{i+1} = r_{i-1} - r_i q_{i+1}$, $g|r_{i+1}$, and g is a divisor of r_i and r_{i+1}. Let c be any other divisor of r_i and r_{i+1}. As $r_{i-1} = r_i q_{i+1} + r_{i+1}$, $c|r_{i-1}$ and therefore $c|g$ because g is the greatest common divisor of r_{i-1} and r_i. Hence $g = \text{GCD}(r_i, r_{i+1})$. $\square$

9.14 EXAMPLE. Find the greatest common divisor of 713 and 253 and find two integers s and t such that

$$713s + 253t = \text{GCD}(713, 253).$$

SOLUTION. By the division algorithm we have

(i) $713 = 2 \cdot 253 + 207$ $a = 713, b = 253, r_1 = 207$
(ii) $253 = 1 \cdot 207 + 46$ $r_2 = 46$
(iii) $207 = 4 \cdot 46 + 23$ $r_3 = 23$
        $46 = 2 \cdot 23 + 0.$ $r_4 = 0$

The last nonzero remainder is the greatest common divisor. Hence $\text{GCD}(713, 253) = 23$.
    We can find the integers s and t by using equations (i)–(iii). We have

$$\begin{aligned}
23 &= 207 - 4 \cdot 46 & &\text{from equation (iii)} \\
&= 207 - 4(253 - 207) & &\text{from equation (ii)} \\
&= 5 \cdot 207 - 4 \cdot 253 & & \\
&= 5 \cdot (713 - 2 \cdot 253) - 4 \cdot 253 & &\text{from equation (i)} \\
&= 5 \cdot 713 - 14 \cdot 253.
\end{aligned}$$

Therefore $s = 5$ and $t = -14$. □

9.15 EXAMPLE. Find a greatest common divisor, $g(x)$, of $a(x) = 2x^4 + 2$ and $b(x) = x^5 + 2$ in $\mathbf{Z}_3[x]$, and find $s(x), t(x) \in \mathbf{Z}_3[x]$ so that

$$g(x) = s(x) \cdot (2x^4 + 2) + t(x) \cdot (x^5 + 2).$$

SOLUTION. By repeated use of the division algorithm we have

(i) $x^5 + 2 = (2x)(2x^4 + 2) + 2x + 2$
(ii) $2x^4 + 2 = (x^3 + 2x^2 + x + 2)(2x + 2) + 1$
(iii) $2x + 2 = (2x + 2) \cdot 1 + 0.$

Hence $\text{GCD}(a(x), b(x)) = 1$.
    From equation (ii) we have

$$1 = 2x^4 + 2 - (x^3 + 2x^2 + x + 2)(2x + 2)$$

$$= 2x^4 + 2 - (x^3 + 2x^2 + x + 2)\{x^5 + 2 - (2x)(2x^4 + 2)\}, \quad \text{from equation (i)}$$

$$= (2x^4 + x^3 + 2x^2 + x + 1)(2x^4 + 2) + (2x^3 + x^2 + 2x + 1)(x^5 + 2).$$

Therefore,

$$s(x) = 2x^4 + x^3 + 2x^2 + x + 1$$

and

$$t(x) = 2x^3 + x^2 + 2x + 1. \quad \square$$

$$
\begin{array}{r}
2x \\
2x^4+2\overline{\smash{\big)}\ x^5+0x^4+0x^3+0x^2+0x+2} \\
\underline{x^5 \qquad\qquad\qquad\ +\ x} \\
2x+2 \\
\end{array}
$$

$$
\begin{array}{r}
x^3+2x^2+x+2 \\
2x+2\overline{\smash{\big)}\ 2x^4+0x^3+\ 0x^2+0x+2} \\
\underline{2x^4+2x^3} \\
x^3 \\
\underline{x^3+\ \ x^2} \\
2x^2 \\
\underline{2x^2+2x} \\
x+2 \\
\underline{x+1} \\
1 \\
\end{array}
$$

9.16 EXAMPLE. Find a greatest common divisor of $a(x) = x^4 + x^3 + 3x - 9$ and $b(x) = 2x^3 - x^2 + 6x - 3$ in $\mathbf{Q}[x]$.

SOLUTION. By the division algorithm we have

$$a(x) = \left(\frac{1}{2}x + \frac{3}{4}\right)b(x) - \frac{9}{4}x^2 - \frac{27}{4}$$

and

$$b(x) = \left(-\frac{8}{9}x + \frac{4}{9}\right)\left(-\frac{9}{4}x^2 - \frac{27}{4}\right).$$

Hence

$$\mathrm{GCD}(a(x), b(x)) = -\frac{9}{4}x^2 - \frac{27}{4}. \quad \square$$

$$
\begin{array}{r}
\frac{1}{2}x + \frac{3}{4} \\
2x^3-x^2+6x-3\overline{\smash{\big)}\ x^4+\ \ x^3+0x^2+3x-9} \\
\underline{x^4-\tfrac{1}{2}x^3+3x^2-\tfrac{3}{2}x} \\
\tfrac{3}{2}x^3-3x^2+\tfrac{9}{2}x-9 \\
\tfrac{3}{2}x^3-\tfrac{3}{4}x^2+\tfrac{9}{2}x-\tfrac{9}{4} \\
\underline{-\tfrac{9}{4}x^2 \qquad\ -\tfrac{27}{4}} \\
\end{array}
$$

$$
\begin{array}{r}
-\frac{8}{9}x + \frac{4}{9} \\
-\tfrac{9}{4}x^2-\tfrac{27}{4}\overline{\smash{\big)}\ 2x^3-x^2+6x-3} \\
\underline{2x^3 \qquad\ +6x} \\
-x^2 \qquad -3 \\
\underline{-x^2 \qquad -3} \\
0 \\
\end{array}
$$

9.17 THEOREM.   Let a and b be integers and p a prime number. If $p|ab$, then $p|a$ or $p|b$.

PROOF.   Suppose p does not divide a. Then $\mathrm{GCD}(a,p)=1$, because the only divisors of p are ± 1 and $\pm p$. By the Euclidean algorithm, there exist integers s and t such that $as+pt=1$. Therefore, $abs+pbt=b$, and since $p|ab$, it follows that $p|b$. $\square$

9.18 THEOREM. $\mathbf{Z}_n$ is a field if and only if n is prime.

PROOF.   Suppose n is prime and that $[a]\cdot[b]=[0]$ in $\mathbf{Z}_n$. Then $n|ab$ and, by Theorem 9.17, $n|a$ or $n|b$. Hence $[a]=[0]$ or $[b]=[0]$, and $\mathbf{Z}_n$ is an integral domain. Since $\mathbf{Z}_n$ is also finite, it follows from Theorem 8.14 that $\mathbf{Z}_n$ is a field.

Suppose n is not prime. Then we can write $n=rs$ where r and s are integers such that $1<r<n$ and $1<s<n$. Now $[r]\neq[0]$ and $[s]\neq[0]$ but $[r]\cdot[s]=[rs]=[0]$. Therefore, $\mathbf{Z}_n$ has zero divisors, and hence is not a field. $\square$

Unique Factorization

One important property of the integers, often known as the Fundamental Theorem of Arithmetic, states that every integer greater than 1 can be written as a finite product of prime numbers, and furthermore, this product is unique up to the ordering of the primes. In this section, we prove a similar result for any Euclidean ring.

9.19 DEFINITION.   Let R be a commutative ring. An element u is called an *invertible element* (or *unit*) of R if there exists an element $v\in R$ such that $uv=1$.

The invertible elements in a ring R are those elements with multiplicative inverses in R. Denote the set of invertible elements of R by R^*

If R is a field, every nonzero element is invertible and $R^*=R-\{0\}$.

The invertible elements in the integers are ± 1. The invertible polynomials in $\mathbf{F}[x]$ are the nonzero constant polynomials, that is, the polynomials of degree 0. The set of invertible elements in the Gaussian integers is $\mathbf{Z}[i]^*=\{\pm 1,\pm i\}$.

9.20 PROPOSITION.   For any commutative ring R, the invertible elements form an Abelian group, $(R^*,\cdot)$, under multiplication.

PROOF.   Let $u_1, u_2 \in R^*$ and let $u_1 v_1 = u_2 v_2 = 1$. Then $(u_1 u_2)(v_1 v_2) = 1$; thus $u_1 u_2 \in R^*$. The group axioms follow immediately. $\square$

Two elements in a Euclidean ring may have many greatest common divisors. For example, in $\mathbf{Q}[x]$, $x + 1$, $2x + 2$, and $\frac{1}{3}x + \frac{1}{3}$ are all greatest common divisors of $x^2 + 2x + 1$ and $x^2 - 1$. However, they can all be obtained from one another by multiplying by invertible elements.

9.21 LEMMA.   If $a|b$ and $b|a$ in an integral domain R, then $a = ub$ where u is an invertible element.

PROOF.   Since $a|b$, $b = va$ for $v \in R$ and, since $b|a$, $a = ub$ for $u \in R$. Therefore, $a = ub = uva$; thus $a(uv - 1) = 0$. If $a = 0$, then $b = 0$. If $a \neq 0$, then, because R has no zero divisors, $uv = 1$, and u is invertible. $\square$

9.22 LEMMA.   If g_2 is a greatest common divisor of a and b in the Euclidean ring R, then g_1 is also a greatest common divisor of a and b if and only if $g_1 = ug_2$, where u is invertible.

PROOF.   If $g_1 = ug_2$ where $uv = 1$, then $g_2 = vg_1$. Hence $g_2|g_1$ and $g_1|g_2$ if and only if $g_1 = ug_2$. The result now follows from the definition of a greatest common divisor. $\square$

9.23 LEMMA.   If a and b are elements in a Euclidean ring R, then $d(a) = d(ab)$ if and only if b is invertible. Otherwise, $d(a) < d(ab)$.

PROOF.   If b is invertible and $bc = 1$, then $d(a) \leqslant d(ab) \leqslant d(abc) = d(a)$. Hence $d(a) = d(ab)$.

If b is not invertible, ab does not divide a and $a = qab + r$ where $d(r) < d(ab)$. Now $r = a(1 - qb)$; thus $d(a) \leqslant d(r)$. Therefore, $d(a) < d(ab)$. $\square$

9.24 DEFINITION.   A noninvertible element p in a Euclidean ring R is said to be *irreducible* if, whenever $p = ab$, either a or b is invertible in R.

The irreducible elements in the integers are the prime numbers together with their negatives.

9.25 LEMMA.   Let R be a Euclidean ring. If $a, b, c \in R$, $\mathrm{GCD}(a, b) = 1$ and $a|bc$ then $a|c$.

PROOF.   By Theorem 9.11, we can write $1 = sa + tb$ where $s, t \in R$. Therefore, $c = sac + tbc$ and $a|c$. $\square$

9.26 PROPOSITION. If p is irreducible in the Euclidean ring R and $p|ab$, then $p|a$ or $p|b$.

PROOF. For any $a \in R$, $GCD(a,p)|p$ and $p = GCD(a,p) \cdot h$. If p is irreducible, either $GCD(a,p)$ or h is invertible, and so either $GCD(a,p) = 1$ or p. Hence if p does not divide a, then $GCD(a,p) = 1$, and it follows from Lemma 9.25 that $p|b$. $\square$

9.27 UNIQUE FACTORIZATION THEOREM. Every nonzero element in a Euclidean ring R is either an invertible element or can be written as the product of a finite number of irreducibles. In such a product, the irreducibles are uniquely determined up to a rearrangement and up to multiplication by invertible elements.

PROOF. We proceed by induction on $d(a)$ for $a \in R$. The least value of $d(a)$ for nonzero a is $d(1)$, because 1 divides any other element. Suppose $d(a) = d(1)$. Then $d(1 \cdot a) = d(1)$ and, by Lemma 9.23, a is invertible.

As the induction hypothesis, suppose that all elements $x \in R$, with $d(x) < d(a)$, are either invertible or can be written as a product of irreducibles. We now prove this for the element a.

If a is irreducible, there is nothing to prove. If not, we can write $a = bc$ where neither b nor c is invertible. By Lemma 9.23, $d(b) < d(bc) = d(a)$ and $d(c) < d(bc) = d(a)$. By the induction hypothesis, b and c can be written as a product of irreducibles, and hence a can also be written as a product of irreducibles.

To prove the uniqueness, suppose that

$$a = p_1 p_2 \cdots p_n = q_1 q_2 \cdots q_m,$$

where each p_i and q_j is irreducible. Now $p_1|a$ and so $p_1|q_1 q_2 \cdots q_m$. By an extension of Proposition 9.26 to m factors, p_1 divides one of the q_is. Rearrange the q_is, if necessary, so that $p_1|q_1$. Therefore, $q_1 = u_1 p_1$ and u_1 is invertible, because p_1 and q_1 are both irreducible.

Now $a = p_1 p_2 \cdots p_n = u_1 p_1 q_2 \cdots q_m$; thus $p_2 \cdots p_n = u_1 q_2 \cdots q_m$. Proceed inductively to show that $p_i = u_i q_i$ for all i, where each u_i is invertible.

If $m < n$, we would obtain the relation $p_{m+1} \cdots p_n = u_1 u_2 \cdots u_m$, which is impossible because irreducibles cannot divide an invertible element. If $m > n$, we would obtain

$$1 = u_1 u_2 \cdots u_n q_{n+1} \cdots q_m,$$

which is again impossible because an irreducible cannot divide 1. Hence

$m = n$, and the primes $p_1, p_2, \ldots, p_n$ are the same as $q_1, q_2, \ldots, q_m$ up to a rearrangement and up to multiplication by invertible elements. $\square$

When the Euclidean ring is the integers, the above theorem yields the Fundamental Theorem of Arithmetic referred to earlier. The ring of polynomials over a field and the Gaussian integers have this unique factorization property enjoyed by the integers. However, the integral domain

$$\mathbf{Z}[\sqrt{-3}\,] = \{a + b\sqrt{-3} \mid a, b \in \mathbf{Z}\},$$

which is a subring of $\mathbf{C}$, does not have the unique factorization property. For example,

$$4 = 2 \cdot 2 = (1 + \sqrt{-3}\,) \cdot (1 - \sqrt{-3}\,),$$

whereas 2, $1 + \sqrt{-3}$, and $1 - \sqrt{-3}$ are all irreducible. Therefore, $\mathbf{Z}[\sqrt{-3}\,]$ cannot be a Euclidean ring.

Factoring Real and Complex Polynomials

The question of whether a polynomial is irreducible or not will be crucial in the next chapter when we extend number fields by adjoining roots of a polynomial. We therefore investigate different methods of factoring polynomials over various coefficient fields.

A polynomial $f(x)$ of positive degree is said to be *reducible over* $\mathbf{F}$ if it can be factored into two polynomials of positive degree in $\mathbf{F}[x]$. If it cannot be so factored, $f(x)$ is called *irreducible over* $\mathbf{F}$, and $f(x)$ is an irreducible element of the ring $\mathbf{F}[x]$. It is important to note that reducibility depends on the field $\mathbf{F}$. The polynomial $x^2 + 1$ is irreducible over $\mathbf{R}$ but reducible over $\mathbf{C}$.

The following basic theorem, first proved by Gauss in his doctoral thesis in 1799, enables us to determine which polynomials are irreducible in $\mathbf{C}[x]$ and in $\mathbf{R}[x]$.

9.28 FUNDAMENTAL THEOREM OF ALGEBRA. If $f(x)$ is a polynomial in $\mathbf{C}[x]$ of positive degree, then $f(x)$ has a root in $\mathbf{C}$.

A proof of this theorem has to use the concept of continuity and therefore requires a knowledge of analysis or topology. A sketch proof is given in Birkhoff and MacLane [1; Ch. V, sect. 3] and in Courant and Robbins [51; Ch. V, sect. 3].

The following useful theorem shows that the complex roots of *real* polynomials occur in conjugate pairs.

9.29 THEOREM. (i) If $z = a + ib$ is a complex root of the real polynomial $f(x) \in \mathbf{R}[x]$, then its conjugate $\bar{z} = a - ib$ is also a root. Furthermore, the real polynomial $x^2 - 2ax + a^2 + b^2$ is a factor of $f(x)$.

(ii) If $a, b, c \in \mathbf{Q}$ and $a + b\sqrt{c}$ is an irrational root of the rational polynomial $f(x) \in \mathbf{Q}[x]$, then $a - b\sqrt{c}$ is also a root, and the rational polynomial $x^2 - 2ax + a^2 - b^2c$ is a factor of $f(x)$.

PROOF. (i) Let $g(x) = x^2 - 2ax + a^2 + b^2 = (x - a - ib)(x - a + ib)$. By the division algorithm in $\mathbf{R}[x]$, there exist real polynomials $q(x)$ and $r(x)$ such that

$$f(x) = q(x)g(x) + r(x) \quad \text{where} \quad r(x) = 0 \quad \text{or} \quad \deg r(x) < 2.$$

Hence $r(x) = r_0 + r_1 x$ where $r_0, r_1 \in \mathbf{R}$. Now $z = a + ib$ is a root of $f(x)$ and of $g(x)$; therefore, it is also a root of $r(x)$, and $0 = r_0 + r_1(a + ib)$. Equating real and imaginary parts, we have $r_0 + r_1 a = 0$ and $r_1 b = 0$. Now

$$r(\bar{z}) = r(a - ib) = r_0 + r_1(a - ib) = r_0 + r_1 a - ir_1 b = 0.$$

Since $\bar{z}$ is a root of $r(x)$ and $g(x)$, it must be a root of $f(x)$.

If z is complex and not real, then $b \neq 0$. In this case $r_1 = 0$ and $r_0 = 0$; thus $g(x) | f(x)$.

(ii) This can be proved in a similar way to part (i). $\square$

9.30 THEOREM. (i) The irreducible polynomials in $\mathbf{C}[x]$ are the polynomials of degree one.

(ii) The irreducible polynomials in $\mathbf{R}[x]$ are the polynomials of degree 1 together with the polynomials of degree 2 of the form $ax^2 + bx + c$, where $b^2 < 4ac$.

PROOF. (i) The polynomials of degree 0 are the invertible elements of $\mathbf{C}[x]$. By the Fundamental Theorem of Algebra, any polynomial of positive degree has a root in $\mathbf{C}$ and hence a linear factor. Therefore, all polynomials of degree greater than 1 are reducible and those of degree 1 are the irreducibles.

(ii) The polynomials of degree 0 are the invertible elements of $\mathbf{R}[x]$. By part (i) and the Unique Factorization Theorem, every real polynomial of positive degree can be factored into linear factors in $\mathbf{C}[x]$. By Theorem 9.29 (i), its nonreal roots fall into conjugate pairs, whose corresponding factors combine to give a quadratic factor in $\mathbf{R}[x]$ of the form $ax^2 + bx + c$, where $b^2 < 4ac$. Hence any real polynomial can be factored into real linear factors and real quadratic factors of the above form. $\square$

9.31 EXAMPLE. Find the kernel and image of the ring morphism $\psi : \mathbf{Q}[x] \rightarrow \mathbf{R}$ defined by $\psi(f(x)) = f(\sqrt{2}\,)$.

SOLUTION. If $p(x) = a_0 + a_1 x + \cdots + a_n x^n \in \mathbf{Q}[x]$, then

$$\psi(p(x)) = a_0 + a_1\sqrt{2} + \cdots + a_n(\sqrt{2}\,)^n$$

$$= (a_0 + 2a_2 + 4a_4 + \cdots) + \sqrt{2}\,(a_1 + 2a_3 + 4a_5 + \cdots)$$

$$\in \mathbf{Q}(\sqrt{2}\,) = \{a + b\sqrt{2} \mid a,b \in \mathbf{Q}\}.$$

where $\mathbf{Q}(\sqrt{2}\,)$ is the subring of $\mathbf{R}$ defined in Example 8.04. Hence $\mathrm{Im}\,\psi \subseteq \mathbf{Q}(\sqrt{2}\,)$ and $\mathrm{Im}\,\psi = \mathbf{Q}(\sqrt{2}\,)$ because $\psi(a + bx) = a + b\sqrt{2}$.

If $p(x) \in \mathrm{Ker}\,\psi$, then $p(\sqrt{2}\,) = 0$; therefore, by Theorem 9.29(ii), $p(-\sqrt{2}\,) = 0$, and $p(x)$ contains a factor $(x^2 - 2)$. Conversely, if $p(x)$ contains a factor $(x^2 - 2)$, then $p(\sqrt{2}\,) = 0$ and $p(x) \in \mathrm{Ker}\,\psi$. Hence $\mathrm{Ker}\,\psi = \{(x^2 - 2)q(x) \mid q(x) \in \mathbf{Q}[x]\}$, that is, the set of all polynomials in $\mathbf{Q}[x]$ with $(x^2 - 2)$ as a factor. $\square$

FACTORING RATIONAL AND INTEGRAL POLYNOMIALS

A rational polynomial can always be reduced to an integer polynomial by multiplying it by the least common multiple of the denominators of its coefficients. We now give various methods for determining whether an integer polynomial has rational roots or is irreducible over $\mathbf{Q}$.

9.32 RATIONAL ROOTS THEOREM. Let $p(x) = a_0 + a_1 x + \cdots + a_n x^n \in \mathbf{Z}[x]$. If r/s is a rational root of $p(x)$ and $\mathrm{GCD}(r,s) = 1$, then

	(i)	$r \mid a_0$
and	(ii)	$s \mid a_n$.

PROOF. If $p(r/s) = 0$, then $a_0 + a_1(r/s) + \cdots + a_{n-1}(r/s)^{n-1} + a_n(r/s)^n = 0$ and $a_0 s^n + a_1 rs^{n-1} + \cdots + a_{n-1} r^{n-1} s + a_n r^n = 0$. Therefore $a_0 s^n = -r(a_1 s^{n-1} + \cdots + a_{n-1} r^{n-2} s + a_n r^{n-1})$; thus $r \mid a_0 s^n$. Since $\mathrm{GCD}(r,s) = 1$, it follows from Lemma 9.25 that $r \mid a_0$. Similarly, $s \mid a_n$. $\square$

9.33 EXAMPLE. Factor $p(x) = 2x^3 + 3x^2 - 1$ in $\mathbf{Q}[x]$.

SOLUTION.   If $p(r/s) = 0$, then, by Theorem 9.32, $r|-1$ and $s|2$. Hence $r = \pm 1$ and $s = \pm 1$ or ± 2, and the only possible values of r/s are $\pm 1, \pm 1/2$.

Instead of testing all these values, we sketch the graph of $p(x)$ to find approximate roots. Differentiating, we have $p'(x) = 6x^2 + 6x = 6x(x+1)$, so $p(x)$ has turning values at 0 and -1.

We see from the graph in Figure 9.02 that -1 is a double root and that there is one more positive root. If it is rational, it can only be $1/2$. Checking this in Table 9.1, we see that $1/2$ is a root; hence $p(x)$ factors as $(x+1)^2(2x-1)$. $\square$

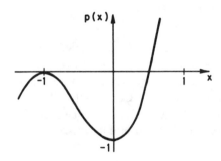

Figure 9.02.   The graph of $p(x) = 2x^3 + 3x^2 - 1$.

Table 9.1

x	-1	0	$1/2$	1	2
$p(x)'$	0	-1	0	4	27

9.34 EXAMPLE.   Prove that $\sqrt[5]{2}$ is irrational.

SOLUTION. $\sqrt[5]{2}$ is a root of $x^5 - 2$. If this polynomial has a rational root r/s, in its lowest terms, it follows from Theorem 9.32 that $r|-2$ and $s|1$. Hence the only possible rational roots are $\pm 1, \pm 2$. We see from Table 9.2 that none of these are roots, and so all the roots of the polynomial $x^5 - 2$ must be irrational. $\square$

Table 9.2

x	-2	-1	1	2
$x^5 - 2$	-34	-3	-1	30

9.35 GAUSS' LEMMA. Let $P(x) = a_0 + \cdots + a_n x^n \in \mathbf{Z}[x]$. If $P(x)$ can be factored in $\mathbf{Q}[x]$ as $P(x) = q(x)r(x)$ with $q(x)$, $r(x) \in \mathbf{Q}[x]$, then $P(x)$ can also be factored in $\mathbf{Z}[x]$.

PROOF. Express the rational coefficients of $q(x)$ in their lowest terms and let u be the least common multiple of their denominators. Then $q(x) = (1/u)\overline{Q}(x)$ where $\overline{Q}(x) \in \mathbf{Z}[x]$. Let s be the greatest common divisor of all the coefficients of $\overline{Q}(x)$; write $q(x) = (s/u)Q(x)$ where $Q(x) \in \mathbf{Z}[x]$, and the greatest common divisor of its coefficients is 1. Write $r(x) = (t/v)R(x)$ in a similar way.

Now $P(x) = q(x)r(x) = \dfrac{s}{u}Q(x)\dfrac{t}{v}R(x) = \dfrac{st}{uv}Q(x)R(x)$ and $uvP(x) = stQ(x)R(x)$.

To prove the theorem, we shown that $uv|st$ by proving that no prime p in uv can divide all the coefficients of $Q(x)R(x)$.

Let $Q(x) = b_0 + \cdots + b_k x^k$ and $R(x) = c_0 + \cdots + c_l x^l$. Choose a prime p and let b_i and c_j be the first coefficients of $Q(x)$ and $R(x)$, respectively, that p fails to divide. These exist because $\mathrm{GCD}(b_0, \ldots, b_k) = 1$ and $\mathrm{GCD}(c_0, \ldots, c_l) = 1$. The coefficient of x^{i+j} in $Q(x)R(x)$ is

$$b_{i+j}c_0 + b_{i+j-1}c_1 + \cdots + b_{i+1}c_{j-1} + b_i c_j + b_{i-1}c_{j+1} + \cdots + b_0 c_{i+j}.$$

Now $p|c_0, p|c_1, \ldots, p|c_{j-1}, p|b_{i-1}, p|b_{i-2}, \ldots, p|b_0$ but $p \nmid b_i c_j$ so this coefficient is not divisible by p. Hence the greatest common divisor of the coefficients of $Q(x)R(x)$ is 1; therefore, $uv|st$ and $P(x)$ can be factored in $\mathbf{Z}[x]$. $\square$

9.36 EXAMPLE. Factor $p(x) = x^4 - 3x^2 + 2x + 1$ into irreducible factors in $\mathbf{Q}[x]$.

SOLUTION. By Theorem 9.32, the only possible rational roots are ± 1. However, these are not roots, so $p(x)$ has no linear factors.

Therefore, if it does factor, it must factor into two quadratics, and by Gauss' Lemma these factors can be chosen to have integral coefficients. Suppose

$$x^4 - 3x^2 + 2x + 1 = (x^2 + ax + b)(x^2 + cx + d)$$

$$= x^4 + (a+c)x^3 + (b+d+ac)x^2 + (bc+ad)x + bd.$$

Thus we have to solve the following system for integer solutions:

$$a + c = 0, \quad b + d + ac = -3, \quad bc + ad = 2 \quad \text{and} \quad bd = 1.$$

Therefore, $b = d = \pm 1$ and $b(a+c) = 2$. Hence $a + c = \pm 2$, which is a contradiction.

The polynomial cannot be factored into two quadratics and therefore is irreducible in $\mathbf{Q}[x]$. $\square$

9.37 EISENSTEIN'S CRITERION. Let $f(x) = a_0 + a_1 x + \cdots + a_n x^n \in \mathbf{Z}[x]$. If, for some prime p,

$\qquad$ (i) $\quad p|a_0, p|a_1, \ldots, p|a_{n-1}$
$\qquad$ (ii) $\quad p \nmid a_n$
and $\qquad$ (iii) $\quad p^2 \nmid a_0$,

then $f(x)$ is irreducible over $\mathbf{Q}$.

PROOF. Suppose $f(x)$ is reducible. By Gauss' Lemma, it factors as two polynomials in $\mathbf{Z}[x]$, that is,

$$f(x) = (b_0 + \cdots + b_r x^r)(c_0 + \cdots + c_s x^s)$$

where $b_i, c_j \in \mathbf{Z}, s > 0$, and $r + s = n$.

Comparing coefficients, we see that $a_0 = b_0 c_0$. Now $p | a_0$, but $p^2 \nmid a_0$, so p must divide b_0 or c_0 but not both. Without loss of generality, suppose that $p | b_0$ and $p \nmid c_0$. Now p cannot divide all of $b_0, b_1, \ldots, b_r$ for then p would divide a_n. Let t be smallest integer for which $p \nmid b_t$; thus $1 \leqslant t \leqslant r < n$. Then $a_t = b_t c_0 + b_{t-1} c_1 + \cdots + b_1 c_{t-1} + b_0 c_t$ and $p | a_t, p | b_0, p | b_1, \ldots, p | b_{t-1}$. Hence $p | b_t c_0$. However, $p \nmid b_t$ and $p \nmid c_0$, so we have a contradiction, and the theorem is proved. $\square$

For example, Eisenstein's Criterion can be used to show that $x^5 - 2$, $x^7 + 2x^3 + 12x^2 - 2$ and $2x^3 + 9x - 3$ are all irreducible over $\mathbf{Q}$.

9.38 EXAMPLE. Show that, for any prime p, $\phi(x) = x^{p-1} + x^{p-2} + \cdots + x + 1$ is irreducible over $\mathbf{Q}$. This is called a *cyclotomic polynomial* and can be written $\phi(x) = (x^p - 1)/(x - 1)$.

SOLUTION. We cannot apply Eisenstein's criterion to $\phi(x)$ as it stands. However, if we put $x = y + 1$, we obtain

$$\phi(y+1) = \frac{(y+1)^p - 1}{y}$$

$$= y^{p-1} + p y^{p-2} + \frac{p(p-1)}{2!} y^{p-3} + \cdots + \frac{p(p-1)}{2!} y + p.$$

Now $\phi(x)$ is irreducible if and only if $\phi(y+1)$ is irreducible, and we can apply Eisenstein's criterion to $\phi(y+1)$ using the prime p.

Hence, for any prime p, $\phi(x)$ is irreducible. $\square$

FACTORING POLYNOMIALS OVER FINITE FIELDS

The roots of a polynomial in $\mathbf{Z}_p[x]$ can be found by trying all the p possible values.

9.39 EXAMPLE. Does $x^4+4\in\mathbf{Z}_7[x]$ have any roots in $\mathbf{Z}_7$?

SOLUTION. We see from Table 9.3 that x^4+4 is never zero and therefore has no roots in $\mathbf{Z}_7$. $\square$

Table 9.3. Values modulo 7

x	0	1	2	3	4	5	6
x^4	0	1	2	4	4	2	1
x^4+4	4	5	6	1	1	6	5

9.40 PROPOSITION. A polynomial in $\mathbf{Z}_2[x]$ has a factor $(x+1)$ if and only if it has an even number of nonzero coefficients.

PROOF. Let $p(x)=a_0+a_1x+\cdots+a_nx^n\in\mathbf{Z}_2[x]$. By the Factor Theorem, $(x+1)$ is a factor of $p(x)$ if and only if $p(1)=0$. (Remember that $x-1=x+1$ in $\mathbf{Z}_2[x]$.) Now $p(1)=a_0+a_1+\cdots+a_n$, which is zero in $\mathbf{Z}_2$ if and only if $p(x)$ has an even number of nonzero coefficients. $\square$

9.41 EXAMPLE. Find all the irreducible polynomials of degree less than or equal to 4 over $\mathbf{Z}_2$.

SOLUTION. Any polynomial of degree 1 is irreducible. So in $\mathbf{Z}_2[x]$ we have x and $x+1$.

Let $p(x)=a_0+\cdots+a_nx^n\in\mathbf{Z}_2[x]$. If $p(x)$ has degree n, then a_n is nonzero, and $a_n=1$. The only possible roots are 0 and 1. The element 0 is a root if and only if $a_0=0$, and 1 is a root if and only if $p(x)$ has an even number of nonzero terms.

The following are the polynomials of degrees 2, 3, and 4 in $\mathbf{Z}_2[x]$ with no linear factors:

$$x^2 + x + 1 \qquad \text{(degree 2)}$$
$$x^3 + x + 1, x^3 + x^2 + 1 \qquad \text{(degree 3)}$$
$$x^4 + x + 1, x^4 + x^2 + 1, x^4 + x^3 + 1, x^4 + x^3 + x^2 + x + 1 \qquad \text{(degree 4)}.$$

If a polynomial of degree 2 or 3 is reducible, it must have a linear factor; hence the above polynomials of degree 2 and 3 are irreducible. If a polynomial of degree 4 is reducible, it either has a linear factor or is the product of two irreducible quadratic factors. Now there is only one irreducible quadratic in $\mathbf{Z}_2[x]$, and its square $(x^2 + x + 1)^2 = x^4 + x^2 + 1$ is reducible.

Hence the irreducible polynomials of degree $\leqslant 4$ over $\mathbf{Z}_2$ are $x, x + 1$, $x^2 + x + 1$, $x^3 + x + 1$, $x^3 + x^2 + 1$, $x^4 + x + 1$, $x^4 + x^3 + 1$, and $x^4 + x^3 + x^2 + x + 1$. $\square$

For example, the polynomials of degree 4 in $\mathbf{Z}_2[x]$ factorize into irreducible factors as follows.

$$
\begin{aligned}
x^4 & = x^4 \\
x^4 + 1 & = (x + 1)^4 \\
x^4 + x & = x(x + 1)(x^2 + x + 1) \\
x^4 + x + 1 & \quad \text{is irreducible} \\
x^4 + x^2 & = x^2(x + 1)^2 \\
x^4 + x^2 + 1 & = (x^2 + x + 1)^2 \\
x^4 + x^2 + x & = x(x^3 + x + 1) \\
x^4 + x^2 + x + 1 & = (x + 1)(x^3 + x^2 + 1) \\
x^4 + x^3 & = x^3(x + 1) \\
x^4 + x^3 + 1 & \quad \text{is irreducible} \\
x^4 + x^3 + x & = x(x^3 + x^2 + 1) \\
x^4 + x^3 + x + 1 & = (x + 1)^2(x^2 + x + 1) \\
x^4 + x^3 + x^2 & = x^2(x^2 + x + 1) \\
x^4 + x^3 + x^2 + 1 & = (x + 1)(x^3 + x + 1) \\
x^4 + x^3 + x^2 + x & = x(x + 1)^3 \\
x^4 + x^3 + x^2 + x + 1 & \quad \text{is irreducible}
\end{aligned}
$$

LINEAR CONGRUENCES AND THE CHINESE REMAINDER THEOREM

The Euclidean algorithm for integers can be used to solve linear congruences. We first find the conditions for a single congruence to have a solution and then show how to find all its solutions, if they exist. We then present the Chinese Remainder Theorem, which gives conditions under which many simultaneous congruences, with coprime moduli, have solutions. These solutions can again be found by using the Euclidean algorithm.

First let us consider a linear congruence of the form

$$ax \equiv b \bmod n.$$

This has a solution if and only if the equation

$$ax + ny = b$$

has integer solutions for x and y. The congruence is also equivalent to the equation $[a][x] = [b]$ in $\mathbf{Z}_n$.

9.42 THEOREM. The equation $ax + ny = b$ has solutions for $x, y \in \mathbf{Z}$ if and only if $\mathrm{GCD}(a, n) | b$.

PROOF. If $ax + ny = b$ has a solution, then $\mathrm{GCD}(a, n)$ divides a and divides n; hence it divides b.

Conversely, if $\mathrm{GCD}(a, n) | b$, then $b = k \cdot \mathrm{GCD}(a, n)$. By Theorem 9.11, there exist $s, t \in \mathbf{Z}$ such that

$$as + nt = \mathrm{GCD}(a, n).$$

Hence $ask + ntk = k \cdot \mathrm{GCD}(a, n)$ and $x = sk, y = tk$ is a solution to $ax + ny = b$. $\square$

The Euclidean algorithm gives a practical way to find the integers s and t in the above theorem. These can then be used to find *one* solution to the equation.

9.43 THEOREM. The congruence $ax \equiv b \bmod n$ has a solution if and only if $\mathrm{GCD}(a, n) | b$. Moreover, if this congruence does have at least one solution, the number of noncongruent solutions modulo n is $\mathrm{GCD}(a, n)$; that is, if $[a][x] = [b]$ has a solution in $\mathbf{Z}_n$, then it has $\mathrm{GCD}(a, n)$ different solutions in $\mathbf{Z}_n$.

PROOF. The condition for the existence of a solution follows immediately from the previous theorem.

Now suppose that x_0 is a solution so that $ax_0 \equiv b \bmod n$. Let $g = \mathrm{GCD}(a, n)$ and $a = ga'$, $n = gn'$. The following statements are all equivalent.

(i) x is a solution to the congruence $ax \equiv b \bmod n$.
(ii) x is a solution to the congruence $a(x - x_0) \equiv 0 \bmod n$.
(iii) $n \mid a(x - x_0)$.
(iv) $n' \mid a'(x - x_0)$.
(v) $n' \mid (x - x_0)$.
(vi) $x = x_0 + kn'$ for some $k \in \mathbf{Z}$.

Now $x_0, x_0 + n', x_0 + 2n', \ldots, x_0 + (g-1)n'$ form a complete set of noncongruent solutions modulo n, and there are g such solutions. $\quad\square$

9.44 EXAMPLE. Find the inverse of $[49]$ in the field $\mathbf{Z}_{53}$.

SOLUTION. Let $[x] = [49]^{-1}$ in $\mathbf{Z}_{53}$. Then $[49] \cdot [x] = [1]$; that is, $49x \equiv 1 \bmod 53$. We can solve this congruence by solving the equation $49x - 1 = 53y$, where $y \in \mathbf{Z}$. By using the Euclidean algorithm we have

$$53 = 1 \cdot 49 + 4 \quad \text{and} \quad 49 = 12 \cdot 4 + 1.$$

Hence $\mathrm{GCD}(49, 53) = 1 = 49 - 12 \cdot 4 = 49 - 12(53 - 49) = 13 \cdot 49 - 12 \cdot 53$. Therefore, $13 \cdot 49 \equiv 1 \bmod 53$ and $[49]^{-1} = [13]$ in $\mathbf{Z}_{53}$. $\quad\square$

9.45 CHINESE REMAINDER THEOREM. Let $m = m_1 m_2 \ldots m_r$, where $\mathrm{GCD}(m_i, m_j) = 1$ if $i \neq j$. Then the system of simultaneous congruences

$$x \equiv a_1 \bmod m_1, \quad x \equiv a_2 \bmod m_2, \ldots, \quad x \equiv a_r \bmod m_r$$

always has an integral solution. Moreover, if b is one solution, the complete solution is the set of integers satisfying $x \equiv b \bmod m$.

PROOF. This result follows from the isomorphism

$$f : \mathbf{Z}_m \longrightarrow \mathbf{Z}_{m_1} \times \mathbf{Z}_{m_2} \times \cdots \times \mathbf{Z}_{m_r}$$

of Theorem 8.26 defined by $f([x]_m) = ([x]_{m_1}, [x]_{m_2}, \ldots, [x]_{m_r})$. The integer x is a solution of the simultaneous congruences if and only if $f([x]_m) = ([a_1]_{m_1}, [a_2]_{m_2}, \ldots, [a_r]_{m_r})$. Therefore, there is always a solution, and the solution set consists of exactly one congruence class modulo m. $\quad\square$

One method of finding the solution to a set of simultaneous congruences is to repeatedly use the Euclidean algorithm.

9.46 EXAMPLE. Solve the simultaneous congruences $\begin{cases} x \equiv 36 \bmod 41 \\ x \equiv 5 \bmod 17 \end{cases}$.

PROOF.   Any solution to the first congruence is of the form $x = 36 + 41t$ where $t \in \mathbf{Z}$. Substituting this into the second congruence, we obtain

$$36 + 41t \equiv 5 \bmod 17 \quad \text{and} \quad 41t \equiv -31 \bmod 17.$$

Reducing modulo 17, we have $7t \equiv 3 \bmod 17$. Solving this by the Euclidean algorithm, we have

$$17 = 2 \cdot 7 + 3 \quad \text{and} \quad 7 = 2 \cdot 3 + 1.$$

Therefore, $1 = 7 - 2(17 - 2 \cdot 7) = 7 \cdot 5 - 17 \cdot 2$ and $7 \cdot 5 \equiv 1 \bmod 17$. Hence $7 \cdot 15 \equiv 3 \bmod 17$ and $t \equiv 15 \bmod 17$ is the solution to $7t \equiv 3 \bmod 17$.

We have shown that, if $x = 36 + 41t$ is a solution to both congruences, then $t = 15 + 17u$ where $u \in \mathbf{Z}$. That is,

$$x = 36 + 41t = 36 + 41(15 + 17u) = 651 + 697u$$

or $x \equiv 651 \bmod 697$ is the complete solution. $\square$

9.47 EXAMPLE.   Find the smallest positive integer that has remainders 4, 3, and 1 when divided by 5, 7, and 9, respectively.

SOLUTION.   We have to solve the three simultaneous congruences

$$. \quad x \equiv 4 \bmod 5, \ x \equiv 3 \bmod 7, \text{ and } x \equiv 1 \bmod 9.$$

The first congruence implies that $x = 4 + 5t$ where $t \in \mathbf{Z}$. Substituting into the second congruence, we have

$$4 + 5t \equiv 3 \bmod 7.$$

Hence $5t \equiv -1 \bmod 7$. Now $5^{-1} = 3$ in $\mathbf{Z}_7$, so $t \equiv 3 \cdot (-1) \equiv 4 \bmod 7$. Therefore, $t = 4 + 7u$, where $u \in \mathbf{Z}$, and any integer satisfying the first two congruences is of the form

$$x = 4 + 5t = 4 + 5(4 + 7u) = 24 + 35u.$$

Substituting this into the third congruence, we have $24 + 35u \equiv 1 \bmod 9$ and $-u \equiv -23 \bmod 9$. Thus $u \equiv 5 \bmod 9$ and $u = 5 + 9v$ for some $v \in \mathbf{Z}$.

Hence any solution of the three congruences is of the form

$$x = 24 + 35u = 24 + 35(5 + 9v) = 199 + 315v.$$

The smallest positive solution is $x = 199$. $\square$

The Chinese Remainder Theorem was known to ancient Chinese astronomers, who used it to date events from observations of various periodic astronomical phenomena. It is used in this computer age as a tool for finding integer solutions to integer equations and for speeding up arithmetic operations in a computer.

Addition of two numbers in conventional representation has to be carried out sequentially on the digits in each position; the digits in the ith position have to be added before the digit to be carried over to the $(i+1)$st position is known. One method of speeding up addition on a computer is to perform addition using residue representation, since this avoids delays due to carry digits.

9.48 DEFINITION. Let $m = m_1 m_2 \ldots m_r$, where the integers m_i are coprime in pairs. The *residue representation* or *modular representation* of any number x in $\mathbf{Z}_m$ is the r-tuple $(a_1, a_2, \ldots, a_r)$ where $x \equiv a_i \bmod m_i$.

For example, every integer from 0 to 29 can be uniquely represented by its residues modulo 2, 3, and 5 in Table 9.4.

Table 9.4.   A residue representation of the integers from 0 to 29

x	Residues Modulo			x	Residues Modulo			x	Residues Modulo		
	2	3	5		2	3	5		2	3	5
0	0	0	0	10	0	1	0	20	0	2	0
1	1	1	1	11	1	2	1	21	1	0	1
2	0	2	2	12	0	0	2	22	0	1	2
3	1	0	3	13	1	1	3	23	1	2	3
4	0	1	4	14	0	2	4	24	0	0	4
5	1	2	0	15	1	0	0	25	1	1	0
6	0	0	1	16	0	1	1	26	0	2	1
7	1	1	2	17	1	2	2	27	1	0	2
8	0	2	3	18	0	0	3	28	0	1	3
9	1	0	4	19	1	1	4	29	1	2	4

This residue representation corresponds exactly to the isomorphism

$$\mathbf{Z}_{30} \longrightarrow \mathbf{Z}_2 \times \mathbf{Z}_3 \times \mathbf{Z}_5.$$

Since this is a ring isomorphism, addition and multiplication are performed by simply adding and multiplying each residue separately.

For example, to add 4 and 7 using residue representation, we have

$$(0, 1, 4) + (1, 1, 2) = (0 + 1, 1 + 1, 4 + 2) = (1, 2, 1).$$

Similarly, multiplying 4 and 7, we have

$$(0,1,4) \cdot (1,1,2) = (0 \cdot 1, 1 \cdot 1, 4 \cdot 2) = (0,1,3).$$

Fast adders can be designed using residue representation, because all the residues can be added simultaneously. Numbers can be converted easily into residue form; however, the reverse procedure of finding a number with a given residue representation requires the Chinese Remainder Theorem. See Szabo and Tanaka [42] for further discussion of the use of residue representations in computers.

Exercises

1–6. Calculate the quotients and remainders in the following divisions.

1. Divide $3x^4 + 4x^3 - x^2 + 5x - 1$ by $2x^2 + x + 1$ in $\mathbf{Q}[x]$.
2. Divide $x^6 + x^4 - 4x^3 + 5x$ by $x^3 + 2x^2 + 1$ in $\mathbf{R}[x]$.
3. Divide $x^7 + x^6 + x^4 + x + 1$ by $x^3 + x + 1$ in $\mathbf{Z}_2[x]$.
4. Divide $2x^5 + x^4 + 2x^3 + x^2 + 2$ by $x^3 + 2x + 2$ in $\mathbf{Z}_3[x]$.
5. Divide $17 + 11i$ by $3 + 4i$ in $\mathbf{Z}[i]$.
6. Divide $20 + 8i$ by $7 - 2i$ in $\mathbf{Z}[i]$.

7–13. Find the greatest common divisors of the following elements a,b in the given Euclidean ring, and find elements s,t in the ring so that $as + bt = \mathrm{GCD}(a,b)$.

7. $a = 33$, $b = 42$ in $\mathbf{Z}$.
8. $a = 2891$, $b = 1589$ in $\mathbf{Z}$.
9. $a = 2x^3 - 4x^2 - 8x + 1$, $b = 2x^3 - 5x^2 - 5x + 2 \in \mathbf{Q}[x]$.
10. $a = x^6 - x^3 - 16x^2 + 12x - 2$, $b = x^5 - 2x^2 - 16x + 8 \in \mathbf{Q}[x]$.
11. $a = x^4 + x + 1$, $b = x^3 + x^2 + x \in \mathbf{Z}_3[x]$.
12. $a = x^4 + 2$, $b = x^3 + 3 \in \mathbf{Z}_5[x]$.
13. $a = 4 - i$, $b = 1 + i \in \mathbf{Z}[i]$.

14–17. Find one solution to each of the following equations with $x,y \in \mathbf{Z}$.

14. $15x + 36y = 3$.   **15.** $24x + 29y = 1$.
16. $24x + 29y = 6$.   **17.** $11x + 31y = 1$.

18–21. Find the inverse to the following elements in the given field.

18. $[4]$ in $\mathbf{Z}_7$.   **19.** $[24]$ in $\mathbf{Z}_{29}$.

20. [35] in $\mathbf{Z}_{101}$.   **21.** [11] in $\mathbf{Z}_{31}$.

22–24.   Find all integral solutions to the following equations.

22. $27x + 15y = 13$.   **23.** $12x + 20y = 14$.
24. $28x + 20y = 16$.

25–36.   Factor the following polynomials into irreducible factors in the given ring.

25. $x^5 - 1$ in $\mathbf{Q}[x]$.   **26.** $x^5 + 1$ in $\mathbf{Z}_2[x]$.
27. $x^4 + 1$ in $\mathbf{Z}_5[x]$.   **28.** $2x^3 + x^2 + 4x + 2$ in $\mathbf{Q}[x]$.
29. $x^4 - 9x + 3$ in $\mathbf{Q}[x]$.   **30.** $2x^3 + x^2 + 4x + 2$ in $\mathbf{C}[x]$.
31. $x^3 - 4x + 1$ in $\mathbf{Q}[x]$.   **32.** $x^4 + 3x^3 + 9x - 9$ in $\mathbf{Q}[x]$.
33. $x^8 - 16$ in $\mathbf{C}[x]$.   **34.** $x^8 - 16 \in \mathbf{R}[x]$.
35. $x^8 - 16$ in $\mathbf{Q}[x]$.   **36.** $x^8 - 16 \in \mathbf{Z}_{17}[x]$.
37. Find all irreducible polynomials of degree 5 over $\mathbf{Z}_2$.
38. Find an irreducible polynomial of degree 2 over $\mathbf{Z}_5$.
39. Find an irreducible polynomial of degree 3 over $\mathbf{Z}_7$.
40. Find the kernel and image of the ring morphism $\psi : \mathbf{R}[x] \to \mathbf{C}$ defined by $\psi(p(x)) = p(i)$ where $i = \sqrt{-1}$.
41. Find the kernel and image of the ring morphism $\psi : \mathbf{Q}[x] \to \mathbf{C}$ defined by $\psi(p(x)) = p(1 + \sqrt{3}\, i)$.

42–47.   Are the following polynomials irreducible in the given ring? Give reasons.

42. $x^3 + x^2 + x + 1$ in $\mathbf{Q}[x]$.
43. $3x^8 - 4x^6 + 8x^5 - 10x + 6$ in $\mathbf{Q}[x]$.
44. $x^4 + x^2 - 6$ in $\mathbf{Q}[x]$.   **45.** $4x^3 + 3x^2 + x + 1$ in $\mathbf{Z}_5[x]$.
46. $x^5 + 15$ in $\mathbf{Q}[x]$.   **47.** $x^4 - 2x^3 + x^2 + 1$ in $\mathbf{R}[x]$.
48. Is $\mathbf{Z}[x]$ a Euclidean ring when $d(f(x)) = \deg f(x)$ for any nonzero polynomial? Is $\mathbf{Z}[x]$ a Euclidean ring with any other definition of $d(f(x))$?
49. Can you define a division algorithm in $\mathbf{R}[x, y]$? How would you divide $x^3 + 3xy + y + 4$ by $xy + y^3 + 2$?
50. Let L_p be the set of all linear functions $f : \mathbf{Z}_p \to \mathbf{Z}_p$ of the form $f(x) = ax + b$, where $a \neq 0$ in $\mathbf{Z}_p$. Show that $(L_p, \circ)$ is a group of order $p(p - 1)$ under composition.
51. If p is a prime, prove that $(x - a)|(x^{p-1} - 1)$ in $\mathbf{Z}_p[x]$ for all nonzero a in $\mathbf{Z}_p$. Hence prove that

$$x^{p-1} - 1 = (x - 1)(x - 2) \cdots (x - p + 1)   \text{in}   \mathbf{Z}_p[x].$$

52. *(Wilson's Theorem)* Prove that $(n - 1)! \equiv -1 \bmod n$ if and only if n is prime.

53. Prove that $\sqrt{2}/\sqrt[3]{5}$ is irrational.

54. Find a polynomial in $\mathbf{Q}[x]$ with $\sqrt{2}+\sqrt{3}$ as a root. Then prove that $\sqrt{2}+\sqrt{3}$ is irrational.

55. Is 5 irreducible in $\mathbf{Z}[i]$?

56. Show that $\mathbf{Z}[\sqrt{-5}\,]=\{a+b\sqrt{-5}\,|a,b\in\mathbf{Z}\}$ does not have the unique factorization property.

57. Prove that a Gaussian integer is irreducible if and only if it is an invertible element times one of the following Gaussian integers:

(i)   any prime p in $\mathbf{Z}$ with $p\equiv3\,\mathrm{mod}\,4$.

(ii) $1+i$.

(iii) $a+bi$ where a is positive and even, and $a^2+b^2=p$, for some prime p in $\mathbf{Z}$ such that $p\equiv1\,\mathrm{mod}\,4$.

58. If r/s is a rational root, in its lowest terms, of a polynomial $p(x)$ with integral coefficients, show that

$$p(x)=(sx-r)g(x)$$

for some polynomial $g(x)$ with *integral* coefficients.

59. Prove that r/s, in its lowest terms, cannot be a root of the integral polynomial $p(x)$ unless $(s-r)|p(1)$. This can be used to shorten the list of possible rational roots of an integral polynomial.

60. Let $m=m_1m_2\ldots m_r$ and $M_i=m/m_i$. If $\mathrm{GCD}(m_i,m_j)=1$ for $i\neq j$, each of the congruences $M_iy\equiv1\,\mathrm{mod}\,m_i$ has a solution $y\equiv b_i\,\mathrm{mod}\,m_i$. Prove that the solution to the simultaneous congruences

$$x\equiv a_1\,\mathrm{mod}\,m_1,\ x\equiv a_2\,\mathrm{mod}\,m_2,\ldots,\ x\equiv a_r\,\mathrm{mod}\,m_r,$$

is

$$x\equiv\sum_{i=1}^{r}M_ib_ia_i\,\mathrm{mod}\,m.$$

61–64. *Solve the following simultaneous congruences.*

61. $x\equiv5\quad\mathrm{mod}\,7$
$\quad\ \ x\equiv4\quad\mathrm{mod}\,6.$

63. $x\equiv0\quad\mathrm{mod}\,2$
$\quad\ \ x\equiv1\quad\mathrm{mod}\,3$
$\quad\ \ x\equiv2\quad\mathrm{mod}\,5.$

62. $x\equiv41\quad\mathrm{mod}\,65$
$\quad\ \ x\equiv35\quad\mathrm{mod}\,72.$

64. $x\equiv9\quad\mathrm{mod}\,12$
$\quad\ \ x\equiv3\quad\mathrm{mod}\,13$
$\quad\ \ x\equiv6\quad\mathrm{mod}\,25.$

65. Prove that $\det\begin{bmatrix}320&461&5264&72\\702&1008&-967&-44\\-91&2333&46&127\\164&-216&1862&469\end{bmatrix}$ is nonzero.

66. Solve the following simultaneous equations

$$26x - 141y = -697$$

$$55x - 112y = 202$$

(i) in $\mathbf{Z}_2$,   (ii) in $\mathbf{Z}_3$,   (iii) in $\mathbf{Z}_5$.

Then use the Chinese Remainder Theorem to solve them in $\mathbf{Z}$ assuming they have a pair of integral solutions between 0 and 29.

67. The value of $\det \begin{bmatrix} 676 & 117 & 522 \\ 375 & 65 & 290 \\ 825 & 143 & 639 \end{bmatrix}$ is positive and less than 100.

Find its value without using a calculator. (If you get tired of doing arithmetic, calculate its value mod 10 and mod 11 and then use the Chinese Remainder Theorem.)

68. The polynomial $x^3 + 5x \in \mathbf{Z}_6[x]$ has six roots. Does this contradict Theorem 9.07?

Quotient Rings | 10

In this chapter, we define a quotient ring in a way similar to our definition of a quotient group. The analogue of a normal subgroup is called an ideal, and a quotient ring consists of the set of cosets of the ring by one of its ideals. As in groups, we have a Morphism Theorem connecting morphisms, ideals, and quotient groups. We discover under what conditions quotient rings are fields. This will enable us to fulfill our long-range goal of extending the number systems by defining new fields using quotient rings of some familiar rings.

IDEALS AND QUOTIENT RINGS

If $(R, +, \cdot)$ is any ring and $(S, +)$ is any subgroup of the Abelian group $(R, +)$, then the quotient *group* $(R/S, +)$ is defined. However, R/S does not have a *ring* structure induced on it by R unless S is a special kind of subset called an ideal.

10.01 DEFINITION. A nonempty subset I of a ring R is called an *ideal* of R if, for all $x,y \in I$ and $r \in R$,

(i) $x - y \in I$

and   (ii) $x \cdot r$ and $r \cdot x \in I$.

Condition (i) implies that $(I, +)$ is a subgroup of $(R, +)$. In any ring R, R itself is an ideal, and $\{0\}$ is an ideal.

10.02 PROPOSITION. Let a be an element of a commutative ring R. The set $\{ar \mid r \in R\}$ of all multiples of a is an ideal of R called the *principal ideal* generated by a. This ideal is denoted by (a).

PROOF. Let $ar, as \in (a)$ and $t \in R$. Then $ar - as = a(r - s) \in (a)$ and $(ar)t = a(rt) \in (a)$. Hence (a) is an ideal of R. $\square$

For example, $(n) = n\mathbf{Z}$, consisting of all integer multiples of n, is the principal ideal generated by n in $\mathbf{Z}$.

The set of all polynomials in $\mathbf{Q}[x]$ that contain $x^2 - 2$ as a factor is the principal ideal $(x^2 - 2) = \{(x^2 - 2) \cdot p(x) \mid p(x) \in \mathbf{Q}[x]\}$ generated by $x^2 - 2$ in $\mathbf{Q}[x]$. The set of all real polynomials that have zero constant term $(x) = \{x \cdot p(x) \mid p(x) \in \mathbf{R}[x]\}$ is the principal ideal generated by x in $\mathbf{R}[x]$. It is also the set of real polynomials with 0 as a root.

The set of all real polynomials, in two variables x and y, that have a zero constant term is an ideal of $\mathbf{R}[x,y]$. However, this ideal is not principal. (See Exercise 30.)

In many rings, every ideal is principal; these are called *principal ideal rings*.

10.03 THEOREM. A Euclidean ring is a principal ideal ring.

PROOF. Let I be any ideal of the Euclidean ring R. If $I = \{0\}$, then $I = (0)$, the principal ideal generated by 0. Otherwise I contains nonzero elements. Let b be a nonzero element of I for which $d(b)$ is minimal. If a is any other element in I, then, by the division algorithm, there exist $q, r \in R$ such that

$$a = q \cdot b + r \quad \text{where} \quad r = 0 \quad \text{or} \quad d(r) < d(b).$$

Now $r = a - q \cdot b \in I$. Since b is an element for which $d(b)$ is minimal, it follows that r must be zero and $a = q \cdot b$. Therefore, $a \in (b)$ and $I \subseteq (b)$.

Conversely, any element of (b) is of the form $q \cdot b$ for some $q \in R$ and $q \cdot b \in I$. Therefore, $I \supseteq (b)$ and $I = (b)$. Hence R is a principal ideal ring. $\square$

10.04 COROLLARY. **Z** is a principal ideal ring and so is **F**[x], if **F** is a field.

PROOF. This follows because **Z** and **F**[x] are Euclidean rings. $\square$

10.05 PROPOSITION. Let I be ideal of the ring R. If I contains the identity 1, then I is the whole ring R.

PROOF. Let $1 \in I$ and $r \in R$. Then $r \cdot 1 = r \in I$ and $I = R$. $\square$

Let I be any ideal in a ring R. Then $(I, +)$ is a normal subgroup of $(R, +)$, and we denote the coset of I in R that contains r by $I + r$. Hence

$$I + r = \{ i + r \in R | i \in I \}.$$

The cosets of I in R are the equivalence classes under the congruence relation modulo I. We have

$$r_1 \equiv r_2 \bmod I \quad \text{if and only if} \quad r_1 - r_2 \in I.$$

By Theorem 4.20, the set of cosets $R / I = \{ I + r | r \in R \}$ is an Abelian group under the operation defined by

$$(I + r_1) + (I + r_2) = I + (r_1 + r_2).$$

10.06 THEOREM. Let I be an ideal in the ring R. Then the set of cosets forms a ring $(R / I, +, \cdot)$ under the operations defined by

$$(I + r_1) + (I + r_2) = I + (r_1 + r_2)$$

and

$$(I + r_1)(I + r_2) = I + (r_1 r_2).$$

10.07 DEFINITION. This ring $(R / I, +, \cdot)$ is called the *quotient ring* (or *factor ring*) of R by I.

PROOF. As mentioned above, $(R / I, +)$ is an Abelian group; thus we only have to verify the axioms related to multiplication.

We first show that multiplication is well defined on cosets. Let $I + r_1' = I + r_1$ and $I + r_2' = I + r_2$ so that $r_1' - r_1 = i_1 \in I$ and $r_2' - r_2 = i_2 \in I$. Then

$$r_1' r_2' = (i_1 + r_1)(i_2 + r_2) = i_1 i_2 + r_1 i_2 + i_1 r_2 + r_1 r_2.$$

Now, since I is an ideal, $i_1 i_2$, $r_1 i_2$ and $i_1 r_2 \in I$. Hence $r_1' r_2' - r_1 r_2 \in I$ and $I + r_1' r_2' = I + r_1 r_2$, which shows that multiplication is well defined on R/I.

Multiplication is associative and distributive over addition. If $r_1, r_2, r_3 \in R$, then

$$(I+r_1)\{(I+r_2)(I+r_3)\} = (I+r_1)(I+r_2 r_3) = I + r_1(r_2 r_3) = I + (r_1 r_2)r_3$$

$$= (I+r_1 r_2)(I+r_3) = \{(I+r_1)(I+r_2)\}(I+r_3).$$

Also

$$(I+r_1)\{(I+r_2)+(I+r_3)\} = (I+r_1)\{I+(r_2+r_3)\} = I + r_1(r_2+r_3)$$

$$= I + (r_1 r_2 + r_1 r_3) = (I+r_1 r_2) + (I+r_1 r_3)$$

$$= \{(I+r_1)(I+r_2)\} + \{(I+r_1)(I+r_3)\}.$$

The other distributive law can be proved similarly. The multiplicative identity is $I+1$. Hence $(R/I, +, \cdot)$ is a ring. $\square$

For example, the quotient ring of $\mathbf{Z}$ by (n) is $\mathbf{Z}/(n) = \mathbf{Z}_n$, the ring of integers modulo n. A coset $(n)+r = \{nz + r \mid z \in \mathbf{Z}\}$ is the equivalent class modulo n containing r.

If R is commutative, so is the quotient ring R/I, because

$$(I+r_1)(I+r_2) = I + r_1 r_2 = I + r_2 r_1 = (I+r_2)(I+r_1).$$

10.08 EXAMPLE. If $I = \{0,2,4\}$ is the ideal generated by 2 in $\mathbf{Z}_6$, find the tables for the quotient ring $\mathbf{Z}_6/I$.

SOLUTION. There are two cosets of $\mathbf{Z}_6$ by I, namely, $I = \{0,2,4\}$ and $I+1 = \{1,3,5\}$. Hence

$$\mathbf{Z}_6/I = \{I, I+1\}.$$

The addition and multiplication tables given in Table 10.1 show that the quotient ring $\mathbf{Z}_6/I$ is isomorphic to $\mathbf{Z}_2$. $\square$

Table 10.1.   The quotient ring $\mathbf{Z}_6/\{0,2,4\}$

+	I	$I+1$	$\cdot$	I	$I+1$
I	I	$I+1$	I	I	I
$I+1$	$I+1$	I	$I+1$	I	$I+1$

COMPUTATIONS IN QUOTIENT RINGS

If **F** is a field, the quotient rings of the polynomial ring **F**[x] form an important class of rings that will be used to construct new fields. Recall that **F**[x] is a principal ideal ring, so that any quotient ring is of the form **F**[x]/(p(x)), for some polynomial $p(x) \in$ **F**[x]. We now look at the structure of such a quotient ring.

The elements of the ring **F**[x]/(p(x)) are equivalence classes under the relation on **F**[x] defined by

$$f(x) \equiv g(x) \bmod (p(x)) \quad \text{if and only if} \quad f(x) - g(x) \in (p(x)).$$

10.09 LEMMA. $f(x) \equiv g(x) \bmod (p(x))$ if and only if $f(x)$ and $g(x)$ have the same remainder when divided by $p(x)$.

PROOF. Let $f(x) = q(x) \cdot p(x) + r(x)$ and $g(x) = s(x) \cdot p(x) + t(x)$, where $r(x)$ and $t(x)$ are zero or have degrees less than that of $p(x)$.

The following statements are equivalent.

  (i) $f(x) \equiv g(x) \bmod (p(x))$.
 (ii) $f(x) - g(x) \in (p(x))$.
(iii) $p(x) | f(x) - g(x)$.
(iv) $p(x) | \{q(x) - s(x)\} \cdot p(x) + r(x) - t(x)$.
 (v) $p(x) | r(x) - t(x)$.
(vi) $r(x) = t(x)$. $\quad\square$

Hence every coset of **F**[x] by $(p(x))$ contains the zero polynomial or a polynomial of degree less than that of $p(x)$.

10.10 THEOREM. Let P be the ideal $(p(x))$, generated by the polynomial $p(x)$ of degree $n > 0$. The different elements of **F**[x]/(p(x)) are precisely those of the form

$$P + a_0 + a_1 x + \cdots + a_{n-1} x^{n-1} \quad \text{where} \quad a_0, a_1, \ldots, a_{n-1} \in \textbf{F}.$$

PROOF. Let $P + f(x)$ be any element of **F**[x]/(p(x)) and let $r(x)$ be the remainder when $f(x)$ is divided by $p(x)$. Then, by Lemma 10.09, $P + f(x) = P + r(x)$, which is of the required form.

Suppose that $P + r(x) = P + t(x)$ where $r(x)$ and $t(x)$ are zero or have degree less than n. Then

$$r(x) \equiv t(x) \bmod (p(x)),$$

and by Lemma 10.09, $r(x) = t(x)$. $\quad\square$

10.11 EXAMPLE. Write down the tables for $Z_2[x]/(x^2+x+1)$.

SOLUTION. Let $P=(x^2+x+1)$ so that

$$Z_2[x]/(x^2+x+1)=\{P+a_0+a_1x|a_0,a_1\in Z_2\}$$
$$=\{P,P+1,P+x,P+x+1\}.$$

The tables for the quotient ring are given in Table 10.2. The addition table is straightforward to calculate. Multiplication is computed as follows.

$$(P+x)^2=P+x^2=P+(x^2+x+1)+x+1=P+x+1$$

and

$$(P+x)(P+x+1)=P+x^2+x=P+(x^2+x+1)+1=P+1. \quad \square$$

Table 10.2.   The ring $Z_2[x]/(x^2+x+1)$

+	P	$P+1$	$P+x$	$P+x+1$
P	P	$P+1$	$P+x$	$P+x+1$
$P+1$	$P+1$	P	$P+x+1$	$P+x$
$P+x$	$P+x$	$P+x+1$	P	$P+1$
$P+x+1$	$P+x+1$	$P+x$	$P+1$	P

·	P	$P+1$	$P+x$	$P+x+1$
P	P	P	P	P
$P+1$	P	$P+1$	$P+x$	$P+x+1$
$P+x$	P	$P+x$	$P+x+1$	$P+1$
$P+x+1$	P	$P+x+1$	$P+1$	$P+x$

10.12 EXAMPLE. Find the sum and product of $P+3x+4$ and $P+5x-6$ in the ring $Q[x]/(x^2-2)=\{P+a_0+a_1x|a_0,a_1\in Q\}$, where P is the principal ideal generated by x^2-2.

SOLUTION. $(P+3x+4)+(P+5x-6)=P+(3x+4)+(5x-6)=P+8x-2$. $(P+3x+4)(P+5x-6)=P+(3x+4)(5x-6)=P+15x^2+2x-24$. By the division algorithm, $15x^2+2x-24=15(x^2-2)+2x+6$. Hence, by Lemma 10.09, $P+15x^2+2x-24=P+2x+6.$ $\square$

There are often easier ways of finding the remainder of $f(x)$ when divided by $p(x)$ than by applying the division algorithm directly. If $\deg p(x)=n$ and $P=(p(x))$, the problem of finding the remainder reduces to the problem of finding a polynomial $r(x)$ of degree less than n such that $f(x) \equiv r(x) \bmod P$. This can often be solved by manipulating congruences, using the fact that $p(x) \equiv 0 \bmod P$.

Consider Example 10.12 in which P is the ideal generated by $x^2 - 2$. Then $x^2 - 2 \equiv 0 \bmod P$ and $x^2 \equiv 2 \bmod P$. Hence, in any congruence modulo P, we can always replace x^2 by 2. For example,

$$15x^2 + 2x - 24 \equiv 15(2) + 2x - 24 \bmod P$$

$$\equiv 2x + 6 \bmod P$$

and so $P + 15x^2 + 2x - 24 = P + 2x + 6$.

In Example 10.11, $P = (x^2 + x + 1)$, so $x^2 + x + 1 \equiv 0 \bmod P$ and $x^2 \equiv x + 1 \bmod P$. (Remember $+1 = -1$ in $\mathbf{Z}_2$.) Therefore, in multiplying two elements in $\mathbf{Z}_2[x]/P$, we can always replace x^2 by $x + 1$. For example,

$$P + x^2 = P + x + 1 \quad \text{and} \quad P + x(x+1) = P + x^2 + x = P + 1.$$

We have usually written the elements of $\mathbf{Z}_n = \mathbf{Z}/(n)$ simply as $0, 1, \ldots,$ $n - 1$ instead of as $[0], [1], \ldots, [n-1]$ or as $(n) + 0, (n) + 1, \ldots, (n) + n - 1$. In a similar way, when there is no confusion, we henceforth write the elements of $\mathbf{F}[x]/(p(x))$ simply as $a_0 + a_1 x + \cdots + a_{n-1} x^{n-1}$ instead of $(p(x)) + a_0 + a_1 x + \cdots + a_{n-1} x^{n-1}$.

Morphism Theorem

10.13 PROPOSITION. If $f: R \to S$ is a ring morphism, then $\mathrm{Ker} f$ is an ideal of R.

PROOF. Since any ring morphism is a group morphism, it follows from Proposition 4.27 that $\mathrm{Ker} f$ is a subgroup of $(R, +)$. If $x \in \mathrm{Ker} f$ and $r \in R$, then $f(xr) = f(x)f(r) = 0 \cdot f(r) = 0$ and $xr \in \mathrm{Ker} f$. Similarly, $rx \in \mathrm{Ker} f$, and $\mathrm{Ker} f$ is an ideal of R. $\quad\square$

Furthermore, any ideal I of a ring R is the kernel of a morphism, for example, the ring morphism $\pi: R \to R/I$ defined by $\pi(r) = I + r$.

The image of $f: R \to S$ can easily be verified to be a subring of S.

10.14 MORPHISM THEOREM FOR RINGS. If $f: R \to S$ is a ring morphism, then $R/\mathrm{Ker} f$ is isomorphic to $\mathrm{Im} f$.

This result is also known as the First Isomorphism Theorem for Rings; the Second and Third Isomorphism Theorems are given in Exercises 19 and 20.

PROOF.   Let $K = \mathrm{Ker} f$. It follows from the Morphism Theorem for Groups 4.29, that $\psi : R/K \to \mathrm{Im} f$, defined by $\psi(K+r) = f(r)$, is a group isomorphism. Hence we need only prove that ψ is a ring morphism. We have

$$\psi\{(K+r)(K+s)\} = \psi\{K+rs\} = f(rs) = f(r)f(s)$$

$$= \psi(K+r)\psi(K+s). \quad \square$$

10.15 EXAMPLE.   Prove that $\mathbf{Q}[x]/(x^2-2) \cong \mathbf{Q}(\sqrt{2}\,)$.

SOLUTION.   Consider the ring morphism $\psi : \mathbf{Q}[x] \to \mathbf{R}$ defined by $\psi(f(x)) = f(\sqrt{2}\,)$ in Example 9.31. The kernel is the set of polynomials containing x^2-2 as a factor, that is, the principal ideal (x^2-2). The image of ψ is $\mathbf{Q}(\sqrt{2}\,)$, and, by the Morphism Theorem for Rings, $\mathbf{Q}[x]/(x^2-2) \cong \mathbf{Q}(\sqrt{2}\,)$. $\square$

In this isomorphism, the element $a_0 + a_1 x \in \mathbf{Q}[x]/(x^2-2)$ is mapped to $a_0 + a_1\sqrt{2} \in \mathbf{Q}(\sqrt{2}\,)$. Addition and multiplication of the elements $a_0 + a_1 x$ and $b_0 + b_1 x$ in $\mathbf{Q}[x]/(x^2-2)$ correspond to the addition and multiplication of the real numbers $a_0 + a_1\sqrt{2}$ and $b_0 + b_1\sqrt{2}$.

10.16 EXAMPLE.   Prove that $\mathbf{R}[x]/(x^2+1) \cong \mathbf{C}$.

SOLUTION.   Define the ring morphism $\psi : \mathbf{R}[x] \to \mathbf{C}$ by $\psi(f(x)) = f(i)$ where $i = \sqrt{-1}$. Any polynomial in $\mathrm{Ker}\,\psi$ has i as a root, and therefore, by Theorem 9.29, also has $-i$ as a root and contains the factor x^2+1. Hence $\mathrm{Ker}\,\psi = (x^2+1)$.

Now $\psi(a+bx) = a+ib$; thus ψ is surjective. By the Morphism Theorem for Rings, $\mathbf{R}[x]/(x^2+1) \cong \mathbf{C}$. $\square$

QUOTIENT POLYNOMIAL RINGS THAT ARE FIELDS.

We now determine when a quotient of a polynomial ring is a field. This result allows us to construct many new fields.

10.17 THEOREM.   Let a be an element of the Euclidean ring R. The quotient ring $R/(a)$ is a field if and only if a is irreducible in R.

PROOF.   Suppose that a is an irreducible element of R and let $(a)+b$ be a nonzero element of $R/(a)$. Then b is not a multiple of a and, since a is irreducible, $\mathrm{GCD}(a,b) = 1$. By Theorem 9.11, there exist $s, t \in R$ such that

$$sa + tb = 1.$$

Now $sa \in (a)$, so $[(a) + t] \cdot [(a) + b] = (a) + 1$, the identity of $R/(a)$. Hence $(a) + t$ is the inverse of $(a) + b$ in $R/(a)$ and $R/(a)$ is a field.

Now suppose that a is not irreducible in R so that there exist elements s and t, which are not invertible, with $st = a$. By Lemma 9.23, $d(s) < d(st) = d(a)$ and $d(t) < d(st) = d(a)$. Hence s is not divisible by a, and $s \notin (a)$. Similarly, $t \notin (a)$, and neither $(a) + s$ nor $(a) + t$ is the zero element of $R/(a)$. However,

$$[(a) + s] \cdot [(a) + t] = (a) + st = (a), \quad \text{the zero element of} \quad R/(a).$$

Therefore, the ring $R/(a)$ has zero divisors and cannot possibly be a field. □

For example, in the quotient ring $\mathbf{Q}[x]/P$, where $P = (x^2 - 1)$, the elements $P + x + 1$ and $P + x - 1$ are zero divisors because

$$(P + x + 1) \cdot (P + x - 1) = P + x^2 - 1 = P, \text{ the zero element.}$$

10.18 COROLLARY. $\mathbf{Z}_p = \mathbf{Z}/(p)$ is a field if and only if p is prime.

PROOF. This result, which we have proved previously in Theorem 9.18, follows from the above theorem because the irreducible elements in $\mathbf{Z}$ are the primes (and their negatives). □

Another particular case of Theorem 10.17 is the following important theorem.

10.19 THEOREM. The ring $\mathbf{F}[x]/(p(x))$ is a field if and only if $p(x)$ is irreducible over the field $\mathbf{F}$. Furthermore, the ring $\mathbf{F}[x]/(p(x))$ always contains a subring isomorphic to the field $\mathbf{F}$.

PROOF. The first part of the theorem is just Theorem 10.17. Let $F = \{(p(x)) + r | r \in \mathbf{F}\}$. This can be verified to be a subring of $\mathbf{F}[x]/(p(x))$, which is isomorphic to the field $\mathbf{F}$ by the isomorphism that takes $r \in \mathbf{F}$ to $(p(x)) + r \in F$. □

10.20 EXAMPLE. Show that $\mathbf{Z}_2[x]/(x^2 + x + 1)$ is a field with four elements.

SOLUTION. We showed in Example 9.41 that $x^2 + x + 1$ is irreducible over $\mathbf{Z}_2$ and in Example 10.11 that the quotient ring has four elements. Hence the quotient ring is a field containing four elements. Its tables are given in Table 10.2. □

10.21 EXAMPLE. Write down the multiplication table for the field $Z_3[x]/(x^2+1)$.

SOLUTION. If $x=0,1$, or 2 in Z_3, then $x^2+1=1,2$, or 2; thus, by the Factor Theorem, x^2+1 has no linear factors. Hence x^2+1 is irreducible over Z_3, and, by Theorem 10.19, the quotient ring $Z_3[x]/(x^2+1)$ is a field. By Theorem 10.10, the elements of this field can be written as

$$Z_3[x]/(x^2+1)=\{a_0+a_1x|a_0,a_1\in Z_3\}.$$

Hence the field contains nine elements. Its multiplication table is given in Table 10.3. This can be calculated by multiplying the polynomials in $Z_3[x]$ and replacing x^2 by -1 or 2, since $x^2\equiv-1\equiv2\bmod(x^2+1)$. $\square$

Table 10.3. Multiplication in $Z_3[x]/(x^2+1)$

·	0	1	2	x	$x+1$	$x+2$	$2x$	$2x+1$	$2x+2$
0	0	0	0	0	0	0	0	0	0
1	0	1	2	x	$x+1$	$x+2$	$2x$	$2x+1$	$2x+2$
2	0	2	1	$2x$	$2x+2$	$2x+1$	x	$x+2$	$x+1$
x	0	x	$2x$	2	$x+2$	$2x+2$	1	$x+1$	$2x+1$
$x+1$	0	$x+1$	$2x+2$	$x+2$	$2x$	1	$2x+1$	2	x
$x+2$	0	$x+2$	$2x+1$	$2x+2$	1	x	$x+1$	$2x$	2
$2x$	0	$2x$	x	1	$2x+1$	$x+1$	2	$2x+2$	$x+2$
$2x+1$	0	$2x+1$	$x+2$	$x+1$	2	$2x$	$2x+2$	x	1
$2x+2$	0	$2x+2$	$x+1$	$2x+1$	x	2	$x+2$	1	$2x$

10.22 EXAMPLE. Show that $Q[x]/(x^3-5)=\{a_0+a_1x+a_2x^2|a_i\in Q\}$ is a field and find the inverse of the element $x+1$.

SOLUTION. By the Rational Roots Theorem 9.32, (x^3-5) has no linear factors and hence is irreducible over Q. Therefore, by Theorem 10.19, $Q[x]/(x^3-5)$ is a field.

If $s(x)$ is the inverse of $x+1$, then $(x+1)s(x)\equiv1\bmod(x^3-5)$; that is, $(x+1)s(x)+(x^3-5)t(x)=1$ for some $t(x)\in Q[x]$.

We can find such polynomials $s(x)$ and $t(x)$ by the Euclidean algorithm. We have

$$x^3-5=(x^2-x+1)(x+1)-6$$

so

$$6\equiv(x^2-x+1)(x+1)\bmod(x^3-5)$$

and

$$1\equiv\frac{(x^2-x+1)}{6}(x+1)\bmod(x^3-5).$$

Hence $(x+1)^{-1} = \dfrac{x^2}{6} - \dfrac{x}{6} + \dfrac{1}{6}$ in $\mathbf{Q}[x]/(x^3-5)$. $\square$

$$
\begin{array}{r}
x^2 - x + 1 \\
x+1\overline{\smash{\big)}\,x^3+0+0 \quad -5} \\
x^3 + x^2 \\
\hline
-x^2 \\
-x^2 - x \\
\hline
x \,-5 \\
x \,+1 \\
\hline
-6
\end{array}
$$

10.23 EXAMPLE. Show that $\mathbf{Z}_3[x]/(x^3+2x+1)$ is a field with 27 elements and find the inverse of the element x^2.

SOLUTION.   If $x=0$, 1, or 2 in $\mathbf{Z}_3$, then $x^3+2x+1=1$; hence x^3+2x+1 has no linear factors and is irreducible. Therefore,

$$\mathbf{Z}_3[x]/(x^3+2x+1) = \{a_0 + a_1 x + a_2 x^2 \,|\, a_i \in \mathbf{Z}_3\}$$

is a field that has $3^3 = 27$ elements.

As in the previous example, to find the inverse of x^2, we apply the Euclidean algorithm to x^3+2x+1 and x^2 in $\mathbf{Z}_3[x]$.

We have $x^3+2x+1 = x(x^2) + (2x+1)$ and $x^2 = (2x+2)(2x+1)+1$. Hence

$$1 = x^2 - (2x+2)\{x^3+2x+1 - x \cdot x^2\}$$

$$= x^2(2x^2+2x+1) - (2x+2)(x^3+2x+1)$$

so

$$1 \equiv x^2(2x^2+2x+1) \bmod (x^3+2x+1)$$

and the inverse of x^2 in $\mathbf{Z}_3[x]/(x^3+2x+1)$ is $2x^2+2x+1$. $\square$

$$
\begin{array}{r}
x \\
x^2\overline{\smash{\big)}\,x^3+0+2x+1} \\
x^3 \\
\hline
2x+1 \\
\hline
\end{array}
\qquad
\begin{array}{r}
2x+2 \\
2x+1\overline{\smash{\big)}\,x^2+0+0} \\
x^2+2x \\
\hline
x \\
x+2 \\
\hline
1 \\
\hline
\end{array}
$$

We cannot use Theorem 10.17 directly on a field to obtain any new quotient fields, because the only ideals of a field are the zero ideal and the whole field. In fact, the following result shows that a field can be characterized by its ideals.

10.24 THEOREM. The nontrivial commutative ring R is a field if and only if (0) and R are its only ideals.

PROOF. Let I be an ideal in the field R. Suppose $I \neq (0)$, so that there is a nonzero element $a \in I$. Since $a^{-1} \in R$, $a \cdot a^{-1} = 1 \in I$. Therefore, by Proposition 10.05, $I = R$. Hence R has only trivial ideals.

Conversely, suppose (0) and R are the only ideals in the ring R. Let a be a nonzero element of R and consider (a) the principal ideal generated by a. Since $1 \cdot a \in (a)$, $(a) \neq (0)$, and hence $(a) = R$. But $1 \in R = (a)$, so there must exist some $b \in R$ such that $a \cdot b = 1$. Therefore, $b = a^{-1}$ and R is a field. $\square$

Exercises

1–6. Find all the ideals in the following rings.

1. $\mathbf{Z}_2 \times \mathbf{Z}_2$.   2. $\mathbf{Z}_{18}$.   3. $\mathbf{Q}$.
4. $\mathbf{Z}_7$.   5. $\mathbf{C}[x]$.   6. $\mathbf{Z}[i]$.

7–10. Construct addition and multiplication tables for the following rings. Find all the zero divisors in each ring. Which of these rings are fields?

7. $\mathbf{Z}_6 / (3)$.   8. $\mathbf{Z}_2[x]/(x^3 + 1)$.
9. $\mathbf{Z}_3 \times \mathbf{Z}_3 / ((1,2))$.   10. $\mathbf{Z}_3[x]/(x^2 + 2x + 2)$.

11–14. Compute the sum and product of the following elements in the given quotient rings.

11. $3x + 4$ and $5x - 2$ in $\mathbf{Q}[x]/(x^2 - 7)$.
12. $x^2 + 3x + 1$ and $-2x^2 + 4$ in $\mathbf{Q}[x]/(x^3 + 2)$.
13. $x^2 + 1$ and $x + 1$ in $\mathbf{Z}_2[x]/(x^3 + x + 1)$.
14. $ax + b$ and $cx + d$ in $\mathbf{R}[x]/(x^2 + 1)$, where $a, b, c, d \in \mathbf{R}$.
15. If U and V are ideals in a ring R, prove that $U \cap V$ is also an ideal in R.
16. Show, by example, that if U and V are ideals in a ring R, then $U \cup V$ is not necessarily an ideal in R. But prove that $U + V = \{u + v | u \in U, v \in V\}$ is always an ideal in R.

17. Find a generator of the following ideals in the given ring and prove a general result for the intersection of two ideals in a principal ideal ring.
(i) $(2) \cap (3)$ in $\mathbf{Z}$.   (ii) $(12) \cap (18)$ in $\mathbf{Z}$.
(iii) $(x^2 - 1) \cap (x + 1)$ in $\mathbf{Q}[x]$.

18. Find a generator of the following ideals in the given ring and prove a general result for the sum of two ideals in a principal ideal ring.
(i) $(2) + (3)$ in $\mathbf{Z}$.   (ii) $(9) + (12)$ in $\mathbf{Z}$.
(iii) $(x^2 + x + 1) + (x^2 + 1)$ in $\mathbf{Z}_2$.

19. *(Second Isomorphism Theorem for Rings)* If I and J are ideals of the ring R, prove that

$$I/(I \cap J) \cong (I + J)/J.$$

20. *(Third Isomorphism Theorem for Rings)* Let I and J be two ideals of the ring R, with $J \subseteq I$. Prove that I/J is an ideal of R/J and that

$$(R/J)/(I/J) \cong R/I.$$

21–29. Prove the following isomorphisms.

21. $\mathbf{R}[x]/(x^2 + 5) \cong \mathbf{C}$.
22. $\mathbf{Z}[x]/(x^2 + 1) \cong \mathbf{Z}[i] = \{a + ib | a, b \in \mathbf{Z}\}$.
23. $\mathbf{Q}[x]/(x^2 - 7) \cong \mathbf{Q}(\sqrt{7}) = \{a + b\sqrt{7} | a, b \in \mathbf{Q}\}$.
24. $\mathbf{Z}[x]/(2x - 1) \cong \{a/b \in \mathbf{Q} | a \in \mathbf{Z}, b = 2^r, r \geqslant 0\}$, a subring of $\mathbf{Q}$.
25. $\mathbf{Z}_{14}/(7) \cong \mathbf{Z}_7$.   **26.** $\mathbf{Z}_{14}/(2) \cong \mathbf{Z}_2$.
27. $\mathbf{R}[x, y]/(x + y) \cong \mathbf{R}[y]$.   **28.** $R \times S/((1, 0)) \cong S$.
29. $\mathcal{P}(X)/\mathcal{P}(X - Y) \cong \mathcal{P}(Y)$, where Y is a subset of X and the operations in these Boolean rings are symmetric difference and intersection.

30. Let I be the set of all polynomials with no constant term in $\mathbf{R}[x, y]$. Find a ring morphism from $\mathbf{R}[x, y]$ to $\mathbf{R}$ whose kernel is the ideal I. Prove that I is not a principal ideal.

31. Let $I = \{p(x) \in \mathbf{Z}[x] | 5 | p(0)\}$. Prove that I is an ideal of $\mathbf{Z}[x]$ by finding a ring morphism from $\mathbf{Z}[x]$ to $\mathbf{Z}_5$ with kernel I. Prove that I is not a principal ideal.

32. Let $I \subseteq \mathcal{P}(X)$ with the property that, if $A \in I$, then all the subsets of A are in I and also if A and B are disjoint sets in I, then $A \cup B \in I$. Prove that I is an ideal in the Boolean ring $(\mathcal{P}(X), \Delta, \cap)$.

33. Is $\{p(x) \in \mathbf{Q}[x] | p(0) = 3\}$ an ideal of $\mathbf{Q}[x]$?

34. Is $\left\{ \begin{pmatrix} a & 0 \\ b & 0 \end{pmatrix} \in \mathfrak{M}(2 \times 2; \mathbf{Z}) | a, b \in \mathbf{Z} \right\}$ an ideal of $\mathfrak{M}(2 \times 2; \mathbf{Z})$?

35. What is the smallest ideal in $\mathfrak{M}(2 \times 2; \mathbf{Z})$ containing $\begin{pmatrix} 1 & 0 \\ 0 & 0 \end{pmatrix}$?

36. Let a, b be elements of a Euclidean ring R. Prove that

$$(a) \subseteq (b) \quad \text{if and only if} \quad b | a.$$

37–38.   Find all the ideals in the following rings and draw the poset diagrams of the ideals under inclusion.

37. $\mathbf{Z}_8$.   **38.** $\mathbf{Z}_{20}$.

39–46.   Which of the following elements are irreducible in the given ring? If an element is irreducible, find the corresponding quotient field modulo the ideal generated by that element.

39. 11 in $\mathbf{Z}$.   **40.** 10 in $\mathbf{Z}$.
41. $x^2 - 2$ in $\mathbf{R}[x]$.   **42.** $x^3 + x^2 + 2$ in $\mathbf{Z}_3[x]$.
43. $x^4 - 2$ in $\mathbf{Q}[x]$.   **44.** $x^7 + 4x^3 - 3ix + 1$ in $\mathbf{C}[x]$.
45. $x^2 - 3$ in $\mathbf{Q}(\sqrt{2})[x]$.   **46.** $3x^5 - 4x^3 + 2$ in $\mathbf{Q}[x]$.

47–56.   Which of the following rings are fields? Give reasons.

47. $\mathbf{Z}_2 \times \mathbf{Z}_2$.   **48.** $\mathbf{Z}_4$.
49. $\mathbf{Z}_{17}$.   **50.** $\mathbf{R}^3$.
51. $\mathbf{Q}[x]/(x^3 - 3)$.   **52.** $\mathbf{Z}_7[x]/(x^2 + 1)$.
53. $\mathbf{Z}_5[x]/(x^2 + 1)$.   **54.** $\mathbf{R}[x]/(x^2 + 7)$.
55. $\mathbf{Q}(\sqrt[4]{11}) = \{a + b11^{\frac{1}{4}} + c11^{\frac{1}{2}} + d11^{\frac{3}{4}} | a, b, c, d \in \mathbf{Q}\}$.
56. $\mathfrak{M}(n \times n; \mathbf{R})$.
57. An ideal $I \neq R$ is said to be a *maximal ideal* in the commutative ring R if, whenever U is an ideal of R such that $I \subseteq U \subseteq R$, then $U = I$ or $U = R$. Show that the nonzero ideal (a) of a Euclidean ring R is maximal if and only if a is irreducible in R.
58. If I is an ideal in a commutative ring R, prove that R/I is a field if and only if I is a maximal ideal of R.
59. Find all the ideals in the ring of formal power series, $\mathbf{R}[[x]]$. Which of the ideals are maximal?
60. Let $C[0, 1] = \{ f : [0, 1] \rightarrow \mathbf{R} | f \text{ is continuous} \}$, the ring of real valued continuous functions on the interval $[0, 1]$. Prove that $I_a = \{ f \in C[0, 1] | f(a) = 0 \}$ is a maximal ideal in $C[0, 1]$ for each $a \in [0, 1]$.
(Every maximal ideal is, in fact, of this form, but this is much harder to prove.)

Field Extensions   11

We proved in the previous chapter that if $p(x)$ is an irreducible polynomial over the field **F**, the quotient ring $\mathbf{K} = \mathbf{F}[x]/(p(x))$ is a field. This field **K** contains a subring isomorphic to **F**; thus **K** can be considered to be an extension of the field **F**. We show that the polynomial $p(x)$ now has a root α in this extension field **K**, even though $p(x)$ was irreducible over **F**. We say that **K** can be obtained from **F** by adjoining the root α. We can construct the complex numbers **C** in such a way, by adjoining a root of $x^2 + 1$ to the real numbers **R**.

Another important achievement is the construction of a finite field with p^n elements for each prime p. Such a field is called a Galois field of order p^n and is denoted by $\mathbf{GF}(p^n)$. We show how this field can be constructed as a quotient ring of the polynomial ring $\mathbf{Z}_p[x]$, by an irreducible polynomial of degree n.

FIELD EXTENSIONS

11.01 DEFINITION.   A *subfield* of a field **K** is a subring **F** that is also a field. In this case, the field **K** is called an *extension of the field* **F**.

For example, **Q** is a subfield of **R**; thus **R** is an extension of the field **Q**.

11.02 EXAMPLE. Let $p(x)$ be a polynomial of degree n irreducible over the field **F**, so that the quotient ring

$$\mathbf{K} = \mathbf{F}[x]/(p(x)) = \{a_0 + a_1 x + \cdots + a_{n-1} x^{n-1} | a_i \in \mathbf{F}\}$$

is a field. Then **K** is an extension field of **F**.

SOLUTION. This follows from Theorem 10.19 when we identify the ideal containing the constant term a_0 with the element a_0 of **F**. $\square$

11.03 PROPOSITION. Let **K** be an extension field of **F**. Then **K** is a vector space over **F**.

PROOF. **K** is an Abelian group under addition. Elements of **K** can be multiplied by elements of **F**. This multiplication satisfies the following properties:

  (i)   if 1 is the identity element of **F** then $1k = k$ for all $k \in \mathbf{K}$.
  (ii) if $\lambda \in \mathbf{F}$ and $k, l \in \mathbf{K}$, then $\lambda(k + l) = \lambda k + \lambda l$.
  (iii) if $\lambda, \mu \in \mathbf{F}$ and $k \in \mathbf{K}$, then $(\lambda + \mu)k = \lambda k + \mu k$.
  (iv) if $\lambda, \mu \in \mathbf{F}$ and $k \in \mathbf{K}$, then $(\lambda \mu)k = \lambda(\mu k)$.

    Hence **K** is a vector space over **F**. $\square$

The fact that a field extension **K** is a vector space over **F** tells us much about the structure of **K**. The elements of **K** can be written as a linear combination of certain elements called basis elements. Furthermore, if the vector space **K** has finite dimension n over the field **F**, there will be n basis elements, and the construction of **K** is particularly simple.

11.04 DEFINITION. The *degree* of the extension **K** of the field **F**, written $[\mathbf{K} : \mathbf{F}]$, is the dimension of **K** as a vector space over **F**. **K** is called a *finite extension* if $[\mathbf{K} : \mathbf{F}]$ is finite.

11.05 EXAMPLE. $[\mathbf{C} : \mathbf{R}] = 2$.

SOLUTION. $\mathbf{C} = \{a + ib | a, b \in \mathbf{R}\}$; therefore, 1 and i span the vector space **C** over **R**. Now 1 and i are linearly independent, since, if $\lambda, \mu \in \mathbf{R}$, then $\lambda 1 + \mu i = 0$ implies that $\lambda = \mu = 0$. Hence $\{1, i\}$ is a basis for **C** over **R** and $[\mathbf{C} : \mathbf{R}] = 2$. $\square$

11.06 EXAMPLE. If $\mathbf{K} = \mathbf{Z}_5[x]/(x^3 + x + 1)$, then $[\mathbf{K} : \mathbf{Z}_5] = 3$.

SOLUTION. $\{1, x, x^2\}$ is a basis for K over Z_5 because, by Theorem 10.10, every element of K can be written uniquely as the coset containing $a_0 + a_1 x + a_2 x^2$ where $a_i \in Z_5$. Hence $[K : Z_5] = 3$. $\square$

The above example is a special case of the following theorem.

11.07 THEOREM. If $p(x)$ is an irreducible polynomial of degree n over the field F and $K = F[x]/(p(x))$, then $[K : F] = n$.

PROOF. By Theorem 10.10, $K = \{a_0 + a_1 x + \cdots + a_{n-1} x^{n-1} | a_i \in F\}$, and such expressions for the elements of K are unique. Hence $\{1, x, x^2, \ldots, x^{n-1}\}$ is a basis for K over F, and $[K : F] = n$. $\square$

11.08 THEOREM. Let L be a finite extension of K and K a finite extension of F. Then L is a finite extension of F and $[L : F] = [L : K][K : F]$.

PROOF. We have three fields, F, K, L, with $L \supseteq K \supseteq F$. We prove the theorem by taking bases for L over K and K over F and constructing a basis for L over F.

Let $[L : K] = m$ and $\{u_1, \ldots, u_m\}$ be a basis for L over K. Let $[K : F] = n$ and $\{v_1, \ldots, v_n\}$ be a basis for K over F. We show that

$$\mathcal{B} = \{v_j u_i | i = 1, \ldots, m, j = 1, \ldots, n,\} \text{ is a basis for } L \text{ over } F.$$

If $x \in L$, then $x = \sum_{i=1}^{m} \lambda_i u_i$, for some $\lambda_i \in K$. Now each element λ_i can be written as $\lambda_i = \sum_{j=1}^{n} \mu_{ij} v_j$, for some $\mu_{ij} \in F$. Hence $x = \sum_{i=1}^{m} \sum_{j=1}^{n} \mu_{ij} v_j u_i$, and $\mathcal{B}$ spans L over F.

Now suppose $\sum_{i=1}^{m} \sum_{j=1}^{n} \mu_{ij} v_j u_i = 0$, where $\mu_{ij} \in F$. Then, since $u_1, \ldots, u_m$ are linearly independent over K, it follows that for each $i = 1, \ldots, m, \sum_{j=1}^{n} \mu_{ij} v_j = 0$. But $v_1, \ldots, v_n$ are linearly independent over F, so, for each i and each j, $\mu_{ij} = 0$.

Hence the elements of $\mathcal{B}$ are linearly independent, and $\mathcal{B}$ is a basis for L over F. Therefore, $[L : F] = m \cdot n = [L : K][K : F]$. $\square$

11.09 EXAMPLE. Show that there is no field lying strictly between Q and $L = Q[x]/(x^3 - 2)$.

SOLUTION. The constant polynomials in L are identified with Q. Suppose K is a field such that $L \supseteq K \supseteq Q$. Then, by Theorem 11.08, $[L : Q] = [L : K][K : Q]$. But, by Theorem 11.07, $[L : Q] = 3$, so $[L : K] = 1$, or $[K : Q] = 1$.

If $[L : K] = 1$, then L is a vector space over K, and $\{1\}$, being linearly independent, is a basis. Hence $L = K$. If $[K : Q] = 1$, then $K = Q$. Hence there is no field lying strictly between L and Q. $\square$

11.10 DEFINITION.   Let **K** be a field extension of **F** and let $a \in$ **K**. The *smallest subfield of* **K** *containing* **F** *and* a is denoted by **F**(a). **F**(a) is called the field obtained by *adjoining* a to **F**.

This smallest subfield exists because the intersection of subfields is a subfield, and **F**(a) is the intersection of all subfields of **K** containing **F** and a.

For example, the smallest field containing **R** and i is the whole of the complex numbers, because this field must contain all elements of the form $a + ib$ where $a, b \in$ **R**. Hence **R**$(i) =$ **C**.

In a similar way, the field obtained by adjoining $a_1, \ldots, a_n \in$ **K** to **F** is denoted by **F**$(a_1, \ldots, a_n)$ and is defined to be the smallest subfield of **K** containing $a_1, \ldots, a_n$ and **F**. It follows that **F**$(a_1, \ldots, a_n) =$ **F**$(a_1, \ldots, a_{n-1})(a_n)$.

11.11 EXAMPLE.   **Q**$(\sqrt{2})$ is equal to the subfield **F** $= \{a + b\sqrt{2} \mid a, b \in$ **Q**$\}$ of **R**.

SOLUTION.   **Q**$(\sqrt{2})$ must contain all rationals and $\sqrt{2}$. Hence **Q**$(\sqrt{2})$ must contain all real numbers of the form $b\sqrt{2}$ for $b \in$ **Q** and also $a + b\sqrt{2}$ for $a, b \in$ **Q**. Therefore, **F** $\subseteq$ **Q**$(\sqrt{2})$. But **Q**$(\sqrt{2})$ is the smallest field containing **Q** and $\sqrt{2}$. Since **F** is another such field, **F** $\supseteq$ **Q**$(\sqrt{2})$ and **F** $=$ **Q**$(\sqrt{2})$. $\square$

If R is an integral domain and x is an indeterminate, then

$$R(x) = \left\{ \frac{a_0 + a_1 x + \cdots + a_n x^n}{b_0 + b_1 x + \cdots + b_m x^m} \,\middle|\, a_i, b_j \in R; \text{ not all the } b_j\text{s are zero} \right\}$$

which is the field of rational functions in R. Any field containing R and x must contain the polynomial ring $R[x]$, and the smallest field containing $R[x]$ is its field of fractions $R(x)$.

ALGEBRAIC NUMBERS

11.12 DEFINITION.   If **K** is a field extension of **F**, the element $k \in$ **K** is called *algebraic* over **F** if there exist $a_0, a_1, \ldots, a_n \in$ **F**, not all zero, such that

$$a_0 + a_1 k + \cdots + a_n k^n = 0.$$

In other words, k is the root of a nonzero polynomial in **F**$[x]$. Elements that are not algebraic over **F** are called *transcendental* over **F**.

For example, $5, \sqrt{3}, i, \sqrt[n]{7} + 3$ are all algebraic over **Q** because they are roots of the polynomials $x - 5, x^2 - 3, x^2 + 1, (x - 3)^n - 7$, respectively.

11.13 EXAMPLE.   Find a polynomial in $\mathbf{Q}[x]$ with $\sqrt[3]{2} + \sqrt{5}$ as a root.

SOLUTION.   Let $x = \sqrt[3]{2} + \sqrt{5}$. We have to eliminate the square and cube roots from this equation. We have $x - \sqrt{5} = \sqrt[3]{2}$, so $(x - \sqrt{5})^3 = 2$ or $x^3 - 3\sqrt{5}\,x^2 + 15x - 5\sqrt{5} = 2$. Hence $x^3 + 15x - 2 = \sqrt{5}\,(3x^2 + 5)$ and $(x^3 + 15x - 2)^2 = 5(3x^2 + 5)^2$. Therefore, $\sqrt[3]{2} + \sqrt{5}$ is a root of $x^6 - 15x^4 - 4x^3 + 75x^2 - 60x - 121 = 0$. $\square$

Not all real and complex numbers are algebraic over $\mathbf{Q}$. The numbers π and e can be proven to be transcendental over $\mathbf{Q}$. (See Stewart [45].) Since π is transcendental,

$$\mathbf{Q}(\pi) = \left\{ \frac{a_0 + a_1\pi + \cdots + a_n\pi^n}{b_0 + b_1\pi + \cdots + b_m\pi^m} \,\middle|\, a_i, b_j \in \mathbf{Q}; \text{ not all the } b_j\text{s are zero} \right\},$$

the field of rational functions in π with coefficients in $\mathbf{Q}$. $\mathbf{Q}(\pi)$ must contain all the powers of π and hence any polynomial in π with rational coefficients. Any nonzero element of $\mathbf{Q}(\pi)$ must have its inverse in $\mathbf{Q}(\pi)$; thus $\mathbf{Q}(\pi)$ contains the set of rational functions in π. The number $b_0 + b_1\pi + \cdots + b_m\pi^m$ is never zero unless $b_0 = b_1 = \cdots = b_m = 0$ because π is not the root of any polynomial with rational coefficients. This set of rational functions in π can be shown to be a subfield of $\mathbf{R}$.

Those readers acquainted with the theory of infinite sets can prove that the set of rational polynomials, $\mathbf{Q}[x]$, is countable. Since each polynomial has only a finite number of roots in $\mathbf{C}$, there are only a countable number of real or complex numbers algebraic over $\mathbf{Q}$. Hence there must be an uncountable number of real and complex numbers transcendental over $\mathbf{Q}$.

11.14 EXAMPLE.   Is $\cos(2\pi/5)$ algebraic or transcendental over $\mathbf{Q}$?

SOLUTION.   We know from De Moivre's Theorem that

$$(\cos 2\pi/5 + i \sin 2\pi/5)^5 = \cos 2\pi + i \sin 2\pi = 1.$$

Taking real parts and writing $c = \cos 2\pi/5$ and $s = \sin 2\pi/5$, we have

$$c^5 - 10s^2c^3 + 5s^4c = 1.$$

Thus $c^5 - 10(1 - c^2)c^3 + 5(1 - c^2)^2 c = 1$, since $s^2 + c^2 = 1$. That is, $16c^5 - 20c^3 + 5c - 1 = 0$ and hence $c = \cos 2\pi/5$ is algebraic over $\mathbf{Q}$. $\square$

11.15 THEOREM.   Let α be algebraic over $\mathbf{F}$ and let $p(x)$ be an irreducible polynomial of degree n over $\mathbf{F}$ with α as a root. Then

$$\mathbf{F}(\alpha) \cong \mathbf{F}[x]/(p(x)),$$

and the elements of $F(\alpha)$ can be written uniquely in the form

$$c_0 + c_1\alpha + c_2\alpha^2 + \cdots + c_{n-1}\alpha^{n-1} \qquad \text{where } c_i \in F.$$

PROOF. Define the ring morphism $f: F[x] \rightarrow F(\alpha)$ by $f(q(x)) = q(\alpha)$. The kernel of f is an ideal of $F[x]$. By Corollary 10.04, all ideals in $F[x]$ are principal; thus $\operatorname{Ker} f = (r(x))$ for some $r(x) \in F[x]$. Since $p(\alpha) = 0$, $p(x) \in \operatorname{Ker} f$, and $r(x) | p(x)$. Since $p(x)$ is irreducible, $p(x) = kr(x)$ for some nonzero element k of F. Therefore, $\operatorname{Ker} f = (r(x)) = (p(x))$.

By the Morphism Theorem,

$$F[x]/(p(x)) \cong \operatorname{Im} f \subseteq F(\alpha).$$

Now, by Theorem 10.19, $F[x]/(p(x))$ is a field; thus $\operatorname{Im} f$ is a subfield of $F(\alpha)$ that contains F and α. Since $\operatorname{Im} f$ cannot be a smaller field than $F(\alpha)$, it follows that $\operatorname{Im} f = F(\alpha)$ and $F[x]/(p(x)) \cong F(\alpha)$.

The unique form for the elements of $F(\alpha)$ follows from the above isomorphism and Theorem 10.10. $\square$

11.16 COROLLARY. If α is a root of the polynomial $p(x)$ of degree n, irreducible over F, then $[F(\alpha) : F] = n$.

PROOF. By Theorems 11.15 and 11.07, $[F(\alpha) : F] = [F[x]/(p(x)) : F] = n$. $\square$

For example, $Q(\sqrt{2}) \cong Q[x]/(x^2 - 2)$ and $[Q(\sqrt{2}) : Q] = 2$. Also, $Q(\sqrt[4]{7}\, i) \cong Q[x]/(x^4 - 7)$ and $[Q(\sqrt[4]{7}\, i) : Q] = 4$ because $\sqrt[4]{7}\, i$ is a root of $x^4 - 7$, which is irreducible over Q, by Eisenstein's Criterion 9.37.

11.17 LEMMA. Let $p(x)$ be an irreducible polynomial over the field F. Then F has a finite extension field K in which $p(x)$ has a root.

PROOF. Let $p(x) = a_0 + a_1 x + a_2 x^2 + \cdots + a_n x^n$ and denote the ideal $(p(x))$ by P. By Theorem 11.07, $K = F[x]/P$ is a field extension of F of degree n whose elements are cosets of the form $P + f(x)$. The element $P + x \in K$ is a root of $p(x)$ because

$$a_0 + a_1(P + x) + a_2(P + x)^2 + \cdots + a_n(P + x)^n$$

$$= a_0 + (P + a_1 x) + (P + a_2 x^2) + \cdots + (P + a_n x^n)$$

$$= P + (a_0 + a_1 x + a_2 x^2 + \cdots + a_n x^n) = P + p(x)$$

$$= P + 0,$$

and this is the zero element of the field K. $\square$

11.18 THEOREM. If $f(x)$ is any polynomial over the field **F**, there is an extension field **K** of **F** over which $f(x)$ splits into linear factors.

PROOF. We prove this by induction on the degree of $f(x)$. If $\deg f(x) \leqslant 1$, there is nothing to prove.

Suppose the result is true for polynomials of degree $n-1$. If $f(x)$ has degree n, we can factor $f(x)$ as $p(x)q(x)$ where $p(x)$ is irreducible over **F**. By the previous lemma, **F** has a finite extension **K'** in which $p(x)$ has a root, say α. Hence, by the Factor Theorem,

$$f(x) = (x - \alpha) g(x) \qquad \text{where } g(x) \text{ is of degree } n-1 \text{ in } \mathbf{K'}[x].$$

By the induction hypothesis, the field **K'** has a finite extension, **K**, over which $g(x)$ splits into linear factors. Hence $f(x)$ also splits into linear factors over **K** and, by Theorem 11.08, **K** is a finite extension of **F**. $\square$

Let us now look at the development of the complex numbers from the real numbers. The reason for constructing the complex numbers is that certain equations, such as $x^2 + 1 = 0$, have no solution in **R**. Since $x^2 + 1$ is a quadratic polynomial in **R**$[x]$ without roots, it is irreducible over **R**. In the above manner, we can extend the real field to

$$\mathbf{R}[x]/(x^2 + 1) = \{a + bx \,|\, a, b \in \mathbf{R}\}.$$

In this field extension

$$(0 + 1x)^2 = -1, \text{ since } x^2 \equiv -1 \mod(x^2 + 1).$$

Denote the element $0 + 1x$ by i, so that $i^2 = -1$ and i is a root of the equation $x^2 + 1 = 0$ in this extension field. The field of complex numbers, **C**, is defined to be **R**(i) and, by Theorem 11.15, there is an isomorphism

$$\psi : \mathbf{R}[x]/(x^2 + 1) \longrightarrow \mathbf{R}(i)$$

defined by $\psi(a + bx) = a + bi$. Since

$$(a + bx) + (c + dx) \equiv (a + c) + (b + d)x \mod(x^2 + 1)$$

and

$$(a + bx)(c + dx) \equiv ac + (ad + bc)x + bdx^2 \mod(x^2 + 1)$$

$$\equiv (ac - bd) + (ad + bc)x \mod(x^2 + 1),$$

addition and multiplication in **C** = **R**(i) are defined in the standard way by

$$(a + bi) + (c + di) = (a + c) + (b + d)i$$

and

$$(a+bi)(c+di) = (ac-bd)+(ad+bc)i.$$

11.19 EXAMPLE. Find $[\mathbf{Q}(\cos 2\pi/5):\mathbf{Q}]$.

SOLUTION. We know, from Example 11.14, that $\cos 2\pi/5$ is algebraic over $\mathbf{Q}$ and is a root of the polynomial $16x^5-20x^3+5x-1$. Using the same methods, we can show that $\cos 2k\pi/5$ is also a root of this equation for each $k \in \mathbf{Z}$. Hence we see from Figure 11.01 that its roots are $1, \cos 2\pi/5 = \cos 8\pi/5$ and $\cos 4\pi/5 = \cos 6\pi/5$. Therefore, $(x-1)$ is a factor of the polynomial and

$$16x^5-20x^3+5x-1 = (x-1)(16x^4+16x^3-4x^2-4x+1)$$

$$= (x-1)(4x^2+2x-1)^2.$$

It follows that $\cos 2\pi/5$ and $\cos 4\pi/5$ are roots of the quadratic $4x^2+2x-1$, and, by the quadratic formula, these roots are $(-1\pm\sqrt{5})/4$. Since $\cos 2\pi/5$ is positive,

$$\cos 2\pi/5 = (\sqrt{5}-1)/4 \qquad \text{and} \qquad \cos 4\pi/5 = (-\sqrt{5}-1)/4.$$

Therefore, $\mathbf{Q}(\cos 2\pi/5) \cong \mathbf{Q}[x]/(4x^2+2x-1)$ because $4x^2+2x-1$ is irreducible over $\mathbf{Q}$. By Corollary 11.16, $[\mathbf{Q}(\cos 2\pi/5):\mathbf{Q}]=2$. $\square$

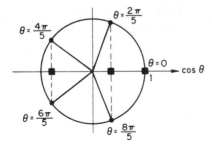

Figure 11.01. The values of $\cos(2k\pi/5)$.

11.20 PROPOSITION. If $[\mathbf{K}:\mathbf{F}]=2$ where $\mathbf{F} \supseteq \mathbf{Q}$, then $\mathbf{K}=\mathbf{F}(\sqrt{\gamma})$ for some $\gamma \in \mathbf{F}$.

PROOF. $\mathbf{K}$ is a vector space of dimension 2 over $\mathbf{F}$. Extend $\{1\}$ to a basis $\{1,\alpha\}$ for $\mathbf{K}$ over $\mathbf{F}$, so that $\mathbf{K}=\{a\alpha+b \mid a,b \in \mathbf{F}\}$.

Now $\mathbf{K}$ is a field, so $\alpha^2 \in \mathbf{K}$ and $\alpha^2 = a\alpha + b$ for some $a, b \in \mathbf{F}$. Hence $(\alpha - (a/2))^2 = b + (a^2/4)$. Put $\beta = \alpha - (a/2)$. Then $\{1, \beta\}$ is also a basis for $\mathbf{K}$ over $\mathbf{F}$, and $\mathbf{K} = \mathbf{F}(\beta)$ where $\beta^2 = b + (a^2/4) = \gamma \in \mathbf{F}$. Hence $\mathbf{K} = \mathbf{F}(\sqrt{\gamma})$. $\square$

11.21 PROPOSITION. If $\mathbf{F}$ is an extension field of $\mathbf{R}$ of finite degree, then $\mathbf{F}$ is isomorphic to $\mathbf{R}$ or $\mathbf{C}$.

PROOF. Let $[\mathbf{F} : \mathbf{R}] = n$. If $n = 1$, then $\mathbf{F}$ is isomorphic to $\mathbf{R}$. Otherwise, $n > 1$ and $\mathbf{F}$ contains some element α not in $\mathbf{R}$. Now $\{1, \alpha, \alpha^2, \ldots, \alpha^n\}$ is a linearly dependent set of elements of $\mathbf{F}$ over $\mathbf{R}$, because it contains more than n elements; hence there exist real numbers $\lambda_0, \lambda_1, \ldots, \lambda_n$, not all zero, such that

$$\lambda_0 + \lambda_1 \alpha + \cdots + \lambda_n \alpha^n = 0.$$

The element α is therefore algebraic over $\mathbf{R}$, and, since the only irreducible polynomials over $\mathbf{R}$ have degree 1 or 2, α must satisfy a linear or quadratic equation over $\mathbf{R}$. If it satisfies a linear equation, then $\alpha \in \mathbf{R}$, contrary to our hypothesis. Therefore, $[\mathbf{R}(\alpha) : \mathbf{R}] = 2$, and, by Proposition 11.20, $\mathbf{R}(\alpha) = \mathbf{R}(\sqrt{\gamma})$. In this case γ must be negative because $\mathbf{R}$ contains all positive square roots; hence $\sqrt{\gamma} = ir$, where $r \in \mathbf{R}$ and $\mathbf{R}(\alpha) = \mathbf{R}(i)$.

Therefore, the field $\mathbf{F}$ contains a subfield isomorphic to the complex numbers and $[\mathbf{F} : \mathbf{C}] = [\mathbf{F} : \mathbf{R}]/2$, which is finite. By an argument similar to the above, any element of $\mathbf{F}$ is the root of an irreducible polynomial over $\mathbf{C}$. However, the only irreducible polynomials over $\mathbf{C}$ are the linear polynomials, and all their roots lie in $\mathbf{C}$. Hence $\mathbf{F}$ is isomorphic to $\mathbf{C}$. $\square$

11.22 EXAMPLE. $[\mathbf{R} : \mathbf{Q}]$ is infinite.

SOLUTION. The real number $\sqrt[n]{2}$ is a root of the polynomial $x^n - 2$, which, by Eisenstein's Criterion, is irreducible over $\mathbf{Q}$. If $[\mathbf{R} : \mathbf{Q}]$ were finite, we could use Theorem 11.08 and Corollary 11.16 to show that

$$[\mathbf{R} : \mathbf{Q}] = [\mathbf{R} : \mathbf{Q}(\sqrt[n]{2})][\mathbf{Q}(\sqrt[n]{2}) : \mathbf{Q}] = [\mathbf{R} : \mathbf{Q}(\sqrt[n]{2})]n.$$

This is a contradiction, because no finite integer is divisible by every integer n. Hence $[\mathbf{R} : \mathbf{Q}]$ must be infinite. $\square$

Galois Fields

In this section, we investigate the structure of finite fields; these fields are called Galois fields in honor of the mathematician Évariste Galois.

We show that the element 1 in any finite field generates a subfield isomorphic to $\mathbf{Z}_p$, for some prime p called the characteristic of the field. Hence a finite field is some finite extension of the field $\mathbf{Z}_p$ and must contain p^m elements, for some integer m.

The characteristic can be defined for any ring, and we give the general definition here, even though we are mainly interested in its application to fields.

For any ring R, define the ring morphism $f : \mathbf{Z} \to R$ by

$$f(n) = \begin{cases} 1 + 1 + \cdots + 1 & (n \text{ copies}) & \text{if } n > 0 \\ 0 & & \text{if } n = 0 \\ -1 - 1 - \cdots - 1 & (|n| \text{ copies}) & \text{if } n < 0. \end{cases}$$

The kernel of f is an ideal of the principal ideal ring $\mathbf{Z}$; hence $\operatorname{Ker} f = (q)$ for some $q \geqslant 0$.

11.23 DEFINITION. The generator $q \geqslant 0$ of $\operatorname{Ker} f$ is called the *characteristic* of the ring R.

The characteristic of R is the least integer $q > 0$ for which $qa = 0$, for all $a \in R$. If no such number exists, the characteristic of R is zero. For example, the characteristic of $\mathbf{Z}$ is 0, and the characteristic of $\mathbf{Z}_n$ is n.

11.24 PROPOSITION. The characteristic of an integral domain is either zero or prime.

PROOF. Let q be the characteristic of an integral domain D. By applying the Morphism Theorem to $f : \mathbf{Z} \to D$, defined by $f(1) = 1$, we see that

$$f(\mathbf{Z}) \cong \begin{cases} \mathbf{Z}_q & \text{if } q \neq 0 \\ \mathbf{Z} & \text{if } q = 0. \end{cases}$$

But $f(\mathbf{Z})$ is a subring of an integral domain; therefore, it has no zero divisors, and, by Theorem 9.18, q must be zero or prime. $\square$

The characteristic of the field $\mathbf{Z}_p$ is p, while $\mathbf{Q}$, $\mathbf{R}$, and $\mathbf{C}$ have zero characteristic.

11.25 PROPOSITION. If the field $\mathbf{F}$ has prime characteristic p, then $\mathbf{F}$ contains a subfield isomorphic to $\mathbf{Z}_p$. If the field $\mathbf{F}$ has zero characteristic, then $\mathbf{F}$ contains a subfield isomorphic to the rational numbers, $\mathbf{Q}$.

PROOF. From the proof of the previous proposition, we see that **F** contains the subring $f(\mathbf{Z})$ which is isomorphic to $\mathbf{Z}_p$ if **F** has prime characteristic p. If the characteristic of **F** is zero, $f:\mathbf{Z}\to f(\mathbf{Z})$ is an isomorphism. We show that **F** contains the field of fractions of $f(\mathbf{Z})$ and that this is isomorphic to **Q**.

Let $Q=\{xy^{-1}\in\mathbf{F}|x,y\in f(\mathbf{Z})\}$, a subring of **F**. Define the function

$$\tilde{f}:\mathbf{Q}\to Q$$

by $\tilde{f}(a/b)=f(a)\cdot f(b)^{-1}$. Since rational numbers are defined as equivalence classes, we have to check that $\tilde{f}$ is well defined. We can show that $\tilde{f}(a/b)=\tilde{f}(c/d)$ if $a/b=c/d$. Furthermore, it can be verified that $\tilde{f}$ is a ring isomorphism. Hence Q is isomorphic to **Q**. $\square$

11.26 COROLLARY. The characteristic of a finite field is nonzero. $\square$

11.27 THEOREM. If **F** is a finite field, it has p^m elements for some prime p and some integer m.

PROOF. By the previous results, **F** has characteristic p, for some prime p, and contains a subfield isomorphic to $\mathbf{Z}_p$. We identify this subfield with $\mathbf{Z}_p$ so that **F** is a field extension of $\mathbf{Z}_p$. The degree of this extension must be finite because **F** is finite. Let $[\mathbf{F}:\mathbf{Z}_p]=m$ and let $\{f_1,\dots,f_m\}$ be a basis of **F** over $\mathbf{Z}_p$, so that

$$\mathbf{F}=\{\lambda_1 f_1+\cdots+\lambda_m f_m|\lambda_i\in\mathbf{Z}_p\}.$$

There are p choices for each λ_i; therefore, **F** contains p^m elements. $\square$

11.28 DEFINITION. A finite field with p^m elements is called a *Galois field* of order p^m and is denoted by $\mathbf{GF}(p^m)$.

It can be shown that for a given prime p and positive integer m, a Galois field $\mathbf{GF}(p^m)$ exists and that all fields of order p^m are isomorphic. See Stewart [45] for a proof of these facts. For $m=1$, the integers modulo p, $\mathbf{Z}_p$, is a Galois field of order p.

From Theorem 11.27 it follows that $\mathbf{GF}(p^m)$ is a field extension of $\mathbf{Z}_p$ of degree m. By Theorem 10.19, each finite field $\mathbf{GF}(p^m)$ can be constructed by finding a polynomial $q(x)$ of degree m, irreducible in $\mathbf{Z}_p[x]$, and defining

$$\mathbf{GF}(p^m)=\mathbf{Z}_p[x]/(q(x)).$$

By Lemma 11.17, there is an element α in $\mathbf{GF}(p^m)$, such that $q(\alpha)=0$, and $\mathbf{GF}(p^m)=\mathbf{Z}_p(\alpha)$, the field obtained by adjoining α to $\mathbf{Z}_p$.

For example, $\mathbf{GF}(4) = \mathbf{Z}_2[x]/(x^2 + x + 1) = \mathbf{Z}_2(\alpha) = \{0, 1, \alpha, \alpha + 1\}$, where $\alpha^2 + \alpha + 1 = 0$. Rewriting Table 10.2, we obtain the Table 11.1 for $\mathbf{GF}(4)$.

Table 11.1.   The Galois field GF(4)

+	0	1	α	$\alpha + 1$
0	0	1	α	$\alpha + 1$
1	1	0	$\alpha + 1$	α
α	α	$\alpha + 1$	0	1
$\alpha + 1$	$\alpha + 1$	α	1	0

$\cdot$	0	1	α	$\alpha + 1$
0	0	0	0	0
1	0	1	α	$\alpha + 1$
α	0	α	$\alpha + 1$	1
$\alpha + 1$	0	$\alpha + 1$	1	α

11.29 EXAMPLE.   Construct a field $\mathbf{GF}(125)$.

SOLUTION.   Since $125 = 5^3$, we can construct such a field if we can find an irreducible polynomial of degree 3 over $\mathbf{Z}_5$.

A reducible polynomial of degree 3 must have a linear factor. Therefore, by the Factor Theorem, $p(x) = x^3 + ax^2 + bx + c$ is irreducible in $\mathbf{Z}_5[x]$ if and only if $p(n) \neq 0$ for $n = 0, 1, 2, 3, 4$ in $\mathbf{Z}_5$.

By trial and error, we find that the polynomial $p(x) = x^3 + x + 1$ is irreducible because $p(0) = 1$, $p(1) = 3$, $p(2) = 11 = 1$, $p(3) = 31 = 1$, and $p(4) = p(-1) = -1 = 4$ in $\mathbf{Z}_5$. Hence

$$\mathbf{GF}(125) = \mathbf{Z}_5[x]/(x^3 + x + 1). \quad \square$$

$(x^3 + x + 1)$ is not the only irreducible polynomial of degree 3 over $\mathbf{Z}_5$. For example, $(x^3 + x^2 + 1)$ is also irreducible. But $\mathbf{Z}_5[x]/(x^3 + x + 1)$ is isomorphic to $\mathbf{Z}_5[x]/(x^3 + x^2 + 1)$.

PRIMITIVE ELEMENTS

The elements of a Galois field $\mathbf{GF}(p^m)$ can be written as

$$\left\{ a_0 + a_1\alpha + \cdots + a_{m-1}\alpha^{m-1} \mid a_i \in \mathbf{Z}_p \right\}$$

where α is a root of a polynomial $q(x)$ of degree m irreducible over $\mathbf{Z}_p$. Addition is easily performed using this representation, because it is simply addition of polynomials in $\mathbf{Z}_p[\alpha]$. However, multiplication is more complicated and requires repeated use of the relation $q(\alpha) = 0$. We show that, by judicious choice of α, the elements of $\mathbf{GF}(p^m)$ can be written as

$$\{0, 1, \alpha, \alpha^2, \alpha^3, \ldots, \alpha^{p^m-2}\} \qquad \text{where } \alpha^{p^m-1} = 1.$$

This element α is called a primitive element of $\mathbf{GF}(p^m)$, and multiplication is easily calculated using powers of α; however, addition is much harder to perform using this representation.

For example, in $\mathbf{GF}(4) = \mathbf{Z}_2(\alpha)$, the element $\alpha + 1 = \alpha^2$ where $\alpha^3 = 1$, and the tables are given in Table 11.2.

Table 11.2. The Galois field GF(4) in terms of a primitive element

+	0	1	α	α^2
0	0	1	α	α^2
1	1	0	α^2	α
α	α	α^2	0	1
α^2	α^2	α	1	0

$\cdot$	0	1	α	α^2
0	0	0	0	0
1	0	1	α	α^2
α	0	α	α^2	1
α^2	0	α^2	1	α

If $\mathbf{F}$ is any field and $\mathbf{F}^* = \mathbf{F} - \{0\}$, we know that $(\mathbf{F}^*, \cdot)$ is an Abelian group under multiplication. We now show that the nonzero elements of a finite field form a cyclic group under multiplication; the generators of this cyclic group are the primitive elements of the field. To prove this theorem, we need some preliminary results about the orders of elements in an Abelian group.

11.30 LEMMA.   If g and h are elements of an Abelian group of orders a and b, respectively, there exists an element of order $\mathrm{LCM}(a,b)$.

PROOF.   Let $p_1, p_2, \ldots, p_s$ be the set of primes occuring in the prime decompositions of a and b and let

$$a = p_1^{\alpha_1} p_2^{\alpha_2} \ldots p_s^{\alpha_s} \quad \text{and} \quad b = p_1^{\beta_1} p_2^{\beta_2} \ldots p_s^{\beta_s}.$$

Now $\mathrm{LCM}(a,b) = p_1^{\gamma_1} p_2^{\gamma_2} \ldots p_s^{\gamma_s}$, where $\gamma_i = \max(\alpha_i, \beta_i)$. Let $d = p_1^{\delta_1} p_2^{\delta_2} \ldots p_s^{\delta_s}$, where

$$\delta_i = \begin{cases} \alpha_i & \text{if} \quad \alpha_i \leqslant \beta_i \\ 0 & \text{if} \quad \alpha_i > \beta_i \end{cases}$$

and $e = p_1^{\epsilon_1} p_2^{\epsilon_2} \ldots p_s^{\epsilon_s}$, where

$$\epsilon_i = \begin{cases} 0 & \text{if} \quad \alpha_i \leqslant \beta_i \\ \beta_i & \text{if} \quad \alpha_i > \beta_i. \end{cases}$$

Then, by Lemma 4.40, g^d and h^e have orders a/d and b/e, respectively. But $\text{GCD}(a/d, b/e) = 1$, since $\alpha_i - \delta_i$ or $\beta_i - \epsilon_i$ is zero, and a/d and b/e have no prime factors in common. Therefore, by Lemma 4.41, $g^d h^e$ is an element of order $ab/de = \text{LCM}(a, b)$, since $\alpha_i + \beta_i - \delta_i - \epsilon_i = \max(\alpha_i, \beta_i) = \gamma_i$. $\square$

11.31 LEMMA. If the maximum order of the elements of an Abelian group G is r, then

$$x^r = e \qquad \text{for all} \qquad x \in G.$$

PROOF. Let $g \in G$ be an element of maximal order r. If h is an element of order t, by the previous lemma, there is an element of order $\text{LCM}(r, t)$. Since $\text{LCM}(r, t) \leqslant r$, t divides r. Therefore, $h^r = e$. $\square$

11.32 THEOREM. Let $\mathbf{GF}(q)^*$ be the set of nonzero elements in the Galois field $\mathbf{GF}(q)$. Then $(\mathbf{GF}(q)^*, \cdot)$ is a cyclic group of order $q - 1$.

PROOF. Let r be the maximal order of elements of $(\mathbf{GF}(q)^*, \cdot)$. Then, by Lemma 11.31,

$$x^r - 1 = 0 \qquad \text{for all} \qquad x \in \mathbf{GF}(q)^*.$$

Hence every nonzero element of the Galois field $\mathbf{GF}(q)$ is a root of the polynomial $x^r - 1$, and, by Theorem 9.07, a polynomial of degree r can have at most r roots over any field; therefore, $r \geqslant q - 1$. But, by Lagrange's Theorem, $r | (q - 1)$; it follows that $r = q - 1$.

$(\mathbf{GF}(q)^*, \cdot)$ is therefore a group of order $q - 1$ containing an element of order $q - 1$ and hence must be cyclic. $\square$

11.33 DEFINITION. A generator of the cyclic group $(\mathbf{GF}(q)^*, \cdot)$ is called a *primitive element* of $\mathbf{GF}(q)$.

For example, in $\mathbf{GF}(4) = \mathbf{Z}_2(\alpha)$, the multiplicative group of nonzero elements, $\mathbf{GF}(4)^*$, is a cyclic group of order 3, and both nonidentity elements α and $\alpha + 1$ are primitive elements.

If α is a primitive element in the Galois field $\mathbf{GF}(q)$, where q is the power of a prime p, then $\mathbf{GF}(q)$ is the field extension $\mathbf{Z}_p(\alpha)$ and

$$\mathbf{GF}(q)^* = \{1, \alpha, \alpha^2, \dots, \alpha^{q-2}\}.$$

Hence

$$\mathbf{GF}(q) = \{0, 1, \alpha, \alpha^2, \dots, \alpha^{q-2}\}.$$

11.34 EXAMPLE. Find all the primitive elements in $GF(9) = Z_3(\alpha)$, where $\alpha^2 + 1 = 0$.

SOLUTION.

$$GF(9) = Z_3[x]/(x^2 + 1) = \{a + bx \,|\, a, b \in Z_3\}.$$

The nonzero elements form a cyclic group $GF(9)^*$ of order 8; hence the multiplicative order of each element is either 1, 2, 4, or 8.

In calculating the powers of each element, we use the relationship $\alpha^2 = -1 = 2$. From Table 11.3, we see that $1 + \alpha$, $2 + \alpha$, $1 + 2\alpha$, and $2 + 2\alpha$ are the primitive elements of $GF(9)$. $\square$

Table 11.3. The nonzero elements of GF(9)

Element x	x^2	x^4	x^8	Order	Primitive
1	1	1	1	1	No
2	1	1	1	2	No
α	2	1	1	4	No
$1 + \alpha$	2α	2	1	8	Yes
$2 + \alpha$	α	2	1	8	Yes
2α	2	1	1	4	No
$1 + 2\alpha$	α	2	1	8	Yes
$2 + 2\alpha$	2α	2	1	8	Yes

11.35 PROPOSITION. (i) $z^{q-1} = 1$ for all elements $z \in GF(q)^*$.
(ii) $z^q = z$ for all elements $z \in GF(q)$.
(iii) If $GF(q) = \{\alpha_1, \alpha_2, \ldots, \alpha_q\}$, then $z^q - z$ factors over $GF(q)$ as

$$(z - \alpha_1)(z - \alpha_2) \cdots (z - \alpha_q).$$

PROOF. We have already shown that (i) is implied by Lemma 11.31. Part (ii) follows immediately because 0 is the only element of $GF(q)$ that is not in $GF(q)^*$. The polynomial $z^q - z$, of degree q, can have at most q roots over any field. By (ii), all elements of $GF(q)$ are roots over $GF(q)$; hence $z^q - z$ factors into q distinct linear factors over $GF(q)$. $\square$

For example, in $GF(4) = Z_2[x]/(x^2 + x + 1) = \{0, 1, \alpha, \alpha + 1\}$ and we have

$$(z + 0)(z + 1)(z + \alpha)(z + \alpha + 1) = (z^2 + z)(z^2 + z + \alpha^2 + \alpha)$$

$$= (z^2 + z)(z^2 + z + 1)$$

$$= z^4 + z = z^4 - z.$$

11.36 DEFINITION. An irreducible polynomial $g(x)$, of degree m over $\mathbf{Z}_p$, is called a *primitive polynomial* if $g(x)|x^k-1$ for $k=p^m-1$ and for no smaller k.

11.37 PROPOSITION. The irreducible polynomial $g(x)\in\mathbf{Z}_p[x]$ is primitive if and only if x is a primitive element in $\mathbf{Z}_p[x]/(g(x))=\mathbf{GF}(p^m)$.

PROOF. The following statements are equivalent.

(i) x is a primitive element in $\mathbf{GF}(p^m)=\mathbf{Z}_p[x]/(g(x))$.
(ii) $x^k=1$ in $\mathbf{GF}(p^m)$ for $k=p^m-1$ and for no smaller k.
(iii) $x^k-1\equiv 0\bmod g(x)$ for $k=p^m-1$ and for no smaller k.
(iv) $g(x)|x^k-1$ for $k=p^m-1$ and for no smaller k. $\square$

For example, x^2+x+1 is primitive in $\mathbf{Z}_2[x]$. From Example 11.34, we see that x^2+1 is not primitive in $\mathbf{Z}_3[x]$. However, $1+\alpha$ and $1+2\alpha=1-\alpha$ are primitive elements, and they are roots of the polynomial

$$(x-1-\alpha)(x-1+\alpha)=(x-1)^2-\alpha^2=x^2+x+2\in\mathbf{Z}_3[x].$$

Hence x^2+x+2 is a primitive polynomial in $\mathbf{Z}_3[x]$. Also, x^2+2x+2 is another primitive polynomial in $\mathbf{Z}_3[x]$ with roots $2+\alpha$ and $2+2\alpha=2-\alpha$.

11.38 EXAMPLE. Let α be a root of the primitive polynomial x^4+x+1 $\in\mathbf{Z}_2[x]$. Show how the nonzero elements of $\mathbf{GF}(16)=\mathbf{Z}_2(\alpha)$ can be represented by the powers of α.

SOLUTION. The representation is given in Table 11.4. $\square$

Arithmetic in $\mathbf{GF}(16)$ can very easily be performed using Table 11.4. Addition is performed by representing elements as polynomials in α of degree less than 4; whereas multiplication is performed using the representation of nonzero elements as powers of α. For example,

$$\frac{1+\alpha+\alpha^3}{1+\alpha^2+\alpha^3}+\alpha+\alpha^2=\frac{\alpha^7}{\alpha^{13}}+\alpha+\alpha^2=\alpha^{-6}+\alpha+\alpha^2$$

$$=\alpha^9+\alpha+\alpha^2,\qquad\text{since}\qquad\alpha^{15}=1$$

$$=\alpha+\alpha^3+\alpha+\alpha^2$$

$$=\alpha^2+\alpha^3.$$

Table 11.4. The representation of GF(16)

Element	α^0	α^1	α^2	α^3
$0 \quad = 0$	0	0	0	0
$\alpha^0 \quad = 1$	1	0	0	0
$\alpha^1 \quad = \quad \alpha$	0	1	0	0
$\alpha^2 \quad = \quad\quad\quad \alpha^2$	0	0	1	0
$\alpha^3 \quad = \quad\quad\quad\quad\quad \alpha^3$	0	0	0	1
$\alpha^4 \quad = 1 + \alpha$	1	1	0	0
$\alpha^5 \quad = \quad\quad \alpha + \alpha^2$	0	1	1	0
$\alpha^6 \quad = \quad\quad\quad\quad\quad \alpha^2 + \alpha^3$	0	0	1	1
$\alpha^7 \quad = 1 + \alpha \quad\quad + \alpha^3$	1	1	0	1
$\alpha^8 \quad = 1 \quad\quad\quad + \alpha^2$	1	0	1	0
$\alpha^9 \quad = \quad\quad \alpha \quad\quad + \alpha^3$	0	1	0	1
$\alpha^{10} = 1 + \alpha + \alpha^2$	1	1	1	0
$\alpha^{11} = \quad\quad \alpha + \alpha^2 + \alpha^3$	0	1	1	1
$\alpha^{12} = 1 + \alpha + \alpha^2 + \alpha^3$	1	1	1	1
$\alpha^{13} = 1 \quad\quad\quad + \alpha^2 + \alpha^3$	1	0	1	1
$\alpha^{14} = 1 \quad\quad\quad\quad\quad + \alpha^3$	1	0	0	1
$\alpha^{15} = 1$				

The concept of primitive polynomials is useful in designing *feedback shift registers* with a long cycle length. Consider the circuit in Figure 11.02 in which the square boxes are delays of one unit of time, and the circle with a cross inside represents a modulo 2 adder.

If the delays are labeled by a representation of the elements of GF(16), a single shift corresponds to multiplying the element of GF(16) by α. Hence, if the contents of the delays are not all zero initially, this shift register will cycle through 15 different states before repeating itself. In general, it is possible to construct a shift register with n delay units that will cycle through $2^n - 1$ different states before repeating itself. The feedback connections have to be derived from a primitive polynomial of degree n over Z_2.

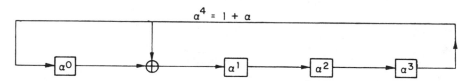

Figure 11.02. A feedback shift register.

Such feedback shift registers are useful in designing error-correcting coders and decoders, random number generators, and radar transmitters. See Chapter 14 of this book and Stone [12; Ch. 9].

Exercises

1–4.   Write out the addition and multiplication tables for the following fields.

1. **GF**(5).   2. **GF**(7).
3. **GF**(9).   4. **GF**(8).

*5–10.   In each of the following cases find, if possible, an irreducible polynomial of degree n over **F**.*

5. $n=3$, $F=Z_{11}$.   6. $n=3$, $F=Q$.
7. $n=4$, $F=R$.   8. $n=3$, $F=GF(4)$.
9. $n=2$, $F=Q(i)$.   10. $n=5$, $F=Z_3$.

*11–13.   In each of the following cases, find a polynomial in **F**[x] with r as a root.*

11. $r=\sqrt{2}+\sqrt{6}$, $F=Q$.
12. $r=\pi+ei$, $F=R$.
13. $r={}^3\sqrt{3}/\sqrt{2}$, $F=Q$.
14. Show that $\theta=2k\pi/7$ satisfies the equation $\cos 4\theta-\cos 3\theta=0$ for each integer k. Hence find an irreducible polynomial over **Q** with $\cos(2\pi/7)$ as a root.
15. Prove that the algebraic numbers

$$A=\{x\in C|\ x \text{ is algebraic over } Q\}$$

form a subfield of **C**.
16. Assuming the Fundamental Theorem of Algebra, prove that every polynomial in **A** has a root in **A**.

17–25.   Calculate the following degrees.

17. $[Q({}^3\sqrt{7}):Q]$.   18. $[C:Q]$.
19. $[Q(i,3i):Q]$.   20. $[C:R(\sqrt{-7})]$.
21. $[Z_3[x]/(x^2+x+2):Z_3]$.   22. $[Q(i,\sqrt{2}):Q]$.
23. $[A:Q]$.   24. $[C:A]$.
25. $[Z_3(t):Z_3]$, where $Z_3(t)$ is the field of rational functions in t over Z_3.
26. Prove that x^2-2 is irreducible over $Q(\sqrt{3})$.

27–32. Find the inverses of the following elements in the given fields. Each field is a finite extension $F(\alpha)$. Express your answers in the form $a_0 + a_1\alpha + \cdots + a_{n-1}\alpha^{n-1}$ where $a_i \in F$ and $[F(\alpha):F] = n$.

27. $1 + \sqrt[3]{2}$ in $Q(\sqrt[3]{2})$.

28. $\sqrt[4]{5} + \sqrt{5}$ in $Q(\sqrt[4]{5})$.

29. $5 + 6\omega$ in $Q(\omega)$, where ω is a complex cube root of 1.

30. $2 - 3i$ in $Q(i)$.

31. α in $GF(32) = Z_2(\alpha)$, where $\alpha^5 + \alpha^2 + 1 = 0$.

32. α in $GF(27) = Z_3(\alpha)$, where $\alpha^3 + \alpha^2 + 2 = 0$.

33–40. Find the characteristic of the following rings. Which of these are fields?

33. $Z_2 \times Z_2$.   **34.** $Z_3 \times Z_4$.

35. $GF(49)$.   **36.** $Z \times Z_2$.

37. $Q(\sqrt[3]{7})$.   **38.** $\mathfrak{M}(2 \times 2; Z_5)$.

39. $Q \times Z_3$.   **40.** $GF(4)[x]$.

41. Let R be any ring and n a positive integer. Prove that $I_n = \{na | a \in R\}$ is an ideal of R and that the characteristic of R/I_n divides n.

42. Let M be a finite subgroup of the multiplicative group F^* of any infinite field F. Prove that M is cyclic, and give an example to show that F^* need not be cyclic.

43. For what values of m is $(Z_m^*, \cdot)$ cyclic? (This is a difficult problem; see Exercises 51–58 of Chapter 4 for other results on Z_m^*.)

44. Let $GF(4) = Z_2(\alpha)$ where $\alpha^2 + \alpha + 1 = 0$. Find an irreducible quadratic in $GF(4)[x]$. If β is the root of such a polynomial show that $GF(4)(\beta)$ is a Galois field of order 16.

45. (i) Show that there are $(p^2 - p)/2$ monic irreducible polynomials of degree 2 over $GF(p)$. (A polynomial is *monic* if the coefficient of the highest power of the variable is 1.)

(ii) Prove that there is a field with p^2 elements for every prime p.

46. (i) How many monic irreducible polynomials of degree 3 are there over $GF(p)$?

(ii) Prove that there is a field with p^3 elements for every prime p.

47. Find an element α such that $Q(\sqrt{2}, \sqrt{-3}) = Q(\alpha)$.

48. Find all the primitive elements in $GF(16) = Z_2(\alpha)$ where $\alpha^4 + \alpha + 1 = 0$.

49. Find all the primitive elements in $GF(32)$.

50–51. Find a primitive polynomial of degree n over the field F.

50. $n = 2$, $F = Z_5$.   **51.** $n = 3$, $F = Z_2$.

52. Let $g(x)$ be a polynomial of degree m over $\mathbf{Z}_p$. If $g(x)|x^k-1$ for $k=p^m-1$ and for no smaller k, show that $g(x)$ is irreducible over $\mathbf{Z}_p$.

53. Prove that $x^8+x\in\mathbf{Z}_2[x]$ will split into linear factors over $\mathbf{GF}(8)$ but not over any smaller field.

54. Let $f(x)=2x^3+5x^2+7x+6\in\mathbf{Q}[x]$. Find a field, smaller than the complex numbers, in which $f(x)$ splits into linear factors.

55. If α and β are roots of x^3+x+1 and $x^3+x^2+1\in\mathbf{Z}_2[x]$, respectively, prove that the Galois fields $\mathbf{Z}_2(\alpha)$ and $\mathbf{Z}_2(\beta)$ are isomorphic.

56. (i) If $p(x)\in\mathbf{Z}_2[x]$, prove that $[p(x)]^2=p(x^2)$.

(ii) If β is a root of $p(x)\in\mathbf{Z}_2[x]$, prove that β^{2^l} is a root for all $l\in\mathbf{N}$.

(iii) Let $\mathbf{GF}(16)=\mathbf{Z}_2(\alpha)$ where $\alpha^4+\alpha+1=0$. Find an irreducible polynomial in $\mathbf{Z}_2[x]$ with α^3 as a root.

57–58.   Solve the following simultaneous linear equations in $\mathbf{GF}(4)=\mathbf{Z}_2(\alpha)$.

57. $\alpha x+(\alpha+1)y=\alpha+1$ **58.** $(\alpha+1)x+y=\alpha$

$\quad\ \ x+\alpha y=1.$ $\quad\ \ x+(\alpha+1)y=\alpha+1.$

59. Solve the quadratic equation $\alpha x^2+(1+\alpha)x+1=0$ over the field $\mathbf{GF}(4)=\mathbf{Z}_2(\alpha)$.

60. Design a feedback shift register using six delays that has a cycle length of 63.

61. What is the cycle length of the feedback shift register in Figure 11.03.

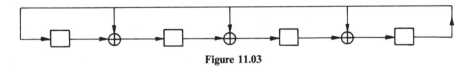

Figure 11.03

62. Design a feedback shift register that has a cycle length of 21.

63. Describe the output sequence of the feedback shift register in Figure 11.04 when the registers initially contain the elements shown.

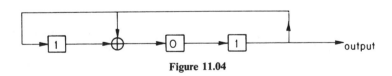

Figure 11.04

64. If a feedback shift register with n delays has a cycle length of 2^n-1, show that the feedback connections must be derived from a primitive irreducible polynomial of degree n over $\mathbf{Z}_2$.

Latin Squares $\boxed{12}$

Latin squares first arose with parlor games such as the problem of arranging the jacks, queens, kings, and aces of a pack of cards in a 4×4 array so that each row and each column contains one card from each suit and one card from each rank. In 1779, Leonard Euler posed the following famous problem of the 36 officers from six ranks and six regiments. He claimed it was impossible to arrange these officers on parade in a 6×6 square so that each row and each column contains one officer from each rank and one from each regiment.

Recently, statisticians have found Latin squares useful in designing experiments, and mathematicians have found close connections between Latin squares and finite geometries.

LATIN SQUARES

12.01 DEFINITION. Let S be a set with n elements. Then a *Latin square* $L = (l_{ij})$, of order n based on S, is an $n \times n$ array of the elements of S such that each element appears exactly once in each row and once in each column.

For example, Table 12.1 illustrates a Latin square of order 3 based on $\{a,b,c\}$.

Table 12.1.
A Latin square

a	b	c
c	a	b
b	c	a

12.02 THEOREM. The table for any finite group $(G, +)$ of order n is a Latin square of order n based on G.

PROOF. We write the operation in G as addition, even though the result still holds if G is not commutative.

Suppose that two elements in one row are equal. Then $x_i + x_j = x_i + x_k$ for some $x_i, x_j, x_k \in G$. Since G is a group, x_i has an inverse $(-x_i)$ such that $(-x_i) + x_i = 0$. Hence $(-x_i) + (x_i + x_j) = (-x_i) + (x_i + x_k)$, and, since the operation is associative, we have $x_j = x_k$. Therefore, an element cannot appear twice in the same row. Similarly, an element cannot appear twice in the same column, and the table is a Latin square. □

Given any Latin square, we can permute the rows among themselves and also the columns among themselves and we still have a Latin square. For example, the addition table for $Z_2 \times Z_2$ is a Latin square of order 4. If we interchange the first and third columns and replace $(0,0)$ by a, $(0,1)$ by b, $(1,0)$ by c, and $(1,1)$ by d, we obtain another Latin square of order 4 based on $\{a,b,c,d\}$. These are illustrated in Table 12.2.

Table 12.2.   Latin squares of order four

(0,0)	(0,1)	(1,0)	(1,1)		c	b	a	d
(0,1)	(0,0)	(1,1)	(1,0)		d	a	b	c
(1,0)	(1,1)	(0,0)	(0,1)		a	d	c	b
(1,1)	(1,0)	(0,1)	(0,0)		b	c	d	a

Latin squares are useful in designing statistical experiments because they can show how an experiment can be arranged so as to reduce the errors without making the experiment too large or too complicated. See Mann [50] for more complete details.

Suppose you wanted to compare the yields of three varieties of hybrid corn. You have a rectangular test plot, but you are not sure that the fertility of the soil is the same everywhere. You could divide up the land into nine rectangular regions and plant the three varieties, a, b, and c, in the form of the Latin square in Table 12.1. Then if one row were more fertile than the others, the Latin square would reduce the error that this might cause. In fact, if the soil fertility was a linear function of the coordinates of the plot, the Latin square arrangement would minimize the error.

Of course the error could be reduced by subdividing the plot into a large number of pieces and planting the varieties at random. But this would make it much more difficult to sow and harvest.

12.03 EXAMPLE.   A smoking machine is used to test the tar content of four brands of cigarettes; the machine has four ports so that four cigarettes can be smoked simultaneously. However, these four ports might not be identical and that might affect the measurements of the tar content. Also, if four runs were made on the machine, testing one brand at a time, the humidity could change, thus affecting the results.

Show how to reduce the errors due to the different ports and the different runs by using a Latin square to design the experiment.

SOLUTION.   If a, b, c, d are the four brands, we can use one of the Latin squares of order 4 that we have constructed. Table 12.3 illustrates which brand should be tested at each port during each of the four runs.   □

Table 12.3.   The design of the smoking experiment

Ports

		1	2	3	4
	1→	c	b	a	d
Runs	2→	d	a	b	c
	3→	a	d	c	b
	4→	b	c	d	a

Not all Latin squares can be obtained from a group table, even if we allow permutations of the rows and columns.

12.04 EXAMPLE.   Show that the Latin square illustrated in Table 12.4 cannot be obtained from a group table.

Table 12.4

$$
\begin{array}{ccccc}
A & B & C & D & E \\
B & A & E & C & D \\
C & D & A & E & B \\
D & E & B & A & C \\
E & C & D & B & A
\end{array}
$$

SOLUTION.   By Corollary 4.14, all groups of order 5 are cyclic and are isomorphic to $(\mathbf{Z}_5, +)$. Suppose that the Latin square in Table 12.4 could be obtained from the addition table of $\mathbf{Z}_5$. Since permutations are reversible, it follows that the rows and columns of this square could be permuted to obtain the table of $\mathbf{Z}_5$. The four elements in the left hand top corner would be taken into four elements forming a rectangle in $\mathbf{Z}_5$, as shown in Table 12.5. Then we would have $p + r = A$, $q + r = B$, $p + s = B$, and $q + s = A$ for some $p, q, r, s \in \mathbf{Z}_5$, where $p \neq q$ and $r \neq s$. Hence $p + r = A = q + s$ and $q + r = B = p + s$. Adding, we have $p + q + 2r = p + q + 2s$ and $2r = 2s$. Therefore, $6r = 6s$, which implies $r = s$ in $\mathbf{Z}_5$, which is a contradiction.   □

Table 12.5

$$
\begin{array}{c|cc}
+ & ..p & ..q.. \\
\hline
. & & \\
. & & \\
r & A & B \\
. & & \\
s & B & A \\
. & &
\end{array}
$$

ORTHOGONAL LATIN SQUARES

Suppose that, in our corn field, besides testing the yields of three varieties of corn, we also wanted to test the effects of three fertilizers on the corn. We could do this in the same experiment by arranging the fertilizers on the nine plots so that each of the three fertilizers was used once on each variety of corn and so that the different fertilizers themselves were arranged in a Latin square of order 3.

Let a, b, c be three varieties of corn and A, B, C be three types of fertilizer. Then the two Latin squares in Table 12.6 could be superimposed to form the design in Table 12.7. In this table, each variety of corn and each type of fertilizer appears exactly once in each row and in each

Table 12.6

a	b	c	A	B	C
c	a	b	B	C	A
b	c	a	C	A	B

Table 12.7

aA	bB	cC
cB	aC	bA
bC	cA	aB

column. Furthermore, each type of fertilizer is used exactly once with each variety of corn. This table could be used to design the experiment. For example, in the top left section of our test plot, we would plant variety "*a*" and use fertilizer "*A*".

12.05 DEFINITION. Two Latin squares of order n are called *orthogonal* if, when the squares are superimposed, each element of the first square occurs exactly once with each element of the second square.

The two Latin squares in Table 12.6 are orthogonal.

Although it is easy to construct Latin squares of any order, the construction of orthogonal Latin squares can be a difficult problem. At this point the reader should try to construct two orthogonal Latin squares of order 4.

Going back to our field of corn and fertilizers, could we use the same trick again to test the effect of three insecticides by choosing another Latin square of order 3 orthogonal to the first two? It can be proved that it is impossible to find such a Latin square. (See Exercise 5.) However, if we have four types of corn, fertilizer, and insecticide, we show, using Theorem 12.08, how they could be distributed on a 4×4 plot using three Latin squares of order 4 orthogonal to each other.

12.06 DEFINITION. If $L_1, \ldots, L_r$ are Latin squares of order n such that L_i is orthogonal to L_j for all $i \neq j$, then $\{L_1, \ldots, L_r\}$ is called a set of r *mutually orthogonal Latin squares* of order n.

We show how to construct $n-1$ mutually orthogonal Latin squares of order n from a finite field with n elements. We know that a finite field has a prime power number of elements, and we are able to construct such squares for $n = 2, 3, 4, 5, 7, 8, 9, 11, 13, 16, 17, \ldots$ etc.

Let $\mathbf{GF}(n) = \{x_0, x_1, x_2, \ldots, x_{n-1}\}$ be a finite field of order $n = p^m$ where $x_0 = 0$ and $x_1 = 1$. Let $L_1 = (a_{ij}^1)$ be the Latin square of order n that is the addition table of $\mathbf{GF}(n)$. Then

$$a_{ij}^1 = x_i + x_j \quad \text{for} \quad 0 \leqslant i \leqslant n-1, \ 0 \leqslant j \leqslant n-1.$$

12.07 PROPOSITION. Define the squares $L_k = (a_{ij}^k)$, for $1 \leqslant k \leqslant n-1$, by

$$a_{ij}^k = x_k \cdot x_i + x_j \quad \text{for} \quad 0 \leqslant i \leqslant n-1, \ 0 \leqslant j \leqslant n-1.$$

Then L_k is a Latin square of order n for $1 \leqslant k \leqslant n-1$ based on $\mathbf{GF}(n)$.

PROOF.   The difference between two elements in the ith row is

$$a_{ij}^k - a_{iq}^k = (x_k \cdot x_i + x_j) - (x_k \cdot x_i + x_q)$$

$$= x_j - x_q \neq 0, \quad \text{if } j \neq q.$$

Hence each row is a permutation of $\mathbf{GF}(n)$.

The difference between two elements in the jth column is

$$a_{ij}^k - a_{rj}^k = (x_k \cdot x_i + x_j) - (x_k \cdot x_r + x_j)$$

$$= x_k \cdot (x_i - x_r) \neq 0 \quad \text{if } i \neq r, \quad \text{since } x_k \neq 0 \quad \text{and} \quad x_i \neq x_r.$$

Hence each column is a permutation of $\mathbf{GF}(n)$ and L_k is a Latin square of order n. $\square$

12.08 THEOREM. $\{L_1, L_2, \ldots, L_{n-1}\}$ is a mutually orthogonal set of Latin squares of order $n = p^m$.

PROOF.   We have to prove that L_k is orthogonal to L_l for all $k \neq l$.

Suppose that when L_k is superimposed on L_l, the pair of elements in the (i,j)th position is the same as the pair in the (r,q)th position. That is, $(a_{ij}^k, a_{ij}^l) = (a_{rq}^k, a_{rq}^l)$ or $a_{ij}^k = a_{rq}^k$ and $a_{ij}^l = a_{rq}^l$. Hence $x_k \cdot x_i + x_j = x_k \cdot x_r + x_q$ and $x_l \cdot x_i + x_j = x_l \cdot x_r + x_q$. Subtracting, we have $(x_k - x_l) \cdot x_i = (x_k - x_l) \cdot x_r$ or $(x_k - x_l) \cdot (x_i - x_r) = 0$. Now the field $\mathbf{GF}(n)$ has no zero divisors; thus either $x_k = x_l$ or $x_i = x_r$. Hence either $k = l$ or $i = r$. But $k \neq l$ and we know from Proposition 12.07 that two elements in the same row of L_k or L_l cannot be equal; therefore, $i \neq r$.

This contradiction proves that when L_k and L_l are superimposed, all the pairs of elements occurring are different. Each element of the first square appears n times and hence must occur with all the n different elements of the second square. Therefore, L_k is orthogonal to L_l, if $k \neq l$. $\square$

If we start with $\mathbf{Z}_3$ and perform the above construction we obtain the two mutually orthogonal Latin squares of order 3 given in Table 12.8.

Table 12.8.   Two orthogonal Latin squares

$$L_1 \begin{array}{|ccc|} \hline 0 & 1 & 2 \\ 1 & 2 & 0 \\ 2 & 0 & 1 \\ \hline \end{array} \qquad L_2 \begin{array}{|ccc|} \hline 0 & 1 & 2 \\ 2 & 0 & 1 \\ 1 & 2 & 0 \\ \hline \end{array}$$

12.09 EXAMPLE.   Construct three mutually orthogonal Latin squares of order 4.

SOLUTION. Apply the method given in Proposition 12.07 to the Galois field $GF(4) = Z_2(\alpha) = \{0, 1, \alpha, \alpha^2\}$, where $\alpha^2 = \alpha + 1$.

L_1 is simply the addition table for $GF(4)$. From the way the square L_k was constructed in Propostion 12.07, we see that its rows are a permutation of the rows of L_1. Hence L_2 can be obtained by multiplying the first column of L_1 by α and then permuting the rows of L_1 so that they start with the correct element. L_3 is also obtained by permuting the rows of L_1 so that the first column is α^2 times the first column of L_1. These are illustrated in Table 12.9. ☐

Table 12.9.   Three mutually orthogonal Latin squares of order 4

$$L_1 \quad \begin{array}{cccc} 0 & 1 & \alpha & \alpha^2 \\ 1 & 0 & \alpha^2 & \alpha \\ \alpha & \alpha^2 & 0 & 1 \\ \alpha^2 & \alpha & 1 & 0 \end{array} \qquad L_2 \quad \begin{array}{cccc} 0 & 1 & \alpha & \alpha^2 \\ \alpha & \alpha^2 & 0 & 1 \\ \alpha^2 & \alpha & 1 & 0 \\ 1 & 0 & \alpha^2 & \alpha \end{array} \qquad L_3 \quad \begin{array}{cccc} 0 & 1 & \alpha & \alpha^2 \\ \alpha^2 & \alpha & 1 & 0 \\ 1 & 0 & \alpha^2 & \alpha \\ \alpha & \alpha^2 & 0 & 1 \end{array}$$

If we write a for 0, b for 1, c for α, and d for α^2 and superimpose the three Latin squares, we obtain Table 12.10. Example 12.09 also allows us to solve the parlor game of laying out the 16 cards that was mentioned at the beginning of the chapter. One solution, using the squares L_2 and L_3 in Table 12.9, is illustrated in Table 12.11.

Table 12.10.   Superimposed Latin squares

aaa	bbb	ccc	ddd
bcd	adc	dab	cba
cdb	dca	abd	bac
dbc	cad	bda	acb

Table 12.11.   The 16 court cards

A ♠	K ◇	Q ♡	J ♣
Q ♣	J ♡	A ◇	K ♠
J ◇	Q ♠	K ♣	A ♡
K ♡	A ♣	J ♠	Q ◇

12.10 EXAMPLE.   A drug company wishes to produce a new cold remedy by combining a decongestant, an antihistamine, and a pain reliever. It plans to test various combinations of three decongestants, three antihistamines, and three pain relievers on four groups of subjects each day from Monday to Thursday. Furthermore, each type of ingredient should also be compared with a placebo. Design this test so as to reduce the effects due to differences between the subject groups and the different days.

SOLUTION. We can use the three mutually orthogonal Latin squares constructed in the previous example to design this experiment.

Make up the drugs given to each group using Table 12.12. The letter in the first position refers to the decongestant, the second to the antihistamine, and the third to the pain reliever. The letter "*a*" refers to a placebo, and "*b*,", "*c*," and "*d*" refer to the three different types of ingredients.   □

Table 12.12.   Testing three different drugs

		Mon	Tue	Wed	Thu
	A	*aaa*	*bbb*	*ccc*	*ddd*
Subject	B	*bcd*	*adc*	*dab*	*cba*
Groups	C	*cdb*	*dca*	*abd*	*bac*
	D	*dbc*	*cad*	*bda*	*acb*

We recognize Euler's problem of the 36 officers on parade, mentioned at the beginning of the chapter, as the problem of constructing two orthogonal Latin squares of order 6. Euler not only conjectured that this problem was impossible to solve, but he also conjectured that it was impossible to find two orthogonal Latin squares of order n, whenever $n \equiv 2 \bmod 4$.

No prime power is congruent to 2 modulo 4; therefore, we cannot use the method of Theorem 12.08 to construct any of these squares. In 1899, G. Tarry, by exhaustive enumeration, proved that the problem of the 36 officers was insoluble. However, in 1959, Euler's general conjecture was shown to be false, and, in fact, Bose, Shrikhande and Parker proved that there exist at least two orthogonal Latin squares of order n, for any $n > 6$. Hence Proposition 12.07 is by no means the only way of constructing orthogonal Latin squares. Denes and Keedwell [48] give a comprehensive survey of all the known results on Latin squares up to the time of its publication.

FINITE GEOMETRIES

The method of constructing $n - 1$ mutually orthogonal Latin squares of order n over **GF**(n) was originally given by R. C. Bose [47]. In the same paper, Bose showed that there is a very close connection between orthogonal Latin squares and geometries with a finite number of points and lines.

The geometries that we consider are called affine planes.

12.11 DEFINITION. An *affine plane* consists of a set, P, of *points*, together with a set, L, of subsets of P called *lines*. The points and lines must satisfy the following incidence axioms.

(i) Any two distinct points lie on exactly one line.
(ii) For each line l and point x not on l, there exists a unique line m containing x and not meeting l.
(iii) There exist three points not lying on a line.

We can define an equivalence relation of *parallelism*, $//$, on the set of lines L, by defining $l//m$ if $l = m$ or l and m contain no common point. Axiom (ii) then states that through each point there is a unique line parallel to any other line.

The points and lines in the Euclidean plane $\mathbf{R}^2$ form such a geometry with an infinite number of points.

If the geometry has only a *finite* number of points, it can be shown that there exists an integer n such that the geometry contains n^2 points and $n^2 + n$ lines, and that each line contains n points, while each point lies on $n + 1$ lines. Such a finite geometry is called an *affine plane of order n*. In an affine plane of order n there are $n + 1$ parallelism classes. (See Exercises 12 and 13.)

The diagram in Figure 12.01 shows an affine plane of order 2 in which $P = \{a, b, c, d\}$ and $L = \{\{a, b\}, \{c, d\}, \{a, c\}, \{b, c\}, \{b, d\}, \{a, d\}\}$.

Bose showed that an affine plane of order n produces a complete set of $n - 1$ mutually orthogonal Latin squares of order n, and conversely, that each set of $n - 1$ mutually orthogonal Latin squares of order n defines an affine plane of order n.

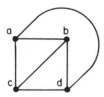

Figure 12.01.   An affine plane with four points.

12.12 THEOREM. There exists an affine plane of order n if and only if there exist $n - 1$ mutually orthogonal Latin squares of order n.

PROOF. Suppose that there exists an affine plane of order n. We coordinatize the points as follows. Take any line and label the n points as $0, 1, 2, \ldots, n-1$. This is called the x-axis, and the point labeled 0 is called the origin. Choose any other line through the origin and label the n points $0, 1, 2, \ldots, n-1$ with 0 at the origin. This line is called the y-axis. A point of the plane is said to have coordinates (a, b) if the unique lines through the point parallel to the y and x-axes meet the axes in points labeled a and b, respectively. This is illustrated in Figure 12.02.

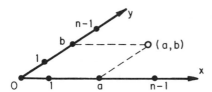

Figure 12.02.   Coordinates in an affine plane.

There are n^2 ordered pairs (a, b) corresponding to the n^2 points of the plane. These points also correspond to the n^2 cells of an $n \times n$ square where (a, b) refers to the cell in the ath row and bth column. We fill these cells with numbers in $n-1$ different ways to produce $n-1$ mutually orthogonal Latin squares of order n.

Consider any complete set of parallel lines that are not parallel to either axis. Label the n parallel lines $0, 1, 2, \ldots, n-1$ in any manner. Through each point, there is exactly one of these lines. In the cell (a, b) place the number of the unique line on which the point (a, b) is found. The numbers in these cells form a Latin square of order n on $\{0, 1, \ldots, n-1\}$. No two numbers in the same row can be the same, because there is only one line through two points in the same row, namely, the line parallel to the x-axis. Hence each number appears exactly once in each row and, similarly, once in each column.

There are $n-1$ sets of parallelism classes that are not parallel to the axes; each of these gives rise to a Latin square. These $n-1$ squares are mutually orthogonal because each line of one parallel system meets all n of the lines of any other system. Hence, when two squares are superimposed, each number of one square occurs once with each number of the second square.

Conversely, suppose that there exists a set of $n-1$ mutually orthogonal Latin squares of order n. We can relabel the elements, if necessary, so that these squares are based on $S = \{0, 1, 2, \ldots, n-1\}$. We define an affine plane with S^2 as the set of points. A set of n points is said to lie on a line if there is a Latin square with the same number in each of the n cells corresponding to these points, or if the n points all have one coordinate the same. It is straightforward to check that this is an affine plane of order n. $\square$

12.13 COROLLARY.   There exists an affine plane of order n whenever n is the power of a prime.

PROOF.   This follows from Theorem 12.08. $\square$

Because of the impossibility of solving Euler's Officer Problem, there are no orthogonal Latin squares of order 6, and hence there is no affine plane of order 6. The only known affine planes have prime power order, although not all of them are derived from the Latin squares produced by Proposition 12.07. (See Horadam [49; pp 295–302].) It is unknown whether a plane of order 10 exists. Many pairs of orthogonal Latin squares of order 10 have been produced, but a large computer search has been unable to find three mutually orthogonal squares. If it could be proved that three mutually orthogonal Latin squares do not exist, this would also prove the nonexistence of an affine plane of order 10. On the other hand, the construction of three such mutually orthogonal squares would help to decide whether there was a plane of order 10.

By the method of Theorem 12.12, we can construct an affine plane of order n from the Galois field $\mathbf{GF}(n)$ whenever n is a prime power. The set of points is

$$P = \mathbf{GF}(n)^2 = \{(x,y) \mid x,y \in \mathbf{GF}(n)\}.$$

It follows from Proposition 12.07 that a line consists of points satisfying a linear equation in x and y with coefficients in $\mathbf{GF}(n)$. The slope of a line is defined in the usual way and is an element of $\mathbf{GF}(n)$ or is infinite. Two lines are parallel if and only if they have the same slope.

For example, if $\mathbf{GF}(4) = \mathbf{Z}_2(\alpha) = \{0, 1, \alpha, \alpha^2\}$, the 16 points of the affine plane of order 4 are shown in Figure 12.03. The horizontal lines are of the form

$$y = \text{constant}$$

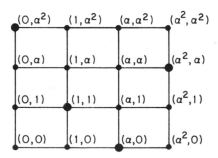

Figure 12.03.   **An affine plane of order 4 with the points of the line** $y = \alpha x + \alpha^2$ **enlarged.**

and have slope 0, whereas the vertical lines are of the form

$$x = \text{constant}$$

and have infinite slope. The line

$$y = \alpha x + \alpha^2$$

has slope α and contains the points $(0, \alpha^2)$, $(1, 1)$, $(\alpha, 0)$ and (α^2, α). This line is parallel to the lines $y = \alpha x$, $y = \alpha x + 1$, and $y = \alpha x + \alpha$.

Given an affine plane of order n, it is possible to construct a *projective plane of order* n by adding a "line at infinity" containing $n + 1$ points corresponding to each parallelism class, so that parallel lines intersect on the line at infinity. The projective plane of order n has $n^2 + n + 1$ points and $n^2 + n + 1$ lines. Furthermore, any projective plane gives rise to an affine plane by taking one line to be the line at infinity. Hence the existence of a projective plane of order n is equivalent to the existence of an affine plane of the same order.

MAGIC SQUARES

Magic squares have been known for thousands of years, and in times when particular numbers were associated with mystical ideas, it was natural that a square that displays such symmetry should have been deemed to have magical properties. Figure 12.04 illustrates an engraving by Dürer, made in 1514, that contains a magic square. Magic squares have no applications, and this section is included for amusement only.

12.14 DEFINITION. A *magic square of order* n consists of the integers 1 to n^2 arranged in an $n \times n$ square array so that the row sums, column sums, and corner diagonal sums are all the same.

The sum of each row must be $n(n^2 + 1)/2$, which is $1/n$ times the sum of all the integers from 1 to n^2. For example, in Dürer's magic square of Figure 12.04, the sum of each row, column, and diagonal is 34.

It is an interesting exercise to try to construct such squares. We show how to construct some magic squares from certain pairs of orthogonal Latin squares. See Ball [46] and Denes and Keedwell [48] for other methods of constructing magic squares.

Let $K = (k_{ij})$ and $L = (l_{ij})$ be two orthogonal Latin squares of order n on the set $S = \{0, 1, \ldots, n-1\}$. Superimpose these two squares to form a square $M = (m_{ij})$ in which the elements of M are numbers in the base n, whose first digit is taken from K and whose second digit is taken from L. That is,

Figure 12.04. "Melancholia," an engraving by Albrecht Dürer. In the upper right there is a magic square of order 4 with the date of the engraving, 1514, in the middle of the bottom row. [Photographed by Walter Steinkopf, courtesy of Staatliche Museen Preussischer Kulturbesitz Kupferstichkabinett Berlin (West)].

$m_{ij} = n \cdot k_{ij} + l_{ij}$. Since K and L are orthogonal, all the possible combinations of two elements from S occur exactly once in M. In other words, all the numbers from 0 to $n^2 - 1$ occur in M.

Now add 1 to every element of M to obtain the square $M' = (m_{ij}')$ where $m_{ij}' = m_{ij} + 1$.

12.15 LEMMA. The square M' contains all the numbers between 1 and n^2 and is row and column magic; that is, the sums of each row and of each column are the same.

PROOF. In any row or column of M, each number from 0 to $n-1$ occurs exactly once as the first digit and exactly once as the second digit. Hence the sum is

$$(0 + 1 + \cdots + n - 1)n + (0 + 1 + \cdots + n - 1) = (n+1)(n-1)n/2 = n(n^2 - 1)/2.$$

Therefore, the sum of any row or column of M' is $n(n^2 - 1)/2 + n = n(n^2 + 1)/2$. $\square$

12.16 EXAMPLE. Construct the square M' from the two orthogonal Latin squares, K and L, in Table 12.13.

Table 12.13.
The construction of a magic square of order 3

1	2	0		0	2	1		10	22	01
0	1	2		2	1	0		02	11	20
2	0	1		1	0	2		21	00	12

K	L	M (in base 3)

3	8	1		4	9	2
2	4	6		3	5	7
7	0	5		8	1	6

M (in base 10)	M'

SOLUTION. Table 12.13 illustrates the superimposed square M in base 3 and in base 10. By adding one to each element, we obtain the magic square M'. $\square$

12.17 THEOREM. If K and L are orthogonal Latin squares of order n on the set $\{0, 1, 2, \ldots, n-1\}$ and the sum of each of the diagonals of K and L is $n(n-1)/2$, then the square M' derived from K and L is a magic square of order n.

PROOF. Lemma 12.15 shows that the sum of each row and each column is $n(n^2+1)/2$. A similar argument shows that the sum of each diagonal is also $n(n^2+1)/2$. ☐

There are two common ways in which the sum of the diagonal elements of K and L can equal $n(n-1)/2$.

(i) The diagonal is a permutation of $\{0,1,\ldots,n-1\}$.
(ii) If n is odd, every diagonal element is $(n-1)/2$.

Both these situations occur in the squares K and L of Table 12.13; thus the square M', which is constructed from these, is a magic square.

12.18 EXAMPLE. Construct a magic square of order 4 from two orthogonal Latin squares in Table 12.9.

SOLUTION. By replacing $0,1,\alpha,\alpha^2$ by $0,1,2,3$, in any order, the squares L_2 and L_3 in Table 12.9 satisfy the conditions of Theorem 12.17, because the diagonal elements are all different. However, L_1 will not satisfy the conditions of Theorem 12.17, whatever substitutions we make. In L_2, replace $0,1,\alpha,\alpha^2$ by $0,1,2,3$, respectively, and in L_3 replace $0,1,\alpha,\alpha^2$ by $3,2,0,1$, respectively, to obtain the squares L_2' and L_3' in Table 12.14. Combine these to obtain the square M with entries in base 4. Add 1 to each entry and convert to base 10 to obtain the magic square M' in Table 12.14. ☐

Table 12.14.
The construction of a magic square of order 4

0	1	2	3
2	3	0	1
3	2	1	0
1	0	3	2

L_2'

3	2	0	1
1	0	2	3
2	3	1	0
0	1	3	2

L_3'

03	12	20	31
21	30	02	13
32	23	11	00
10	01	33	22

M (in base 4)

4	7	9	14
10	13	3	8
15	12	6	1
5	2	16	11

M'

Exercises

1. Construct a Latin square of order 7 on $\{a,b,c,d,e,f,g\}$.
2. Construct four mutually orthogonal Latin squares of order 5.

3. Construct four mutually orthogonal Latin squares of order 8.

4. Construct two mutually orthogonal Latin squares of order 9.

5. Prove that there are at most $(n-1)$ mutually orthogonal Latin squares of order n. (You can always relabel each square so that the first rows are the same.)

6. Let $L = (l_{ij})$ be a Latin square of order l on $\{1, 2, \ldots, l\}$ and $M = (m_{ij})$ be a Latin square of order m on $\{1, 2, \ldots, m\}$. Describe how to construct a Latin square of order lm on $\{1, 2, \ldots, l\} \times \{1, 2, \ldots, m\}$ from L and M.

7. Is the Latin square of Table 12.15 the multiplication table for a group of order 6 with identity A?

Table 12.15

A	B	C	D	E	F
B	A	F	E	C	D
C	F	B	A	D	E
D	C	E	B	F	A
E	D	A	F	B	C
F	E	D	C	A	B

8. A chemical company wants to test a chemical reaction using seven different levels of catalyst, a, b, c, d, e, f, g. In the manufacturing process, the raw material comes from the previous stage in batches, and the catalyst must be added immediately. If there are seven reactors, A, B, C, D, E, F, G, in which the catalytic reaction can take place, show how to design the experiment using seven batches of raw material so as to minimize the effect of the different batches and of the different reactors.

9. A supermarket wishes to test the effect of putting cereal on four shelves at different heights. Show how to design such an experiment lasting four weeks and using four brands of cereal.

10. A manufacturer has five types of toothpaste. He would like to test these on five subjects by giving each subject a different type each week for five weeks. Each type of toothpaste is identified by a different color, red, blue, green, white, or purple, and the manufacturer changes the color code each week to reduce the psychological effect of the color. Show how to design this experiment.

11. Quality control would like to find the best type of music to play to its assembly line workers in order to reduce the number of faulty products. As an experiment, a different type of music is played on four days in a week, and on the fifth day no music at all is played. Design such an experiment to last five weeks that will reduce the effect of the different days of the week.

12. The relation of parallelism, $//$, on the set of lines of an affine plane is defined by $l//m$ if and only if $l=m$ or $l \cap m = \varnothing$. Prove that $//$ is an equivalence relation.

13. Let P be the set of points and L be the set of lines of a finite affine plane.

(i) Show that the number of points on a line l equals the number of lines in any parallelism class not containing l.

(ii) Deduce that all the lines contain the same number of points.

(iii) If each line contains n points, show that the plane contains n^2 points and $n^2 + n$ lines, each point lying on $n+1$ lines. Show also that there are $n+1$ parallelism classes.

14. Find all the lines in the affine plane of order 3 whose point set is $\mathbf{Z}_3^2$.

15–17.   The following examples refer to the affine plane of order 9 obtained from $\mathbf{GF}(9) = \mathbf{Z}_3(\alpha)$, *where* $\alpha^2 + 1 = 0$.

15. Find the line through $(2\alpha, 1)$ that is parallel to the line $y = \alpha x + 2 + \alpha$.

16. Find the point of intersection of the lines $y = x + \alpha$ and $y = (\alpha + 1)x + 2\alpha$.

17. Find the equation of the line through $(0, 2\alpha)$ and $(2, \alpha + 1)$.

18. Prove that a magic square of order 3 must have 5 at its center.

19. Prove that 1 cannot be a corner element of a magic square of order 3.

20. How many different magic squares of order 3 are there?

21. How many essentially different magic squares of order 3 are there? That is, ones that cannot be obtained from each other by a symmetry of the square.

22. Is there a magic square of order 2?

23. Find two magic squares of order 4, different from that in Example 12.18.

24. Find a magic square of order 5.

25. Find a magic square of order 8.

26. Can you construct a magic square of order 9 or 11 with the present year in the last two squares of the bottom row?

Geometrical
Constructions $\boxed{13}$

The only geometric instruments used by the ancient Greeks were a straight-edge and a compass. They did not possess reliable graduated rulers or protractors. However, with these two instruments, they could still perform a wide variety of constructions; they could divide a line into any number of equal parts, and they could bisect angles and construct parallel and perpendicular lines. There were three famous problems that the Greeks could not solve using these methods. These problems were (i) *the duplication of the cube*; that is, given one edge of a cube, construct the edge of a cube whose volume is double that of the given cube; (ii) *the trisection of any given angle*; and (iii) *the squaring of the circle*; that is, given any circle, construct a square whose area is the same as that of the circle. For centuries, the solution to these problems eluded mathematicians, despite the fact that large prizes were offered for their discovery.

It was not until the nineteenth century that mathematicians suspected and, in fact, proved that these constructions were impossible. In the beginning of that century, nonexistence proofs began appearing in algebra; it was proved that the general polynomial equation of degree 5 could not be solved in terms of nth roots and rational operations. Similar algebraic methods were then applied to these geometric problems. The geometric problems could be converted into algebraic problems by determining which multiples of a given length could be constructed using only straight-edge and compass.

Some of the classical constructions are illustrated in Figure 13.01.

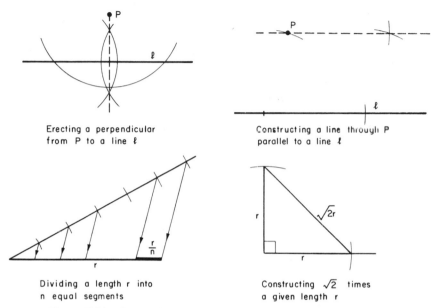

Erecting a perpendicular from P to a line ℓ

Constructing a line through P parallel to a line ℓ

Dividing a length r into n equal segments

Constructing $\sqrt{2}$ times a given length r

Figure 13.01. Geometrical constructions using straight-edge and compass.

CONSTRUCTIBLE NUMBERS

We are interested in those lengths which can be constructed from a given length. For convenience, we choose our unit of length to be this given length. We see that we can divide a length into any number of equal parts, and hence we can construct any rational multiple. However, we can do more than this; we can construct irrational multiples such as $\sqrt{2}$ by using right-angled triangles.

13.01 DEFINITION. Given any line segment in the plane, choose rectangular coordinates so that its end points are $(0,0)$ and $(1,0)$. Any point in the plane that can be constructed from this line segment by using a straight-edge and compass is called a *constructible point*. A real number is called *constructible* if it occurs as one coordinate of a constructible point.

Points can be constructed by performing the following allowable operations a finite number of times. We can

(i)   draw a line through two previously constructed points;
(ii) draw a circle with center at a previously constructed point and with radius equal to the distance between two previously constructed points;
(iii) mark the point of intersection of two straight lines;
(iv) mark the points of intersection of a straight line and a circle;
(v)   mark the points of intersection of two circles.

13.02 THEOREM.   The set of constructible numbers, **K**, is a subfield of **R**.

PROOF.   **K** is a subset of **R**, so we have to show that it is a field. That is, if $a, b \in$ **K**, we have to show that $a + b$, $a - b$, ab, and, if $b \neq 0$, $a/b \in$ **K**.

If $a, b \in$ **K**, we can mark off lengths a and b on a line to construct lengths $a + b$ and $a - b$.

If $a, b, c \in$ **K**, mark off a segment OA of length a on one line and mark off segments OB and OC of length b and c on another-line through O as shown in Figure 13.02. Draw a line through B parallel to CA and let it meet OA in X. The triangles OAC and OXB are similar, and if $OX = x$, then $x/a = b/c$ and $x = ab/c$.

By taking $c = 1$, we can construct ab, and by taking $b = 1$, we can construct a/c. Hence **K** is a subfield of **R**.   □

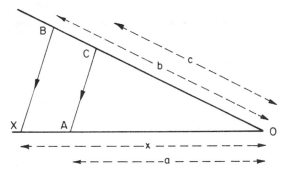

Figure 13.02.   Constructing products and quotients.

13.03 COROLLARY.   **K** is an extension field of **Q**.

PROOF.   Since $1 \in$ **K** and sums and differences of constructible numbers are constructible, it follows that **Z** $\subseteq$ **K**. Since quotients of constructible numbers are constructible, **Q** $\subseteq$ **K**.   □

13.04 PROPOSITION. If $k \in \mathbf{K}$ and $k > 0$, then $\sqrt{k} \in \mathbf{K}$.

PROOF. Mark off segments AB and BC of lengths k and 1 on a line. Draw the circle with diameter AC and construct the perpendicular to AC at B as shown in Figure 13.03. Let it meet the circle at D and E. Then, by a standard theorem in geometry, $AB \cdot BC = DB \cdot BE$: thus $BD = BE = \sqrt{k}$. $\square$

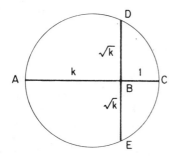

Figure 13.03. Constructing square roots.

13.05 EXAMPLE. $\sqrt[4]{2}$ is constructible.

SOLUTION. we apply the construction of Proposition 13.04 twice to construct $\sqrt{2}$ and then $\sqrt{\sqrt{2}} = \sqrt[4]{2}$. $\square$

We can construct any number that can be written in terms of rational numbers, $+$, $-$, $\cdot$, $\div$ and $\sqrt{}$ signs. For example, the numbers $1 + 4\sqrt{5}$, $\sqrt{2} + \sqrt{4/5}$, and $\sqrt{3 - \sqrt{7}}$ are all constructible. If k_1 is a positive rational number, all the elements of the extension field $\mathbf{K}_1 = \mathbf{Q}(\sqrt{k_1})$ are constructible. $\mathbf{K}_1$ has degree 1 or 2 over $\mathbf{Q}$ depending on whether $\sqrt{k_1}$ is rational or irrational. If k_2 is a positive element of $\mathbf{K}_1$, all the elements of $\mathbf{K}_2 = \mathbf{K}_1(\sqrt{k_2}) = \mathbf{Q}(\sqrt{k_1}, \sqrt{k_2})$ are constructible, and $[\mathbf{K}_2 : \mathbf{K}_1] = 1$ or 2 depending on whether $\sqrt{k_2}$ is an element of $\mathbf{K}_1$ or not. We now show that every constructible number lies in a field obtained by repeating the above extensions.

13.06 THEOREM. The number α is constructible if and only if there exists a sequence of real fields $\mathbf{K}_0, \mathbf{K}_1, \ldots, \mathbf{K}_n$ such that $\alpha \in \mathbf{K}_n \supseteq \mathbf{K}_{n-1} \supseteq \cdots \supseteq \mathbf{K}_0 = \mathbf{Q}$ and $[\mathbf{K}_i : \mathbf{K}_{i-1}] = 2$ for $1 \leqslant i \leqslant n$.

PROOF. Suppose that $\alpha \in \mathbf{K}_n \supseteq \mathbf{K}_{n-1} \supseteq \cdots \supseteq \mathbf{K}_0 = \mathbf{Q}$ where $[\mathbf{K}_i : \mathbf{K}_{i-1}] = 2$. By Proposition 11.20, $\mathbf{K}_i = \mathbf{K}_{i-1}(\sqrt{\gamma_{i-1}})$ for $\gamma_{i-1} \in \mathbf{K}_{i-1}$, and, since $\mathbf{K}_i$ is real,

$\gamma_{i-1} > 0$. Therefore, by repeated application of Proposition 13.04, it can be shown that every element of $\mathbf{K}_n$ is constructible.

Conversely, suppose that $\alpha \in \mathbf{K}$; thus α appears as the coordinate of a point constructible from $(0,0)$ and $(1,0)$ by a finite number of the operations (i) to (v) in Definition 13.01. We prove the result by induction on m, the number of constructible numbers used in reaching α.

Suppose $X_k = \{x_1, \ldots, x_k\}$ is a set of numbers that have already been constructed, that is, have appeared as coordinates of constructible points. When the next number x_{k+1} is constructed, we show that $[\mathbf{Q}(X_{k+1}) : \mathbf{Q}(X_k)] = 1$ or 2, where $\mathbf{Q}(X_{k+1}) = \mathbf{Q}(x_1, \ldots, x_k, x_{k+1})$.

We first show that if we perform either of the operations (i) or (ii) in Definition 13.01 using previously constructed numbers in X_k, the coefficients in the equation of the line or circle remain in $\mathbf{Q}(X_k)$. If we perform the operation (iii), the newly constructed numbers remain in $\mathbf{Q}(X_k)$, and if we perform the operations (iv) or (v), the newly constructed numbers are either in $\mathbf{Q}(X_k)$ or an extension field of degree 2 over $\mathbf{Q}(X_k)$.

Operation (i).   The line through (α_1, β_1) and (α_2, β_2) is $(y - \beta_1)/(\beta_2 - \beta_1) = (x - \alpha_1)/(\alpha_2 - \alpha_1)$, and, if $\alpha_1, \alpha_2, \beta_1, \beta_2 \in X_k$, the coefficients in the equation of this line lie in $\mathbf{Q}(X_k)$.

Operation (ii).   The circle with center (α_1, β_1) and radius equal to the distance from (α_2, β_2) to (α_3, β_3) is $(x - \alpha_1)^2 + (y - \beta_1)^2 = (\alpha_2 - \alpha_3)^2 + (\beta_2 - \beta_3)^2$, and all the coefficients in this equation lie in $\mathbf{Q}(X_k)$.

Operation (iii).   Let $\alpha_{ij}, \beta_j \in \mathbf{Q}(X_k)$. Then the lines

$$\alpha_{11}x + \alpha_{12}y = \beta_1$$

$$\alpha_{21}x + \alpha_{22}y = \beta_2$$

meet in the point (x, y) where, using Cramer's rule,

$$x = \det\begin{pmatrix} \beta_1 & \alpha_{12} \\ \beta_2 & \alpha_{22} \end{pmatrix} \Big/ \det\begin{pmatrix} \alpha_{11} & \alpha_{12} \\ \alpha_{21} & \alpha_{22} \end{pmatrix}$$

and

$$y = \det\begin{pmatrix} \alpha_{11} & \beta_1 \\ \alpha_{21} & \beta_2 \end{pmatrix} \Big/ \det\begin{pmatrix} \alpha_{11} & \alpha_{12} \\ \alpha_{21} & \alpha_{22} \end{pmatrix}$$

as long as they are not parallel. Both of these coordinates are in $\mathbf{Q}(X_k)$.

Operation (iv).   To obtain the points of intersection of a circle and line with coefficients in $\mathbf{Q}(X_k)$, we eliminate y from the equations to obtain an equation of the form

$$\alpha x^2 + \beta x + \gamma = 0 \quad \text{where} \quad \alpha, \beta, \gamma \in \mathbf{Q}(X_k).$$

The line and circle intersect if $\beta^2 - 4\alpha\gamma \geq 0$ and the x coordinates of the intersection points are $x = \dfrac{-\beta \pm \sqrt{\beta^2 - 4\alpha\gamma}}{2\alpha}$, which are in $\mathbf{Q}(X_k)$ or $\mathbf{Q}(X_k)(\sqrt{\beta^2 - 4\alpha\gamma})$. Similarly, the y coordinates are in $\mathbf{Q}(X_k)$ or in an extension field of degree 2 over $\mathbf{Q}(X_k)$.

Operation (v). The intersection of the two circles

$$x^2 + y^2 + \alpha_1 x + \beta_1 y + \gamma_1 = 0$$

$$x^2 + y^2 + \alpha_2 x + \beta_2 y + \gamma_2 = 0$$

is the same as the intersection of one of them with the line

$$(\alpha_1 - \alpha_2)x + (\beta_1 - \beta_2)y + \gamma_1 - \gamma_2 = 0.$$

This is now the same situation as in operation (iv).

Initially, $m = 2$, $X_2 = \{0, 1\}$, and $\mathbf{Q}(X_2) = \mathbf{Q}$. It follows by induction on m, the number of constructible points used, that

$$\alpha \in \mathbf{Q}(X_m) \supseteq \mathbf{Q}(X_{m-1}) \supseteq \cdots \supseteq \mathbf{Q}(X_3) \supseteq \mathbf{Q}(X_2) = \mathbf{Q}$$

where $[\mathbf{Q}(X_{k+1}) : \mathbf{Q}(X_k)] = 1$ or 2 for $2 \leq k \leq m - 1$. Furthermore, each extension field $\mathbf{Q}(X_k)$ is a subfield of $\mathbf{R}$ because $\mathbf{Q}$ and X_k are sets of real numbers. By dropping each field $\mathbf{Q}(X_i)$ that is a trivial extension of $\mathbf{Q}(X_{i-1})$, it follows that

$$\alpha \in \mathbf{K}_n \supseteq \mathbf{K}_{n-1} \supseteq \cdots \supseteq \mathbf{K}_0 = \mathbf{Q}$$

where $[\mathbf{K}_i : \mathbf{K}_{i-1}] = 2$ for $1 \leq i \leq n$. $\quad \square$

13.07 COROLLARY. If α is constructible, then $[\mathbf{Q}(\alpha) : \mathbf{Q}] = 2^r$ for some $r \geq 0$.

PROOF. If α is constructible, then $\alpha \in \mathbf{K}_n \supseteq \mathbf{K}_{n-1} \supseteq \cdots \supseteq \mathbf{K}_0 = \mathbf{Q}$ where $\mathbf{K}_i$ is an extension field of degree 2 over $\mathbf{K}_{i-1}$. By Theorem 11.08,

$$\left[\mathbf{K}_n : \mathbf{Q}(\alpha)\right]\left[\mathbf{Q}(\alpha) : \mathbf{Q}\right] = \left[\mathbf{K}_n : \mathbf{Q}\right] = \left[\mathbf{K}_n : \mathbf{K}_{n-1}\right]\left[\mathbf{K}_{n-1} : \mathbf{K}_{n-2}\right] \cdots \left[\mathbf{K}_1 : \mathbf{Q}\right] = 2^n.$$

Hence $[\mathbf{Q}(\alpha) : \mathbf{Q}] | 2^n$; thus $[\mathbf{Q}(\alpha) : \mathbf{Q}] = 2^r$ for some $r \geq 0$. $\quad \square$

13.08 COROLLARY. If $[\mathbf{Q}(\alpha):\mathbf{Q}]\neq 2^r$ for some $r\geqslant 0$, then α is not constructible. $\square$

Corollary 13.07 does not give a sufficient condition for α to be constructible, as Example 13.18 shows.

13.09 EXAMPLE. Can a root of the polynomial x^5+4x+2 be constructed using straight-edge and compass?

SOLUTION. Let α be a root of x^5+4x+2. By Eisenstein's criterion, x^5+4x+2 is irreducible over $\mathbf{Q}$; thus, by Corollary 11.16, $[\mathbf{Q}(\alpha):\mathbf{Q}]=5$. Since 5 is not a power of 2, it follows from Corollary 13.08 that α is not constructible. $\square$

13.10 EXAMPLE. Can a root of the polynomial x^4-3x^2+1 be constructed using straight-edge and compass?

SOLUTION. Solving the equation $x^4-3x^2+1=0$, we obtain $x^2=(3\pm\sqrt{5})/2$ and $x=\pm\sqrt{(3\pm\sqrt{5})/2}$. It follows from Theorem 13.06 that all these roots can be constructed. $\square$

DUPLICATING THE CUBE

Let l be the length of the sides of a given cube so that its volume is l^3. A cube with double the volume will have sides of length $\sqrt[3]{2}\, l$.

13.11 PROPOSITION. $\sqrt[3]{2}$ is not constructible.

PROOF. $\sqrt[3]{2}$ is a root of x^3-2 which, by the Rational Roots Theorem 9.32, is irreducible over $\mathbf{Q}$. Hence, by Corollary 11.16, $[\mathbf{Q}(\sqrt[3]{2}):\mathbf{Q}]=3$, and, by Corollary 13.08, $\sqrt[3]{2}$ is not constructible. $\square$

Since we cannot construct a length of $\sqrt[3]{2}\, l$ starting with a length l, the ancient problem of duplicating the cube is insoluble.

TRISECTING AN ANGLE

Certain angles can be trisected using straight-edge and compass. For example, $\pi,\pi/2,3\pi/4$ can be trisected because $\pi/3$, $\pi/6$, and $\pi/4$ can be constructed. However, we show that not all angles are trisectable by proving that $\pi/3$ cannot be trisected.

If we are given the angle ϕ, we can drop a perpendicular, from a point at a unit distance from the angle, to construct the lengths $\cos\phi$ and $\sin\phi$, as shown in Figure 13.04. Conversely, if either $\cos\phi$ or $\sin\phi$ is constructible, it is possible to construct the angle ϕ. Hence, if we are given an angle ϕ, we can construct all numbers in the extension field $\mathbf{Q}(\cos\phi)$. Of course, if $\cos\phi \in \mathbf{Q}$, then $\mathbf{Q}(\cos\phi) = \mathbf{Q}$.

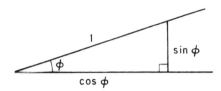

Figure 13.04. Constructing $\sin\phi$ and $\cos\phi$ from the angle ϕ.

We can now consider those numbers which are constructible from $\mathbf{Q}(\cos\phi)$. This notion of constructibility is similar to our previous notion, and similar results hold, except that the starting field is $\mathbf{Q}(\cos\phi)$ instead of $\mathbf{Q}$.

13.12 THEOREM. The angle ϕ can be trisected if and only if the polynomial $4x^3 - 3x - \cos\phi$ is reducible over $\mathbf{Q}(\cos\phi)$.

PROOF. Let $\theta = \phi/3$. The angle θ can be constructed from ϕ if and only if $\cos\theta$ can be constructed from $\cos\phi$. It follows from De Moivre's Theorem and the Binomial Theorem that

$$\cos\phi = \cos 3\theta = 4\cos^3\theta - 3\cos\theta.$$

Hence $\cos\theta$ is a root of $f(x) = 4x^3 - 3x - \cos\phi$.

If $f(x)$ is reducible over $\mathbf{Q}(\cos\phi)$, then $\cos\theta$ is a root of a polynomial of degree 1 or 2 over $\mathbf{Q}(\cos\phi)$; thus $[\mathbf{Q}(\cos\phi, \cos\theta) : \mathbf{Q}(\cos\phi)] = 1$ or 2. Hence, by Propositions 11.20 and 13.04, $\cos\theta$ is constructible from $\mathbf{Q}(\cos\phi)$.

If $f(x)$ is irreducible over $\mathbf{Q}(\cos\phi)$, then $[\mathbf{Q}(\cos\phi, \cos\theta) : \mathbf{Q}(\cos\phi)] = 3$, and it follows, by a proof similar to that of Theorem 13.06, that $\cos\theta$ cannot be constructed from $\mathbf{Q}(\cos\phi)$ by using straight-edge and compass. $\square$

13.13 COROLLARY. If $\cos\phi \in \mathbf{Q}$, then the angle ϕ can be trisected if and only if $4x^3 - 3x - \cos\phi$ is reducible over $\mathbf{Q}$. $\square$

For example, if $\phi = \pi/2$, then ϕ can be trisected because the polynomial $4x^3 - 3x + 0$ is reducible over $\mathbf{Q}$.

13.14 PROPOSITION. $\pi/3$ cannot be trisected by straight-edge and compass.

PROOF. The polynomial $f(x) = 4x^3 - 3x - \cos(\pi/3) = 4x^3 - 3x - (1/2)$. Now, by the Rational Roots Theorem 9.32, the only possible roots of $2f(x) = 8x^3 - 6x - 1$ are ± 1, $\pm 1/2$, $\pm 1/4$, or $\pm 1/8$. We see from the graph of $f(x)$ in Figure 13.05 that none of these are roots, except possibly $-1/4$ or $-1/8$. However, $f(-1/4) = 3/16$ and $f(-1/8) = -17/128$; thus $f(x)$ has no rational roots. Hence $f(x)$ is irreducible over **Q**, and, by Corollary 13.13, $\pi/3$ cannot be trisected by straight-edge and compass. □

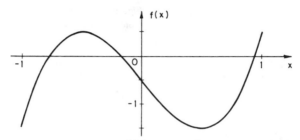

Figure 13.05.   The graph of $f(x) = 4x^3 - 3x - (1/2)$.

13.15 EXAMPLE. Archimedes showed that, if we are allowed to mark our straight-edge, it *is* possible to trisect any angle.

CONSTRUCTION. Let AOB be the angle ϕ we are to trisect. Draw a circle with center O and any radius r and let this circle meet OA and OB in P and Q. Mark two points X and Y on our straight-edge of distance r apart. Now move this straight-edge through Q, keeping X on OA until Y lies on the circle, as shown in Figure 13.06. Then we claim that the angle OXY is $\phi/3$, and hence the angle AOB is trisected.

SOLUTION. Let angle $OXY = \theta$. Since triangle XYO is isosceles, the angle $XOY = \theta$. Now

$$\text{angle } OYQ = \text{angle } OXY + \text{angle } XOY = 2\theta.$$

Triangle YOQ is isosceles, so angle $OQY = 2\theta$. Also,

$$\phi = \text{angle } AOB = \text{angle } OXQ + \text{angle } OQX = \theta + 2\theta = 3\theta.$$

Hence $\theta = \phi/3$.   □

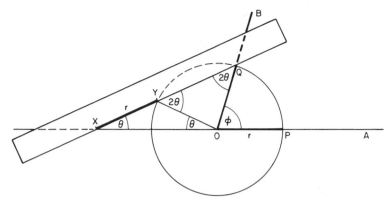

Figure 13.06.   The trisection of the angle ϕ using a marked ruler.

SQUARING THE CIRCLE

Given any circle of radius r, its area is πr^2, so that a square with the same area has sides of length $\sqrt{\pi}\, r$. We can square the circle if and only if $\sqrt{\pi}$ is constructible.

13.16 PROPOSITION. $[\mathbf{Q}(\sqrt{\pi}):\mathbf{Q}]$ is infinite, and hence $\sqrt{\pi}$ is not constructible.

PROOF.   The proof of this depends on the fact that π is transcendental over $\mathbf{Q}$; that is, π does not satisfy any polynomial equation with rational coefficients. This was mentioned in Chapter 11, and a proof is given in Stewart [45].

$\mathbf{Q}(\pi)$ is a subfield of $\mathbf{Q}(\sqrt{\pi})$ because $\pi=(\sqrt{\pi})^2\in\mathbf{Q}(\sqrt{\pi})$. Since π is transcendental, $\pi,\pi^2,\pi^3,\dots$ are linearly independent over $\mathbf{Q}$, and $[\mathbf{Q}(\pi):\mathbf{Q}]$ is infinite. Therefore,

$$\left[\mathbf{Q}(\sqrt{\pi}):\mathbf{Q}\right]=\left[\mathbf{Q}(\sqrt{\pi}):\mathbf{Q}(\pi)\right]\left[\mathbf{Q}(\pi):\mathbf{Q}\right]$$

is also infinite. Hence, by Corollary 13.08, $\sqrt{\pi}$ is not constructible. $\square$

Hence the circle cannot be squared by straight-edge and compass.

CONSTRUCTING REGULAR POLYGONS

Another problem that has been of great interest to mathematicians from the time of the ancient Greeks is that of constructing a regular n-gon, that is, a regular polygon with n sides. This is equivalent to constructing the angle $(2\pi/n)$ or the number $\cos(2\pi/n)$. The Greeks knew how to construct regular polygons with three, four, five, and six sides, but were unable to construct a regular 7-gon.

It is well known how to construct an equilateral triangle and a square using straight-edge and compass. We proved in Example 11.19 that $\cos(2\pi/5) = (\sqrt{5}-1)/4$; thus a regular pentagon can be constructed. Furthermore, if a regular n-gon is constructible, so is a regular $2n$-gon, because angles can be bisected using straight-edge and compass. Proposition 13.14 shows that $\pi/9$ cannot be constructed; hence $2\pi/9$ and a regular 9-gon cannot be constructed.

In 1796, at the age of nineteen, Gauss discovered that a regular 17-gon could be constructed and later showed the only regular n-gons that are constructible are the ones for which $n = 2^k p_1 \cdots p_r$, where $k \geqslant 0$ and $p_1, \ldots, p_r$ are distinct primes of the form $2^{2^m} + 1$. Prime numbers of the form $2^{2^m} + 1$ are called *Fermat primes*. Pierre de Fermat (1601–1665) conjectured that all numbers of the form $2^{2^m} + 1$ are prime. When $m = 0$, 1, 2, 3, and 4, the numbers are 3, 5, 17, 257, and 65537, respectively, and they are all prime. However, in 1732, Euler discovered that $2^{2^5} + 1$ is divisible by 641. Computers have checked many of these numbers for $m > 5$, and they have all been composite. In fact, no more Fermat primes are known today.

A complete proof of Gauss' result is beyond the scope of this book, since it requires more group and field theory than we have covered. (See Stewart [45] or Adamson [43].) However, we can prove the following.

13.17 THEOREM. If p is a prime for which a regular p-gon is constructible, then p is a Fermat prime.

PROOF. Let $\xi_p = \cos(2\pi/p) + i\sin(2\pi/p)$, a pth root of unity. If a regular p-gon can be constructed, $\cos(2\pi/p)$ and $\sin(2\pi/p)$ are constructible numbers and $[\mathbf{Q}(\cos(2\pi/p), \sin(2\pi/p)) : \mathbf{Q}] = 2^r$ for some integer r. Hence

$$\left[\mathbf{Q}(\cos(2\pi/p), \sin(2\pi/p), i) : \mathbf{Q} \right] = 2^{r+1}.$$

Now $\mathbf{Q}(\xi_p) \subseteq \mathbf{Q}(\cos(2\pi/p), \sin(2\pi/p), i)$ and so, by Theorem 11.08, $[\mathbf{Q}(\xi_p) : \mathbf{Q}] = 2^k$ for some integer $k \leqslant r+1$.

The pth root of unity, ξ_p, is a root of the cyclotomic polynomial $\phi(x) = x^{p-1} + x^{p-2} + \cdots + x + 1$ which, by Example 9.38, is irreducible over $\mathbf{Q}$. Hence $[\mathbf{Q}(\xi_p) : \mathbf{Q}] = p - 1$, and therefore $p - 1 = 2^k$.

The number $p = 2^k + 1$ is a prime only if $k = 0$ or k is a power of 2. Suppose that k contains an odd factor $b > 1$ and that $k = a \cdot b$. Then $2^a + 1 | (2^a)^b + 1$, since $x + 1 | x^b + 1$ if b is odd. Hence $2^{ab} + 1$ cannot be prime.

The case $p=2$ gives rise to the degenerate 2-gon. Otherwise p is a Fermat prime, $2^{2^m}+1$, for some integer $m \geqslant 0$.   □

A Nonconstructible Number of Degree Four

This next example shows that Corollary 13.07 does not give a sufficient condition for a number to be constructible.

13.18 EXAMPLE.   There is a real root r_i, of the irreducible polynomial $x^4 - 4x + 2$, that is not constructible, even though $[\mathbf{Q}(r_i):\mathbf{Q}]=2^2$.

SOLUTION.   By Eisenstein's criterion, $x^4 - 4x + 2$ is irreducible over $\mathbf{Q}$, so that $[\mathbf{Q}(r_i):\mathbf{Q}]=4$ for each root r_i. However, we can factor this polynomial into two quadratics over $\mathbf{R}$, say

$$x^4 - 4x + 2 = (x^2 + ax + b)(x^2 + cx + d).$$

Comparing coefficients, and then using equation (i), we have

$$
\begin{array}{lll}
0 = a + c & \text{and} \quad c = -a & \text{(i)} \\
0 = b + d + ac & \text{and} \quad b + d = a^2 & \text{(ii)} \\
-4 = bc + ad & \text{and} \quad -4 = a(d - b) & \text{(iii)} \\
2 = bd & & \text{(iv)}.
\end{array}
$$

Let $t = b + d$, so that $16 = a^2\{(b+d)^2 - 4bd\} = t(t^2 - 8)$. This number t satisfies the equation

$$t^3 - 8t - 16 = 0 \qquad \text{(v)}.$$

By Theorem 9.32, this equation (v) has no rational roots; thus $t^3 - 8t - 16$ is irreducible over $\mathbf{Q}$. We see from Figure 13.07 that the equation does

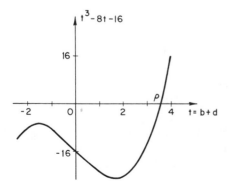

Figure 13.07.   The graph of $t^3 - 8t - 16$.

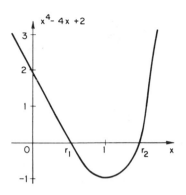

Figure 13.08. The graph of $x^4 - 4x + 2$.

have a real root ρ between 3 and 4, and the coefficients a, b, c, d can be expressed in terms of ρ.

Either a or c is positive. Without loss of generality suppose that $c > 0$; thus $b + d = t = \rho$, $a = -c = -\sqrt{\rho}$, and $d - b = 4/\sqrt{\rho}$. Therefore we have $b = \rho/2 - 2/\sqrt{\rho}$ and $d = \rho/2 + 2/\sqrt{\rho}$, and the roots of $x^2 + ax + b$ are

$$\frac{-a \pm \sqrt{a^2 - 4b}}{2} = \frac{1}{2}\left[\sqrt{\rho} \pm \sqrt{-\rho + \frac{8}{\sqrt{\rho}}}\,\right],$$

which are real, since $\rho < 4$. These are the roots r_1 and r_2 in Figure 13.08.

If both these roots of $x^4 - 4x + 2$ are constructible, then $(r_1 + r_2)^2 = \rho$ is also constructible. But this is impossible, since ρ is a root of the irreducible polynomial $t^3 - 8t - 16$ and $[\mathbf{Q}(\rho) : \mathbf{Q}] = 3$.

Hence $x^4 - 4x + 2$ has a real root that is not constructible. $\square$

This example was adapted from the article by Kalmanson [52].

Exercises

1–6.   Which of the following numbers are constructible?

1. $\sqrt[4]{5 + \sqrt{2}}$.

2. $\sqrt[6]{2}$.

3. $\dfrac{2}{1 + \sqrt{7}}$.

4. $(1 - 4\sqrt{7}\,)^3$.

5. $1 - \sqrt[5]{27}$.

6. $\sqrt[3]{7 - 5\sqrt{2}}$.

7. Is $\mathbf{Q}(\cos\phi) = \mathbf{Q}(\sin\phi)$ for every angle ϕ?

8. If $\tan\phi$ is constructible, show how to construct the angle ϕ.

9. Prove that all the constructible numbers are algebraic over **Q**.

10–13.   For each of the following values of $\cos \phi$, determine whether you can trisect the angle ϕ by straight-edge and compass.

10. $\cos \phi = 1/4$.   **11.** $\cos \phi = -9/16$.
12. $\cos \phi = 1/\sqrt{2}$.   **13.** $\cos \phi = \sqrt{2} /8$.
14. By writing $\pi/15$ in terms of $\pi/5$ and $\pi/3$, show that it is possible to trisect $\pi/5$ and also possible to construct a regular 15-gon.
15. Can $\pi/7$ be trisected?
16. Construct a regular pentagon using straight-edge and compass only.
17. Prove that $\cos 2\pi/7$ is a root of $8x^3 + 4x^2 - 4x - 1$ and that $2\cos 2\pi/7$ is a root of $x^3 + x^2 - 2x - 1$. Hence show that a regular septagon is not constructible.
18. If the regular n-gon is constructible and $n = qr$, show that the regular q-gon is also constructible.
19. Let $\xi = \cos(2\pi/p^2) + i\sin(2\pi/p^2)$. Show that ξ is a root of

$$f(x) = 1 + x^p + x^{2p} + \cdots + x^{(p-1)p}.$$

Prove that $f(x)$ is irreducible over **Q** by applying Eisenstein's criterion to $f(1+x)$.
20. Using the two previous exercises, prove that, if a regular n-gon is constructible, then $n = 2^k p_1 \cdots p_r$ where $p_1, \ldots, p_r$ are distinct Fermat primes.
21. Prove that a regular 17-gon is constructible.

22–33.   Can you construct a root of the following polynomials?

22. $x^2 - 7x - 13$.   **23.** $x^4 - 5$.
24. $x^8 - 16$.   **25.** $x^3 - 10x^2 + 2$.
26. $x^4 + x^3 - 12x^2 + 7x - 1$.   **27.** $x^5 - 9x^3 + 3$.
28. $x^6 + x^3 - 1$.   **29.** $3x^6 - 8x^4 + 1$.
30. $4x^4 - x^2 + 2x + 1$.   **31.** $x^4 + x - 1$.
32. $x^{48} - 1$.   **33.** $x^4 - 4x^3 + 4x^2 - 2$.

Error-Correcting Codes

14

With the increased usage of electronic instrumentation and computers, there is a growing need for methods of transmitting information quickly and *accurately* over radio and telephone lines and to and from magnetic tape. Over any transmission line, there is liable to be noise, that is, extraneous signals that can alter the transmitted information. This is not very noticeable in listening to the radio or even in reading a telegram, because normal English is about 20% redundant. However, in transmissions from satellites and in computer link-ups, the redundancy is usually zero; thus we would like to detect, and possibly correct, any errors in the transmitted message. We can do this by putting the message into a code that will detect and correct most of the errors. (These are not the sorts of codes useful to a spy. Secret codes are made deliberately hard to break, whereas error-correcting codes are designed to be easily decoded.)

One familiar code is the parity check digit that is usually attached to each number inside a computer. A number is written in binary form and a check digit is added that is the sum modulo 2 of the other digits. The sum of the digits of any number and its check digit is always even, unless an error has occurred. This check digit will detect any odd number of errors but not an even number of errors. This is useful if the probability of two errors occurring in the same word is very small. When a parity check failure occurs in reading words from a computer memory, the computer automatically rereads the faulty word. If a parity check failure occurs inside the arithmetic unit, the program usually has the be rerun.

All the codes we construct are obtained by adding a certain number of check digits to each block of information. Codes can either be used to simply *detect* errors or can be used to *correct* errors. A code that will detect $2t$ or fewer errors can be used to correct t or fewer errors.

Error-detecting codes are used when it is relatively easy to send the original message again, whenever an error is detected. The single parity check code in a computer is an example of an error-detecting code.

Sometimes it is impossible or too expensive to retransmit the original message when an error is detected. Error-correcting codes then have to be employed. These are used, for example, in transmissions from satellites and space probes (see Figure 14.01). The extra equipment needed to store and retransmit messages from a satellite would add unnecessary weight to the payload. Error-correcting codes are also used when transmitting data from computer memories to magnetic tape units. If a message containing an error is stored on magnetic tape, it may be weeks before it is read and the error detected; by this time the original word would be lost.

The Coding Problem

In most digital computers and many communication systems, information is handled in binary form; that is, messages are formed from the symbols 0 and 1. Therefore, in this chapter, we only discuss *binary codes*. However, most of the results generalize to codes whose symbols come from any finite field.

We assume that, when a message is transmitted over a channel, the probability of the digit 1 being changed into 0 is the same as that of 0 being changed into 1. Such channels are called *binary symmetric*. Figure 14.02 illustrates what might happen to a message over a noisy channel.

In order to transmit a message over a noisy channel, we break up the message into blocks of k digits and we *encode* each block by attaching $n - k$ *check digits* to obtain a *code word* consisting of n digits, as shown in Figure 14.03. Such a code is referred to as an *(n, k)-code*.

The code words can now be transmitted over the noisy channel, and, after being received, they can be processed in one of two ways. The code can be used to *detect errors* by checking whether the received word is a code word or not. If the received word is a code word, it is assumed to be

Figure 14.01 In 1969 the Mariners 6 and 7 space probes sent back over 200 close-up photographs of Mars. Each photograph was divided into 658,240 dots and each dot was given a brightness level ranging from 1 to 2^8. Therefore, each photograph required about five million bits of information. These bits were encoded, using an error-correcting code, and transmitted back to Earth at a rate of 16,200 bits per second, where they were received and decoded into photographs. (Photographs from JPL/NASA)

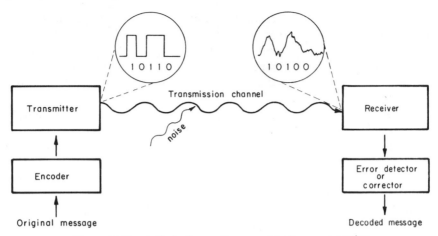

Figure 14.02. A block diagram for error detection or correction.

the transmitted word. If the received word is not a code word, an error must have occurred during transmission, and the receiver can request that the word be retransmitted. However, the code could also be used to *correct errors*. In this case, the decoder chooses the transmitted code word that is most likely to produce each received word.

Whether a code is used as an error-detecting or as an error-correcting code depends on each individual situation. More equipment is required to correct errors, and fewer errors can be corrected than could be detected; on the other hand, when a code only detects errors, there is the trouble of stopping the decoding process and requesting retransmission every time an error occurs.

In an (n,k)-code, the original message is k digits long and there are 2^k different possible messages and hence 2^k code words. The received words have n digits; hence there are 2^n possible words that could be received, only 2^k of which are code words.

The extra $n-k$ check digits that are added to produce the code word are called *redundant digits* because they carry no new information but only allow the existing information to be transmitted more accurately. The ratio $R = k/n$ is called the *code rate* or *information rate*.

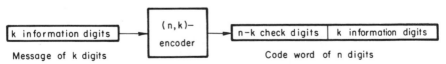

Figure 14.03. Encoding a block of k digits.

For each particular communications channel, it is a major problem to design a code that will transmit useful information as fast as possible and, at the same time, as reliably as possible.

It was proved by C. E. Shannon in 1948 that each channel has a definite capacity C, and for any rate $R < C$, there exist codes of rate R such that the probability of erroneous decoding is arbitrarily small. In other words, by increasing the code length n and keeping the code rate R below the channel capacity C, it is possible to make the probability of erroneous decoding as small as we please. However, this theory provides no useful method of finding such codes.

For codes to be efficient, they usually have to be very long; they may contain 2^{100} messages and many times that number of possible received words. In order to be able to effectively encode and decode such long codes, we look at codes that have a strong algebraic structure.

SIMPLE CODES

We now compare two very simple codes of length 3. The first is the $(3, 2)$-code that attaches a single parity check to a message of length 2. The parity check is the sum modulo 2 of the digits in the message. Hence a received word is a code word if and only if it contains an even number of 1s. The code words are given in Table 14.1. The second code is the $(3, 1)$-code that repeats a message, consisting of a single digit, three times. Its two code words are illustrated in Table 14.2.

If one error occurs in the $(3, 2)$ parity check code during transmission, say 101 is changed to 100, then this would be detected because there would be an odd number of 1s in the received word. However, this code will not correct any errors; the received word 100 is just as likely to have come from 110 or 000 as from 101. This code will not detect two errors either. If 101 was the transmitted code word and errors occurred in the first two

Table 14.1.
The $(3, 2)$ parity check code

Message	Code Word
00	000
01	101
10	110
11	011

↑
parity check

Table 14.2.
The $(3, 1)$ repeating code

Message	Code Word
0	000
1	111

positions, the received word would be 011, and this would be erroneously decoded as 11.

The decoder first performs a parity check on the received word. If there are an even number of 1s in the word, the word passes the parity check, and the message is the last two digits of the word. If there are an odd number of ones in the received work, it fails the parity check, and the decoder registers an error. Examples of this decoding are shown in Table 14.3.

Table 14.3.   The $(3,2)$ parity check code used to detect errors

Received Word:	101	111	100	000	110
Parity check:	Passes	Fails	Fails	Passes	Passes
Received Message:	01	Error	Error	00	10

The $(3,1)$ repeating code can be used as an error-detecting code, and it will detect one or two transmission errors but, of course, not three errors. This same code can also be used as an error-correcting code. If the received word contains more 1s than 0s, the decoder assumes that the message is 1, otherwise it assumes the message is 0. This will correctly decode messages containing one error, but will erroneously decode messages containing more than one error. Examples of this decoding are shown in Table 14.4.

Table 14.4.   The $(3,1)$ repeating code used to correct errors

Received Word:	111	010	011	000
Decoded Message:	1	0	1	0

One useful way to discover the error-detecting and error-correcting capabilities of a code is by means of the Hamming distance. The *Hamming distance* between two words u and v of the same length is defined to be the number of positions in which they differ. This distance is denoted by $d(u,v)$. For example, $d(101,100)=1$, $d(101,010)=3$, and $d(010,010)=0$.

The Hamming distance between two words is the number of single errors needed to change one word into the other. In an (n,k)-code, the 2^n received words can be thought of as placed at the vertices of an n-dimensional cube with unit sides. The Hamming distance between two words is the shortest distance between their corresponding vertices along the edges of the n-cube. The 2^k code words form a subset of the 2^n vertices, and the

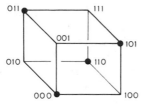

Figure 14.04. The code words of the (3,2) parity check code are shown as large dots.

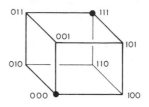

Figure 14.05. The code words of the (3, 1) repeating code are shown as large dots.

code has better error-correcting and detecting capabilities the further these code words are apart. Figure 14.04 illustrates the (3,2) parity check code whose code words are at Hamming distance 2 apart. Figure 14.05 illustrates the (3, 1) repeating code whose code words are at Hamming distance 3 apart.

14.01 PROPOSITION. A code will detect all sets of t or fewer errors if and only if the minimum Hamming distance between code words is at least $t+1$.

PROOF. If r errors occur when the code word u is transmitted, the received word v is at Hamming distance r from u. These transmission errors will be detected if and only if v is not another code word. Hence all sets of t or fewer errors in the code word u will be detected if and only if the Hamming distance of u from all the other code words is at least $t+1$. ☐

14.02 PROPOSITION. A code is capable of correcting all sets of t or fewer errors if and only if the minimum Hamming distance between code words is at least $2t+1$.

PROOF. Suppose the code contains two code words u_1 and u_2 at Hamming distance $2t$ or closer. Then there exists a received word v that differs from u_1 and u_2 in t or fewer positions. This received word v could have originated from u_1 or u_2 with t or fewer errors and hence would not be correctly decoded in both these situations.

Conversely, any code whose code words are at least $2t+1$ apart is capable of correcting up to t errors. This can be achieved in decoding by choosing the code word that is closest to each received word. ☐

Table 14.5 summarizes these results.

Table 14.5. The detection capabilities of various codes

Code	Minimum distance between code words	Number of errors detectable	Number of errors correctable	Information Rate
(3, 2) parity check code	2	1	0	2/3
(3, 1) repeating code	3	2	1	1/3
General (n, k) code	d	$d - 1$	$\leqslant (d-1)/2$	k/n

POLYNOMIAL REPRESENTATION

There are various ways that a word of n binary digits can be represented algebraically. One convenient way is by means of a polynomial in $\mathbf{Z}_2[x]$ of degree less than n. The word $a_0 a_1 \ldots a_{n-1}$ can be represented by the polynomial

$$a_0 + a_1 x + \cdots + a_{n-1} x^{n-1} \in \mathbf{Z}_2[x].$$

We now use this representation to show how codes can be constructed.

14.03 DEFINITION. Let $p(x) \in \mathbf{Z}_2[x]$ be a polynomial of degree $n - k$. The *polynomial code generated by* $p(x)$ is an (n, k)-code whose code words are precisely those polynomials, of degree less than n, which are divisible by $p(x)$.

A message of length k is represented by a polynomial $m(x)$, of degree less than k. In order that the higher order coefficients in a code polynomial carry the message digits, we multiply $m(x)$ by x^{n-k}. This has the effect of shifting the message $n - k$ places to the right. To *encode* the message polynomial $m(x)$, we divide $x^{n-k}m(x)$ by $p(x)$ and add the remainder, $r(x)$, to $x^{n-k}m(x)$ to form the code polynomial

$$v(x) = r(x) + x^{n-k}m(x).$$

This code polynomial is always a multiple of $p(x)$ because, by the division algorithm,

$$x^{n-k}m(x) = q(x) \cdot p(x) + r(x) \quad \text{where} \quad \deg r(x) < n - k \quad \text{or} \quad r(x) = 0;$$

thus

$$v(x) = r(x) + x^{n-k}m(x) = -r(x) + x^{n-k}m(x) = q(x) \cdot p(x).$$

(Remember $r(x) = -r(x)$ in $\mathbf{Z}_2[x]$.) The polynomial $x^{n-k}m(x)$ has zeros in the $n-k$ lowest order terms, whereas the polynomial $r(x)$ is of degree less than $n-k$; hence the k highest order coefficients of the code polynomial $v(x)$ are the message digits, and the $n-k$ lowest order coefficients are the check digits. These check digits are precisely the coefficients of the remainder $r(x)$.

For example, let $p(x) = 1 + x^2 + x^3 + x^4$ be the generator polynomial of a $(7,3)$-code. We encode the message 101 as follows.

$$
\begin{aligned}
\text{Message} &= 1 \quad 0 \quad 1 \\
m(x) &= 1 \quad\quad + x^2 \\
x^4 m(x) &= \quad\quad\quad\quad\quad\quad x^4 \quad + x^6 \\
r(x) &= 1 + x \\
v(x) = r(x) + x^4 m(x) &= 1 + x \quad\quad\quad + x^4 \quad + x^6 \\
\text{Code word} &= 1 \quad 1 \quad 0 \quad 0 \quad 1 \quad 0 \quad 1
\end{aligned}
$$

$$\underbrace{}_{\text{Check Digits}} \quad \underbrace{}_{\text{Message}}$$

$$
\begin{array}{r}
x^2 + x + 1 \\
x^4 + x^3 + x^2 + 0 + 1 \overline{\big)\, x^6 \quad\quad + x^4 } \\
\underline{x^6 + x^5 + x^4 \quad\quad + x^2 } \\
x^5 \quad\quad\quad + x^2 \\
\underline{x^5 + x^4 + x^3 \quad\quad + x } \\
x^4 + x^3 + x^2 + x \\
\underline{x^4 + x^3 + x^2 \quad\quad + 1 } \\
x + 1
\end{array}
$$

The generator polynomial $p(x) = a_0 + a_1 x + \cdots + a_{n-k} x^{n-k}$ is always chosen so that $a_0 = 1$ and $a_{n-k} = 1$, since this avoids wasting check digits. If $a_0 = 0$, any code polynomial would be divisible by x and the first digit of the code word would always be 0; if $a_{n-k} = 0$, the coefficient of x^{n-k-1} in the code polynomial would always be 0.

14.04 EXAMPLE. Write down all the code words for the code generated by the polynomial $p(x) = 1 + x + x^3$ when the message length k is 3.

SOLUTION. Since $\deg p(x) = 3$, there will be three check digits and, since the message length k is 3, the code word length n will be 6. The number of messages is $2^k = 8$.

Consider the message 110, which is represented by the polynomial $m(x) = 1 + x$. Its check digits are the coefficients of the remainder $r(x) = 1 + x^2$, obtained by dividing $x^3 m(x) = x^3 + x^4$ by $p(x)$. Hence the code polynomial is $v(x) = r(x) + x^3 m(x) = 1 + x^2 + x^3 + x^4$, and the code word is 101110. Table 14.6 shows all the code words. $\square$

Table 14.6.   The (6,3)-code generated by $1 + x + x^3$

Message			Code Word					
			Check Digits			Message Digits		
0	0	0	0	0	0	0	0	0
1	0	0	1	1	0	1	0	0
0	1	0	0	1	1	0	1	0
0	0	1	1	1	1	0	0	1
1	1	0	1	0	1	1	1	0
1	0	1	0	0	1	1	0	1
0	1	1	1	0	0	0	1	1
1	1	1	0	1	0	1	1	1
↑	↑	↑	↑	↑	↑	↑	↑	↑
1	x	x^2	1	x	x^2	x^3	x^4	x^5

$$
\begin{array}{r}
x + 1 \\
x^3 + 0 + x + 1 \,\overline{\smash{\big)}\, x^4 + x^3 } \\
x^4 + x^2 + x \\
\hline
x^3 + x^2 + x \\
x^3 x + 1 \\
\hline
x^2 + 1
\end{array}
$$

A received message can be checked for errors by testing whether it is divisible by the generator polynomial $p(x)$. If the remainder, when the received polynomial $u(x)$ is divided by $p(x)$, is nonzero, an error must have occurred during transmission. If the remainder is zero, the received polynomial $u(x)$ is a code word, and either no error has occurred or an undetectable error has occurred.

14.05 EXAMPLE.   If the generator polynomial is $p(x) = 1 + x + x^3$, test whether the following received words contain detectable errors.
(i)   100011,   (ii)   100110,   (iii)   101000.

SOLUTION.   The received polynomials are $1+x^4+x^5$, $1+x^3+x^4$ and $1+x^2$, respectively. These contain detectable errors if and only if they have nonzero remainders when divided by $p(x)=1+x+x^3$.

$$
\begin{array}{r}
x^2+x\ +1 \\
x^3+x+1\ \overline{\smash{\big)}\ x^5+x^4+0\ +0\ +0+1} \\
x^5\quad\ +x^3+x^2 \\
\hline
x^4+x^3+x^2 \\
x^4\quad\ +x^2+x \\
\hline
x^3\quad\ +x+1 \\
x^3\quad\ +x+1 \\
\hline
0
\end{array}
$$

$$
\begin{array}{r}
x\ +1 \\
x^3+x+1\ \overline{\smash{\big)}\ x^4+x^3+0\ +0+1} \\
x^4\quad\ +x^2+x \\
\hline
x^3+x^2+x+1 \\
x^3\quad\ +x+1 \\
\hline
x^2
\end{array}
$$

$$
\begin{array}{r}
0 \\
x^3+x+1\ \overline{\smash{\big)}\ x^2+0+1}
\end{array}
$$

Hence $1+x^4+x^5$ is divisible by $p(x)$, but $1+x^3+x^4$ and $1+x^2$ are not. Therefore, errors have occurred in the latter two words but are unlikely to have occurred in the first.   □

Table 14.6 lists all the code words for this code. Hence, in the above example, we can tell at a glance whether a word is a code word simply by noting whether it is on this list. However, in practice, the list of code words is usually so large that it is easier to calculate the remainder when the received polynomial is divided by the generator polynomial.

Furthermore, this remainder can easily be computed using shift registers. Figure 14.06 shows a shift register for dividing by $1+x+x^3$. The square boxes represent unit delays, and the circle with a cross inside denotes a modulo 2 adder (or Exclusive OR gate).

The delays are initiaĺly zero, and a polynomial $u(x)$ is fed into this shift register with the high-order coefficients first. When all the coefficients of $u(x)$ have been fed in, the delays contain the remainder of $u(x)$ when divided by $1+x+x^3$. If these are all zero, the polynomial $u(x)$ is a code

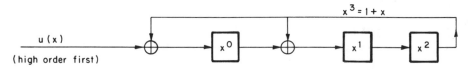

Figure 14.06.   A shift register for dividing by $1+x+x^3$.

Table 14.7. The contents of the shift register when $1+x^3+x^4$ is divided by $1+x+x^3$

Stage	Received Polynomial waiting to enter register						Register Contents x^0	x^1	x^2	
0	1	0	0	1	1	0	0	0	0	←Register initially zero
1		1	0	0	1	1	0	0	0	
2			1	0	0	1	1	0	0	
3				1	0	0	1	1	0	
4					1	0	0	1	1	
5						1	1	1	1	
6							0	0	1	←Remainder is x^2

word; otherwise a detectable error has occurred. Table 14.7 illustrates this shift register in operation.

This same register could be modified to encode messages, because the check digits for $m(x)$ are the coefficients of the remainder when $x^3 m(x)$ is divided by $1+x+x^3$. However, the circuit in Figure 14.07 is more efficient for encoding.

The message $m(x)$ is fed directly into the high end of the shift register. This has the effect of multiplying $m(x)$ by x^3. While $m(x)$ is being fed in, the switch is in position 1 and the remainder is calculated by the register. Then the switch is changed to position 2, and the check digits are let out to immediately follow the message.

This encoding circuit could also be used for error detection. When $u(x)$ is fed into the encoding circuit with the switch in position 1, the register calculates the remainder of $x^3 u(x)$ when divided by $p(x)$. However, $u(x)$ is divisible by $p(x)$ if and only if $x^3 u(x)$ is divisible by $p(x)$, assuming that $p(x)$ does not contain a factor x.

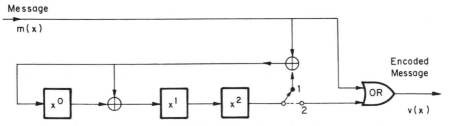

Figure 14.07. An encoding circuit for a code generated by $1+x+x^3$.

How is the generator polynomial chosen so that the code has useful properties without adding too many check digits? We now give some examples.

14.06 PROPOSITION.   The polynomial $p(x) = 1 + x$ generates the $(n, n-1)$ parity check code.

PROOF.   By Proposition 9.40, a polynomial in $\mathbf{Z}_2[x]$ is divisible by $1 + x$ if and only if it contains an even number of nonzero coefficients. Hence the code words of a code generated by $1 + x$ are those words containing an even number of 1s. The check digit for the message polynomial $m(x)$ is the remainder when $xm(x)$ is divided by $1 + x$. Therefore, by the Remainder Theorem 9.05, the check digit is $m(1)$, the parity of the number of 1s in the message. This code is the parity check code. $\square$

The $(3, 1)$ code that repeats the single message digit three times has code words 000 and 111, and is generated by the polynomial $1 + x + x^2$.

We now give one method, using primitive polynomials, of finding a generator for a code that will always detect single, double, or triple errors. Furthermore, the degree of the generator polynomial will be as small as possible so that the check digits are reduced to a minimum. Recall from Definition 11.36 that an irreducible polynomial $p(x)$ of degree m over $\mathbf{Z}_2$ is primitive if $p(x)|1 + x^k$ for $k = 2^m - 1$ and for no smaller k.

14.07 THEOREM.   If $p(x)$ is a primitive polynomial of degree m, then the $(n, n-m)$-code generated by $p(x)$ detects all single and double errors whenever $n \leqslant 2^m - 1$.

PROOF.   Let $v(x)$ be a transmitted code word and $u(x) = v(x) + e(x)$ be the received word. The polynomial $e(x)$ is called the *error polynomial*. An error is detectable if and only if $p(x) \nmid u(x)$. Since $p(x)$ does divide the code word $v(x)$, an error $e(x)$ will be detectable if and only if $p(x) \nmid e(x)$.

If a single error occurs, the error polynomial contains a single term, say x^i, where $0 \leqslant i < n$. Since $p(x)$ is irreducible, it does not have 0 as a root; therefore, $p(x) \nmid x^i$, and the error x^i is detectable.

If a double error occurs, the error polynomial $e(x)$ is of the form $x^i + x^j$ where $0 \leqslant i < j < n$. Hence $e(x) = x^i(1 + x^{j-i})$ where $0 < j - i < n$. Now $p(x) \nmid x^i$ and, since $p(x)$ is primitive, $p(x) \nmid 1 + x^{j-i}$ if $j - i < 2^m - 1$. Since $p(x)$ is irreducible, $p(x) \nmid x^i(1 + x^{j-i})$ whenever $n \leqslant 2^m - 1$, and all double errors are detectable. $\square$

14.08 COROLLARY.   If $p_1(x)$ is a primitive polynomial of degree m, the

$(n, n - m - 1)$-code generated by $p(x) = (1 + x)p_1(x)$ detects all double errors and any odd number of errors whenever $n \le 2^m - 1$.

PROOF.   The code words in the code generated by $p(x)$ must be divisible by $p_1(x)$ and by $(1 + x)$. The factor $(1 + x)$ has the effect of adding an overall parity check digit to the code. By Proposition 9.40, all the code words have an even number of terms, and the code will detect any odd number of errors. Since the code words are divisible by the primitive polynomial $p_1(x)$, the code will detect all double errors if $n \le 2^m - 1$. $\square$

Some primitive polynomials of low degree are given in Table 14.8. For example, by adding 11 check digits to a message of length 1012 or less, using the generator polynomial $(1 + x)(1 + x^3 + x^{10}) = 1 + x + x^3 + x^4 + x^{10} + x^{11}$, we can detect single, double, triple, and any odd number of errors. Furthermore, the encoding and detecting can be done by a small shift register using only 11 delay units. The number of different messages of length 1012 is 2^{1012}, an enormous figure! When written out in base 10, it would contain 305 digits.

Table 14.8.   A short table of primitive polynomials in $Z_2[x]$

Primitive Polynomial	Degree m	$2^m - 1$
$1 + x$	1	1
$1 + x + x^2$	2	3
$1 + x + x^3$	3	7
$1 + x + x^4$	4	15
$1 + x^2 + x^5$	5	31
$1 + x + x^6$	6	63
$1 + x^3 + x^7$	7	127
$1 + x^2 + x^3 + x^4 + x^8$	8	255
$1 + x^4 + x^9$	9	511
$1 + x^3 + x^{10}$	10	1023

MATRIX REPRESENTATION

Another natural way to represent a word $a_1 a_2 \ldots a_n$ of length n is by the element $(a_1, a_2, \ldots, a_n)^T$ of the vector space $Z_2^n = Z_2 \times Z_2 \times \ldots \times Z_2$ of dimension n over Z_2. We denote the elements of our vector spaces as column vectors, and $(a_1, a_2, \ldots, a_n)^T$ denotes the transpose of $(a_1, a_2, \ldots, a_n)$. In an

(n, k)-code, the 2^k possible messages of length k are all the elements of the vector space $\mathbf{Z}_2^k$, whereas the 2^n possible recieved words of length n form the vector space $\mathbf{Z}_2^n$. An encoder is an injective function

$$\gamma : \mathbf{Z}_2^k \to \mathbf{Z}_2^n$$

that assigns to each k digit message an n digit code word.

14.09 DEFINITION. An (n, k)-code is called a *linear code* if the encoding function is a linear transformation from $\mathbf{Z}_2^k$ to $\mathbf{Z}_2^n$.

Nearly all block codes in use are linear codes, and, in particular, all polynomial codes are linear.

14.10 PROPOSITION. Let $p(x)$ be a polynomial of degree $n - k$ that generates an (n, k)-code. Then this code is linear.

PROOF. Let $\gamma : \mathbf{Z}_2^k \to \mathbf{Z}_2^n$ be the encoding function defined by the generator polynomial $p(x)$. Let $m_1(x)$ and $m_2(x)$ be two message polynomials of degree less than k and let $\mathbf{m}_1$ and $\mathbf{m}_2$ be the same messages considered as vectors in $\mathbf{Z}_2^k$. The code vector $\gamma(\mathbf{m}_i)$ corresponds to the code polynomial $v_i(x) = r_i(x) + x^{n-k} m_i(x)$, where $r_i(x)$ is the remainder when $x^{n-k} m_i(x)$ is divided by $p(x)$. Now

$$v_1(x) + v_2(x) = r_1(x) + r_2(x) + x^{n-k} \big[m_1(x) + m_2(x) \big],$$

and $r_1(x) + r_2(x)$ has degree less than $n - k$; therefore, $r_1(x) + r_2(x)$ is the remainder when $x^{n-k} m_1(x) + x^{n-k} m_2(x)$ is divided by $p(x)$. Hence $v_1(x) + v_2(x)$ corresponds to the code vector $\gamma(\mathbf{m}_1 + \mathbf{m}_2)$ and

$$\gamma(\mathbf{m}_1 + \mathbf{m}_2) = \gamma(\mathbf{m}_1) + \gamma(\mathbf{m}_2).$$

Since the only scalars are 0 and 1, this implies that γ is a linear transformation. $\square$

Let $\{\mathbf{e}_1, \mathbf{e}_2, \ldots, \mathbf{e}_n\}$ be the standard basis of the vector space $\mathbf{Z}_2^n$, where $\mathbf{e}_i$ contains a 1 in the ith position and 0s elsewhere.

14.11 DEFINITION. Let G be the $n \times k$ matrix that represents, with respect to the standard basis, the transformation $\gamma : \mathbf{Z}_2^k \to \mathbf{Z}_2^n$, defined by an (n, k) linear code. This matrix G is called the *generator matrix* or *encoding matrix* of the code.

If $\mathbf{m}$ is a message vector, its code word is $\mathbf{v} = G\mathbf{m}$. The code vectors are the vectors in the image of γ, and they form a vector subspace of $\mathbf{Z}_2^n$ of dimension k. The columns of G are a basis for this subspace, and therefore, a vector is a code vector if and only if it is a linear combination of the columns of the generator matrix G.

(Most coding theorists write the elements of their vector spaces as row vectors instead of column vectors, as used here. In this case, their generator matrix is the transpose of ours, and it operates on the right of the message vector.)

In the $(3,2)$ parity check code, a vector $\mathbf{m}=(m_1,m_2)^T$ is encoded as $\mathbf{v}=(c,m_1,m_2)^T$, where the parity check $c=m_1+m_2$. Hence the generator matrix is

$$G = \begin{bmatrix} 1 & 1 \\ 1 & 0 \\ 0 & 1 \end{bmatrix} \quad \text{because} \quad \begin{bmatrix} 1 & 1 \\ 1 & 0 \\ 0 & 1 \end{bmatrix}\begin{pmatrix} m_1 \\ m_2 \end{pmatrix} = \begin{pmatrix} c \\ m_1 \\ m_2 \end{pmatrix}.$$

If the code word is to contain the message digits in its last k positions, the generator matrix must be of the form $G = \left(\dfrac{P}{I_k}\right)$ where P is an $(n-k)\times k$ matrix and I_k is the $k\times k$ identity matrix.

14.12 EXAMPLE. Find the generator matrix for the $(6,3)$-code of Example 14.04 that is generated by the polynomial $1+x+x^3$.

SOLUTION. The columns of the generator matrix G are the code vectors corresponding to messages consisting of basis elements $\mathbf{e}_1=(1,0,0)^T$, $\mathbf{e}_2=(0,1,0)^T$, and $\mathbf{e}_3=(0,0,1)^T$. We see from Table 14.6 that the generator matrix is

$$G = \begin{bmatrix} 1 & 0 & 1 \\ 1 & 1 & 1 \\ 0 & 1 & 1 \\ 1 & 0 & 0 \\ 0 & 1 & 0 \\ 0 & 0 & 1 \end{bmatrix}. \quad \square$$

Any message vector, $\mathbf{m}$, in the $(6,3)$-code of the above example can be encoded by calculating $G\mathbf{m}$. However, given any received vector $\mathbf{u}$ it is not easy to determine, from the generator matrix G, whether $\mathbf{u}$ is a code vector or not. The code vectors form a subspace, $\operatorname{Im}\gamma$, of dimension k in $\mathbf{Z}_2^n$, generated by the columns of G. We now find a linear transformation $\eta:\mathbf{Z}_2^n\to\mathbf{Z}_2^{n-k}$, represented by a matrix H, whose kernel is precisely $\operatorname{Im}\gamma$. Hence a vector $\mathbf{u}$ will be a code vector if and only if $H\mathbf{u}=\mathbf{0}$.

14.13 THEOREM. Let $\gamma : \mathbf{Z}_2^k \to \mathbf{Z}_2^n$ be the encoding function for a linear (n, k)-code with generator matrix $G = \begin{pmatrix} P \\ I_k \end{pmatrix}$, where P is an $(n-k) \times k$ matrix and I_k is the $k \times k$ identity matrix. Then the linear transformation

$$\eta : \mathbf{Z}_2^n \to \mathbf{Z}_2^{n-k}$$

defined by the $(n-k) \times n$ matrix $H = (I_{n-k} | P)$ has the following properties.

(i) $\operatorname{Ker} \eta = \operatorname{Im} \gamma$.
(ii) A received vector $\mathbf{u}$ is a code vector if and only if $H\mathbf{u} = \mathbf{0}$.

14.14 DEFINITION. This $(n-k) \times n$ matrix H is called the *parity check matrix* of the (n, k)-code.

PROOF OF THE THEOREM. The composition $\eta \circ \gamma : \mathbf{Z}_2^k \to \mathbf{Z}_2^{n-k}$ is the zero transformation because

$$HG = (I_{n-k} | P) \begin{pmatrix} P \\ I_k \end{pmatrix} = (I_{n-k} P + PI_k) = P + P = 0$$

using block multiplication of matrices over the field $\mathbf{Z}_2$. Hence $\operatorname{Im} \gamma \subseteq \operatorname{Ker} \eta$.

Since the first $n-k$ columns of H consist of the standard basis vectors in $\mathbf{Z}_2^{n-k}$, $\operatorname{Im} \eta$ spans $\mathbf{Z}_2^{n-k}$ and contains 2^{n-k} elements. By the Morphism Theorem for groups

$$\# \operatorname{Ker} \eta = \frac{\# \mathbf{Z}_2^n}{\# \operatorname{Im} \eta} = \frac{2^n}{2^{n-k}} = 2^k.$$

But $\operatorname{Im} \gamma$ also contains 2^k elements, and therefore $\operatorname{Im} \gamma$ must equal $\operatorname{Ker} \eta$. □

The parity check matrix of the $(3, 2)$ parity check code is the 1×3 matrix $H = (1 \quad 1 \quad 1)$. A received vector $\mathbf{u} = (u_1, u_2, u_3)^T$ is a code vector if and only if

$$H\mathbf{u} = (1 \quad 1 \quad 1) \begin{bmatrix} u_1 \\ u_2 \\ u_3 \end{bmatrix} = u_1 + u_2 + u_3 = 0.$$

The parity check matrix of the $(3, 1)$-code that repeats the message three times is the 2×3 matrix $H = \begin{pmatrix} 1 & 0 & 1 \\ 0 & 1 & 1 \end{pmatrix}$. A received vector $\mathbf{u} = (u_1, u_2, u_3)^T$

is a code vector if and only if $H\mathbf{u}=\mathbf{0}$; that is, if and only if $u_1+u_3=0$ and $u_2+u_3=0$. In $\mathbf{Z}_2$, this is equivalent to $u_1=u_2=u_3$.

The parity check matrix for the $(6,3)$-code of Examples 14.04 and 14.12 is

$$H = \begin{bmatrix} 1 & 0 & 0 & 1 & 0 & 1 \\ 0 & 1 & 0 & 1 & 1 & 1 \\ 0 & 0 & 1 & 0 & 1 & 1 \end{bmatrix}.$$

The received vector $\mathbf{u}=(u_1,\ldots,u_6)^T$ is a code vector if and only if

$$\begin{aligned} u_1 \quad\;\; + u_4 \quad\;\;\; + u_6 &= 0 \\ u_2 \;+ u_4 + u_5 + u_6 &= 0 \\ u_3 \quad\;\;\; + u_5 + u_6 &= 0. \end{aligned}$$

That is, if and only if

$$\begin{aligned} u_1 &= u_4 \quad\;\; + u_6 \\ u_2 &= u_4 + u_5 + u_6 \\ u_3 &= \quad\;\; u_5 + u_6. \end{aligned}$$

In this code, the three digits on the right, u_4, u_5, and u_6, are the message digits, whereas u_1, u_2, and u_3 are the check digits. For each code vector $\mathbf{u}$, the equation $H\mathbf{u}=\mathbf{0}$ expresses each check digit in terms of the message digits. This is why H is called the parity check matrix.

14.15 EXAMPLE. Find the generator matrix and parity check matrix for the $(9,4)$-code generated by $p(x)=(1+x)(1+x+x^4)=1+x^2+x^4+x^5$. Then use the parity check matrix to determine whether the word 110110111 is a code word.

SOLUTION. The check digits attached to a message polynomial $m(x)$ are the coefficients of the remainder when $x^5 m(x)$ is divided by $p(x)$. The message polynomials are linear combinations of $1, x, x^2$ and x^3. We can calculate the remainders when x^5, x^6, x^7, and x^8 are divided by $p(x)$ as follows. (This is just like the action of a shift register that divides by $p(x)$.)

$$x^5 \equiv 1 + x^2 + x^4 \bmod p(x)$$

$$x^6 \equiv x + x^3 + x^5 \equiv 1 + x + x^2 + x^3 + x^4 \bmod p(x)$$

$$x^7 \equiv x + x^2 + x^3 + x^4 + x^5 \equiv 1 + x + x^3 \bmod p(x)$$

$$x^8 \equiv x + x^2 + x^4 \bmod p(x).$$

Therefore, every code polynomial is a linear combination of the following basis polynomials:

$$1 + x^2 + x^4 + x^5$$

$$1 + x + x^2 + x^3 + x^4 + x^6$$

$$1 + x + x^3 + x^7$$

$$x + x^2 + x^4 + x^8.$$

The generator matrix G is obtained from the coefficients of the above polynomials, and the parity check matrix H is obtained from G. Hence

$$G = \begin{bmatrix} 1 & 1 & 1 & 0 \\ 0 & 1 & 1 & 1 \\ 1 & 1 & 0 & 1 \\ 0 & 1 & 1 & 0 \\ 1 & 1 & 0 & 1 \\ 1 & 0 & 0 & 0 \\ 0 & 1 & 0 & 0 \\ 0 & 0 & 1 & 0 \\ 0 & 0 & 0 & 1 \end{bmatrix} \quad \text{and} \quad H = \begin{bmatrix} 1 & 0 & 0 & 0 & 0 & 1 & 1 & 1 & 0 \\ 0 & 1 & 0 & 0 & 0 & 0 & 1 & 1 & 1 \\ 0 & 0 & 1 & 0 & 0 & 1 & 1 & 0 & 1 \\ 0 & 0 & 0 & 1 & 0 & 0 & 1 & 1 & 0 \\ 0 & 0 & 0 & 0 & 1 & 1 & 1 & 0 & 1 \end{bmatrix}.$$

If the received vector is $\mathbf{u} = (1\ 1\ 0\ 1\ 1\ 0\ 1\ 1\ 1)^T$, $H\mathbf{u} = (1\ 0\ 0\ 1\ 1)^T$ and hence $\mathbf{u}$ is not a code vector. $\square$

Summing up, if $G = \left(\dfrac{P}{I_k} \right)$ is the generator matrix of an (n, k)-code, then $H = (I_{n-k} | P)$ is the parity check matrix. We *encode* a message $\mathbf{m}$ by calculating $G\mathbf{m}$, and we can *detect errors* in a received vector $\mathbf{u}$ by calculating $H\mathbf{u}$.

A linear code is determined by either giving its generator matrix or by giving its parity check matrix.

ERROR CORRECTING AND DECODING

We would like to find an efficient method for *correcting* errors and decoding. One crude method would be to calculate the Hamming distance between a received word and each code word. The code word closest to the received word would be assumed to be the most likely transmitted word. However, the magnitude of this task becomes enormous as soon as the message length is quite large.

Consider an (n,k) linear code with encoding function $\gamma : \mathbf{Z}_2^k \rightarrow \mathbf{Z}_2^n$. Let $V = \mathrm{Im}\,\gamma$ be the subspace of code vectors. If the code vector $\mathbf{v} \in V$ is sent through a channel and an error $\mathbf{e} \in \mathbf{Z}_2^n$ occurs during transmission, the received vector will be $\mathbf{u} = \mathbf{v} + \mathbf{e}$. The decoder receives the vector $\mathbf{u}$ and has to determine the most likely transmitted code vector $\mathbf{v}$ by finding the most likely error pattern $\mathbf{e}$. This error is $\mathbf{e} = -\mathbf{v} + \mathbf{u} = \mathbf{v} + \mathbf{u}$. The decoder does not know what the code vector $\mathbf{v}$ is, but knows that the error $\mathbf{e}$ lies in the coset $V + \mathbf{u}$.

14.16 DEFINITION. The most likely error pattern in each coset of $\mathbf{Z}_2^n$ by V is called the *coset leader*.

The coset leader will usually be the element of the coset containing the fewest number of 1s. If two or more error patterns are equally likely, one is chosen arbitrarily. In many transmission channels, errors such as those caused by a stroke of lightning tend to come in bursts that affect several adjacent digits. In these cases, the coset leaders are chosen so that the 1s in each error pattern are bunched together as much as possible.

The cosets of $\mathbf{Z}_2^n$ by the subspace V can be characterized by means of the parity check matrix H. The subspace V is the kernel of the transformation $\eta : \mathbf{Z}_2^n \rightarrow \mathbf{Z}_2^{n-k}$; therefore, by the Morphism Theorem, the set of cosets $\mathbf{Z}_2^n / V$ is isomorphic to $\mathrm{Im}\,\eta$ where the isomorphism sends the coset $V + \mathbf{u}$ to $\eta(\mathbf{u}) = H\mathbf{u}$. Hence the coset $V + \mathbf{u}$ is characterized by the vector $H\mathbf{u}$.

14.17 DEFINITION. If H is an $(n-k) \times n$ parity check matrix and $\mathbf{u} \in \mathbf{Z}_2^n$, then the $(n-k)$ dimensional vector $H\mathbf{u}$ is called the *syndrome* of $\mathbf{u}$.

(Syndrome is a medical term meaning a pattern of symptoms that characterizes a condition or disease.)

Every element of $\mathbf{Z}_2^{n-k}$ is a syndrome; thus there are 2^{n-k} different cosets and 2^{n-k} different syndromes.

14.18 THEOREM. Two vectors are in the same coset of $\mathbf{Z}_2^n$ by V if and only if they have the same syndrome.

PROOF. If $\mathbf{u}_1, \mathbf{u}_2 \in \mathbf{Z}_2^n$, then the following statements are equivalent:

(i) $V + \mathbf{u}_1 = V + \mathbf{u}_2$,   (ii) $\mathbf{u}_1 - \mathbf{u}_2 \in V$,

(iii) $H(\mathbf{u}_1 - \mathbf{u}_2) = 0$,   (iv) $H\mathbf{u}_1 = H\mathbf{u}_2$. $\square$

14.19 METHOD. We can decode received words to correct errors by using the following procedure.

(i)   Calculate the syndrome of the received word.

(ii)   Find the coset leader in the coset corresponding to this syndrome.

(iii)   Subtract the coset leader from the received word to obtain the most likely transmitted word.

(iv)   Drop the check digits to obtain the most likely message.

For a polynomial code generated by $p(x)$, the syndrome of a received polynomial $u(x)$ is the remainder obtained by dividing $u(x)$ by $p(x)$. This is because the jth column of H is the remainder obtained by dividing x^{j-1} by $p(x)$. Hence the syndrome of elements in a polynomial code can be easily calculated by means of a shift register that divides by the generator polynomial.

14.20 EXAMPLE.   Write out the cosets and syndromes for the $(6,3)$-code with parity check matrix

$$H = \begin{bmatrix} 1 & 0 & 0 & 1 & 0 & 1 \\ 0 & 1 & 0 & 1 & 1 & 1 \\ 0 & 0 & 1 & 0 & 1 & 1 \end{bmatrix}.$$

SOLUTION.   Each of the rows in Table 14.9 forms a coset with its corresponding syndrome. The top row is the set of code words.

Table 14.9.   **The syndromes and all the words of a $(6,3)$-code**

Syn-drome	Coset leader			Words				
000	000000	110100	011010	111001	101110	001101	100011	010111
100	100000	010100	111010	011001	001110	101101	000011	110111
010	010000	100100	001010	101001	111110	011101	110011	000111
001	001000	111100	010010	110001	100110	000101	101011	011111
110	000100	110000	011110	111101	101010	001001	100111	010011
011	000010	110110	011000	111011	101100	001111	100001	010101
111	000001	110101	011011	111000	101111	001100	100010	010110
101	000110	110010	011100	111111	101000	001011	100101	010001

The element in each coset that is most likely to occur as an error pattern is chosen as coset leader and placed at the front of each row. In the top row 000000 is clearly the most likely error pattern to occur. This means that any received word in this row is assumed to contain no errors. In each of the next six rows, there is one element containing precisely one nonzero digit; these are chosen as coset leaders. Any received word in one of these rows is assumed to have one error corresponding to the nonzero digit in its coset leader. In the last row, every word contains at least two nonzero digits. We choose 000110 as coset leader. We could have chosen 101000 or 010001, since these also contain two nonzero digits; however, if the errors occur in bursts, then 000110 is a more likely error pattern. Any received word in this last row must contain at least two errors. In decoding with 000110 as coset leader, we are assuming that the two errors occur in the fourth and fifth digits.

Each word in Table 14.9 can be constructed by adding its coset leader to the code word at the top of its column. $\square$

A word could be decoded by looking it up in the table and taking the code word at the top of the column in which it appears. When the code is large, this decoding table is enormous, and it would be impossible to store it in a computer. However, in order to decode, all we really need is the parity check matrix, to calculate the syndromes, and the coset leaders corresponding to each syndrome.

14.21 EXAMPLE. Decode 111001, 011100, 000001, 100011, and 101011 using Table 14.10 which contains the syndromes and coset leaders. The parity check matrix is

$$H = \begin{bmatrix} 1 & 0 & 0 & 1 & 0 & 1 \\ 0 & 1 & 0 & 1 & 1 & 1 \\ 0 & 0 & 1 & 0 & 1 & 1 \end{bmatrix}.$$

Table 14.10. Syndromes and coset leaders for a (6,3)-code

Syndrome	Coset leader
000	000000
100	100000
010	010000
001	001000
110	000100
011	000010
111	000001
101	000110

SOLUTION. Table 14.11 shows the calculation of the syndromes and the decoding of the received words. $\square$

Table 14.11. Decoding using syndromes and coset leaders

Received word **u**:	111001	011100	000001	100011	101011
Syndrome H**u**:	000	101	111	000	001
Coset leader **e**:	000000	000110	000001	000000	001000
Code word **u** + **e**:	111001	011010	000000	100011	100011
Message:	001	010	000	011	011

14.22 EXAMPLE. Calculate the table of coset leaders and syndromes for the $(9,4)$ polynomial code of Example 14.15, which is generated by $p(x) = 1 + x^2 + x^4 + x^5$.

SOLUTION. There is no simple algorithm for finding all the coset leaders. One method of finding them is as follows.

We write down, in Table 14.12, the 2^5 possible syndromes and try to find their corresponding coset leaders. We start filling in the table by first entering the error patterns, with zero or one errors, next to their syndromes. These will be the most likely errors to occur. The error pattern with one error in the jth position is the jth standard basis vector in $\mathbf{Z}_2^9$ and its syndrome is the jth column of the parity check matrix H, given in Example 14.15. So, for instance, $H(000000001) = 01101$, the last column of H.

The next most likely errors to occur are those with two adjacent errors. We enter all these in the table. For example,

$$H(000000011) = H(000000010) + H(000000001)$$
$$= 11010 + 01101, \text{ the last two columns of } H$$
$$= 10111.$$

This still does not fill the table. We now look at each syndrome without a coset leader and find the simplest way the syndrome can be constructed from the columns of H. Most of them come from adding two columns, but some have to be obtained by adding three columns. □

The $(9,4)$-code in the above example will, by Corollary 14.08, *detect* single, double, and triple errors. Hence it will *correct* any single error. It will not detect all errors involving four digits or correct all double errors, because 000000000 and 100001110 are two code words of Hamming distance 4 apart. For example, if the received word is 100001000, whose syndrome is 00101, the above table would decode this as 100001110 rather than 000000000; both these code words differ from the received word by a double error.

Table 14.12.   Syndromes and their coset leaders for a (9,4)-code

Syndrome	Coset leader	Syndrome	Coset leader	Syndrome	Coset leader
00000	000000000	01011	000011100	10110	000111000
00001	000010000	01100	011000000	10111	000000011
00010	000100000	01101	000000001	11000	110000000
00011	000110000	01110	011100000	11001	110010000
00100	001000000	01111	000001010	11010	000000010
00101	000000110	10000	100000000	11011	000010010
00110	001100000	10001	001001000	11100	111000000
00111	001110000	10010	000000101	11101	000100100
01000	010000000	10011	001101000	11110	000010100
01001	010010000	10100	000011000	11111	000000100
01010	000001100	10101	000001000		

14.23 EXAMPLE. Decode 100110010, 100100101, 111101100, and 000111110 using the parity check matrix in Example 14.15 and the coset leaders in Table 14.12.

SOLUTION.   Table 14.13 illustrates the decoding process.   □

Table 14.13.   Decoding using syndromes and coset leaders

Received word $\mathbf{u}$:	100110010	100100101	111101100	000111110
Syndrome $\mathbf{Hu}$:	01000	00000	10111	10011
Coset leader $\mathbf{e}$:	010000000	000000000	000000011	001101000
Code word $\mathbf{u}+\mathbf{e}$:	110110010	100100101	111101111	001010110
Message:	0010	0101	1111	0110

BCH CODES

The most powerful class of error-correcting codes known to date was discovered around 1960 by Hocquenghem and independently by Bose and Chaudhari. For any positive integers m and t, with $t < 2^{m-1}$, there exists a Bose–Chaudhuri–Hocquenghem (BCH) code of length $n = 2^m - 1$ that will correct any combination of t or fewer errors. These codes are polynomial codes with a generator $p(x)$ of degree $\leqslant mt$ and have message length at least $n - mt$.

14.24 DEFINITION.   A *t-error-correcting BCH code* of length $n = 2^m - 1$ has a generator polynomial $p(x)$ that is constructed as follows. Take a primitive element α in the Galois field **GF**(2^m). Let $p_i(x) \in \mathbf{Z}_2[x]$ be the irreducible polynomial with α^i as a root. Then

$$p(x) = \mathrm{LCM}(p_1(x), p_2(x), \ldots, p_{2t}(x)).$$

It is clear that $\alpha, \alpha^2, \alpha^3, \ldots, \alpha^{2t}$ are all roots of $p(x)$. By Exercise 56 of Chapter 11, $[p_i(x)]^2 = p_i(x^2)$ and hence α^{2i} is a root of $p_i(x)$. Therefore

$$p(x) = \mathrm{LCM}(p_1(x), p_3(x), \ldots, p_{2t-1}(x)).$$

Since **GF**(2^m) is a vector space of degree m over $\mathbf{Z}_2$, for any $\beta = \alpha^i$, the elements $1, \beta, \beta^2, \ldots, \beta^m$ are linearly dependent. Hence β satisfies a polynomial of degree at most m in $\mathbf{Z}_2[x]$, and the irreducible polynomial $p_i(x)$ must also have degree at most m. Therefore,

$$\deg p(x) \leqslant \deg p_1(x) \cdot \deg p_3(x) \cdots \deg p_{2t-1}(x) \leqslant mt.$$

14.25 EXAMPLE. Find the generator polynomials of the t-error-correcting BCH codes of length $n = 15$ for each value of t less than 8.

SOLUTION. Let α be a primitive element of **GF**(16), where $\alpha^4 + \alpha + 1 = 0$. We repeatedly refer back to the elements of **GF**(16) given in Table 11.4 on page 251 when performing arithmetic operations in **GF**$(16) = \mathbf{Z}_2(\alpha)$.

We first calculate the irreducible polynomials $p_i(x)$ that have α^i as roots. We only need to look at the odd powers of α. The element α itself is the root of $x^4 + x + 1$. Therefore, $p_1(x) = x^4 + x + 1$.

If the polynomial $p_3(x)$ contains α^3 as a root, it also contains

$$\left(\alpha^3\right)^2 = \alpha^6, \quad \left(\alpha^6\right)^2 = \alpha^{12}, \quad \left(\alpha^{12}\right)^2 = \alpha^{24} = \alpha^9 \quad \text{and} \quad \left(\alpha^9\right)^2 = \alpha^{18} = \alpha^3.$$

Hence

$$p_3(x) = (x - \alpha^3)(x - \alpha^6)(x - \alpha^{12})(x - \alpha^9)$$

$$= \left(x^2 + (\alpha^3 + \alpha^6)x + \alpha^9\right)\left(x^2 + (\alpha^{12} + \alpha^9)x + \alpha^{21}\right)$$

$$= (x^2 + \alpha^2 x + \alpha^9)(x^2 + \alpha^8 x + \alpha^6)$$

$$= x^4 + (\alpha^2 + \alpha^8)x^3 + (\alpha^9 + \alpha^{10} + \alpha^6)x^2 + (\alpha^{17} + \alpha^8)x + \alpha^{15}$$

$$= x^4 + x^3 + x^2 + x + 1.$$

The polynomial $p_5(x)$ has roots α^5, α^{10}, and $\alpha^{20} = \alpha^5$. Hence

$$p_5(x) = (x - \alpha^5)(x - \alpha^{10})$$

$$= x^2 + x + 1.$$

The polynomial $p_7(x)$ has roots $\alpha^7, \alpha^{14}, \alpha^{28} = \alpha^{13}$, $\alpha^{26} = \alpha^{11}$, and $\alpha^{22} = \alpha^7$. Hence

$$p_7(x) = (x - \alpha^7)(x - \alpha^{14})(x - \alpha^{13})(x - \alpha^{11})$$

$$= (x^2 + \alpha x + \alpha^6)(x^2 + \alpha^4 x + \alpha^9)$$

$$= x^4 + x^3 + 1.$$

Now every power of α is a root of one of the polynomials $p_1(x)$, $p_3(x)$, $p_5(x)$, or $p_7(x)$. For example, $p_9(x)$ contains α^9 as a root, and therefore, $p_9(x) = p_3(x)$.

The BCH code that corrects one error is generated by $p(x) = p_1(x) = x^4 + x + 1$.

The BCH code that corrects two errors is generated by

$$p(x) = \text{LCM}(p_1(x), p_3(x)) = (x^4 + x + 1)(x^4 + x^3 + x^2 + x + 1).$$

This least common multiple is the product because $p_1(x)$ and $p_3(x)$ are different irreducible polynomials. Hence $p(x) = x^8 + x^7 + x^6 + x^4 + 1$.

The BCH code that corrects three errors is generated by

$$p(x) = \text{LCM}(p_1(x), p_3(x), p_5(x))$$

$$= (x^4 + x + 1)(x^4 + x^3 + x^2 + x + 1)(x^2 + x + 1)$$

$$= x^{10} + x^8 + x^5 + x^4 + x^2 + x + 1.$$

The BCH code that corrects four errors is generated by

$$p(x) = \text{LCM}(p_1(x), p_3(x), p_5(x), p_7(x))$$

$$= p_1(x) \cdot p_3(x) \cdot p_5(x) \cdot p_7(x)$$

$$= \frac{x^{15} + 1}{x + 1} = \sum_{i=0}^{14} x^i.$$

This polynomial contains all the elements of **GF**(16) as roots, except for 0 and 1.

Since $p_9(x) = p_3(x)$, the five-error-correcting BCH code is generated by

$$p(x) = \text{LCM}(p_1(x), p_3(x), p_5(x), p_7(x), p_9(x))$$

$$= (x^{15} + 1)/(x + 1),$$

and this is also the generator of the six- and seven-error-correcting BCH codes.

These results are summarized in Table 14.14. □

Table 14.14.   The construction of t-error-correcting BCH codes of length 15

t	Roots of $p_{2t-1}(x)$	Degree $p_{2t-1}(x)$	$p(x)$	Deg$p(x)$ $= 15 - k$	Message length $= k$
1	$\alpha, \alpha^2, \alpha^4, \alpha^8$	4	$p_1(x)$	4	11
2	$\alpha^3, \alpha^6, \alpha^{12}, \alpha^9$	4	$p_1(x)p_3(x)$	8	7
3	α^5, α^{10}	2	$p_1(x)p_3(x)p_5(x)$	10	5
4	$\alpha^7, \alpha^{14}, \alpha^{13}, \alpha^{11}$	4	$(x^{15}+1)/(x+1)$	14	1
5	$\alpha^9, \alpha^3, \alpha^6, \alpha^{12}$	4	$(x^{15}+1)/(x+1)$	14	1
6	$\alpha^{11}, \alpha^7, \alpha^{14}, \alpha^{13}$	4	$(x^{15}+1)/(x+1)$	14	1
7	$\alpha^{13}, \alpha^{11}, \alpha^7, \alpha^{14}$	4	$(x^{15}+1)/(x+1)$	14	1

For example, the two-error-correcting BCH code is a $(15,7)$-code with generator polynomial $x^8 + x^7 + x^6 + x^4 + 1$. It contains seven message digits and eight check digits.

The seven-error-correcting code generated by $(x^{15}+1)/(x+1)$ has message length 1, and the two code words are the sequence of 15 zeros and the sequence of 15 ones. Each received word can be decoded by majority rule to give the message 1, if the word contains more 1s than 0s, and to give the message 0 otherwise. It is clear that this will correct up to seven errors.

We now show that the BCH code given in Definition 14.24 does indeed correct t errors.

14.26 LEMMA.   The minimum Hamming distance between code words of a linear code is the minimum number of ones in the nonzero code words.

PROOF.   If v_1 and v_2 are code words, then, since the code is linear, $v_1 - v_2$ is also a code word. The Hamming distance between v_1 and v_2 is equal to the number of ones in $v_1 - v_2$. The result now follows because the zero word is always a code word, and its Hamming distance from any other word is the number of ones in that word. $\square$

14.27 THEOREM. If $t < 2^{m-1}$, the minimum distance between code words in the BCH code given in Definition 14.24 is at least $2t+1$, and hence this code corrects t or fewer errors.

PROOF. Suppose the code contains a code polynomial with fewer than $2t+1$ nonzero terms,

$$v(x) = v_1 x^{r_1} + \cdots + v_{2t} x^{r_{2t}} \quad \text{where} \quad r_1 < \cdots < r_{2t}.$$

This code polynomial is divisible by the generator polynomial $p(x)$ and hence has roots $\alpha, \alpha^2, \alpha^3, \ldots, \alpha^{2t}$. Therefore, if $1 \leqslant i \leqslant 2t$,

$$v(\alpha^i) = v_1 \alpha^{ir_1} + \cdots + v_{2t} \alpha^{ir_{2t}}$$

$$= \alpha^{ir_1} \left(v_1 + \cdots + v_{2t} \alpha^{ir_{2t} - ir_1} \right).$$

Put $s_i = r_i - r_1$ so that the elements $v_1, \ldots, v_{2t}$ satisfy the following linear equations:

$$v_1 + v_2 \alpha^{s_2} + \cdots + v_{2t} \alpha^{s_{2t}} = 0$$
$$v_1 + v_2 \alpha^{2s_2} + \cdots + v_{2t} \alpha^{2s_{2t}} = 0$$
$$\vdots \qquad\qquad \vdots \qquad\qquad \vdots$$
$$v_1 + v_2 \alpha^{2ts_2} + \cdots + v_{2t} \alpha^{2ts_{2t}} = 0.$$

The coefficient matrix is nonsingular because its determinant is the Vandermonde determinant

$$\det \begin{pmatrix} 1 & \alpha^{s_2} & \cdots & \alpha^{s_{2t}} \\ 1 & \alpha^{2s_2} & & \alpha^{2s_{2t}} \\ \vdots & & & \vdots \\ 1 & \alpha^{2ts_2} & \cdots & \alpha^{2ts_{2t}} \end{pmatrix} = \prod_{2t \geqslant i > j \geqslant 2} (\alpha^{s_i} - \alpha^{s_j}) \neq 0.$$

This determinant is nonzero because $\alpha, \alpha^2, \ldots, \alpha^{2t}$ are all different if $t < 2^{m-1}$. (The expression for the Vandermonde determinant can be verified as follows. When the jth column is subtracted from the ith column, each term contains a factor $(\alpha^{s_i} - \alpha^{s_j})$; hence the determinant contains this factor. Both sides are polynomials in $\alpha^{s_2}, \ldots, \alpha^{s_{2t}}$ of the same degree and hence must differ by a multiplicative constant. By looking at the leading diagonal, we see that this constant is 1.)

The above linear equations must have the unique solution $v_1 = v_2 = \cdots = v_{2t} = 0$. Therefore, there are no nonzero code words with fewer than $2t + 1$ ones, and, by Lemma 14.26 and Proposition 14.02, the code will correct t or fewer errors. $\square$

There is, for example, a BCH $(127, 92)$-code which will correct up to five errors. This code adds 35 check digits to the 92 information digits and hence contains 2^{35} syndromes. It would be impossible to store all these syndromes and their coset leaders in a computer, so decoding has to be done by other methods. The errors in BCH codes can be found by algebraic means without listing the table of syndromes and coset leaders.

In fact, any code with a relatively high information rate must be long and consequently, to be useful, must possess a simple algebraic decoding algorithm. Further details of the BCH and other codes can be found in references [55] to [61].

Exercises

1. Which of the following received words contain detectable errors when using the $(3, 2)$ parity check code?

$$110, 010, 001, 111, 101, 000.$$

2. Decode the following words using the $(3, 1)$ repeating code to correct errors:

$$111, 011, 101, 010, 000, 001.$$

Which of the words contain detectable errors?

3. An ancient method of detecting errors when performing the arithmetical operations of addition, multiplication, and subtraction is the method known as *casting out nines*.

For each number occurring in a calculation, a check digit is found by adding together the digits in the number and casting out any multiples of nine. The original calculation is then performed on these check digits instead of on the original numbers. The answer obtained, after casting out nines, should equal the check digit of the original answer. If not, an error has occurred.

For example, check the following:

$$9642 \times (425 - 163) = 2526204.$$

Add the digits of each number; $9 + 6 + 4 + 2 = 21 = 2 \times 9 + 3$, $4 + 2 + 5 = 9 + 2$, $1 + 6 + 3 = 9 + 1$. Cast out the nines and perform the calculation on these check digits:

$$3 \times (2-1) = 3.$$

Now 3 is the check digit for the answer because $2+5+2+6+2+0+4 = 2 \times 9 + 3$; hence this calculation checks. Why does this method work?

4. Find the redundancy of the English language. Copy a paragraph from a book leaving out every nth letter and ask a friend to try to read the paragraph. (Try $n = 2, 3, 4, 5, 6$. If a passage with every fifth letter missing can usually be read, then the redundancy is at least $1/5$ or 20%.)

5. Each recent book, when published, is given an International Standard Book Number (ISBN) consisting of ten digits, for example, 0-471-29891-3. The first digit is a code for the language group, the second set of digits is a code for the publisher, and the third group is the publisher's number for the book. The last digit is one of $0, 1, 2, \ldots, 9, X$ and is a check digit. Have a look at some recent books and discover how this check digit is calculated. What is the 1×10 parity check matrix? How many errors does this code detect? Will it correct any?

6. Is $1 + x^3 + x^4 + x^6 + x^7$ or $x + x^2 + x^3 + x^6$ a code word in the $(8, 4)$ polynomial code generated by $p(x) = 1 + x^2 + x^3 + x^4$?

7. Write down all the code words in the $(6, 3)$-code generated by $p(x) = 1 + x^2 + x^3$.

8. Design a code for messages of length 20, by adding as few check digits as possible, that will detect single, double, and triple errors. Also give a shift register encoding circuit for your code.

9. Decode the following, using the $(6, 3)$-code given in Table 14.9:

$$000101, \ 011001, \ 110000.$$

10. A $(7, 4)$ linear code is defined by the equations

$$u_1 = u_4 + u_5 + u_7, \ u_2 = u_4 + u_6 + u_7, \ u_3 = u_4 + u_5 + u_6,$$

where u_4, u_5, u_6, u_7 are the message digits and u_1, u_2, u_3 are the check digits. Write down the generator and parity check matrices for this code. Decode the received words 0000111 and 0001111 to correct any errors.

11. Find the minimum Hamming distance between the code words of the code with generator matrix G where

$$G^T = \begin{bmatrix} 0 & 0 & 1 & 0 & 1 & 1 & 0 & 0 & 0 \\ 0 & 1 & 0 & 1 & 0 & 0 & 1 & 0 & 0 \\ 1 & 0 & 1 & 0 & 0 & 0 & 0 & 1 & 0 \\ 0 & 1 & 1 & 0 & 1 & 0 & 0 & 0 & 1 \end{bmatrix}.$$

Discuss the error-detecting and correcting capabilities of this code, and write down the parity check matrix.

12. Encode the following messages using the generator matrix of the $(9, 4)$-code of Example 14.15:

$$1101, 0111, 0000, 1000.$$

13–15. Find the generator and parity check matrices for the following polynomial codes.

13. The $(4, 1)$-code generated by $1 + x + x^2 + x^3$.

14. The $(7, 3)$-code generated by $(1 + x)(1 + x + x^3)$.

15. The $(9, 4)$-code generated by $1 + x^2 + x^5$.

16. Find the syndromes of all the received words in the $(3, 2)$ parity check code.

17. Using the parity check matrix in Example 14.21 and the syndromes in Table 14.10, decode the following words:

$$101110, 011000, 001011, 111111, 110011.$$

18. Using the parity check matrix in Example 14.15 and the syndromes in Table 14.12, decode the following words:

$$110110110, 001001101, 111111111, 000000111.$$

19–22. Construct a table of coset leaders and syndromes for each of the following codes.

19. The $(3, 1)$-code generated by $1 + x + x^2$.

20. The $(7, 4)$-code with parity check matrix

$$H = \begin{bmatrix} 1 & 0 & 0 & 1 & 1 & 1 & 0 \\ 0 & 1 & 0 & 1 & 1 & 0 & 1 \\ 0 & 0 & 1 & 1 & 0 & 1 & 1 \end{bmatrix}.$$

21. The $(9, 4)$-code generated by $1 + x^2 + x^5$.

22. The $(7, 3)$-code generated by $(1 + x)(1 + x + x^3)$.

23. Consider the $(63, 56)$-code generated by $(1 + x)(1 + x + x^6)$.

(i)   What is the number of digits in the message before coding?
(ii)   What is the number of check digits?
(iii) How many different syndromes are there?
(iv). What is the information rate?

(v)   What sort of errors will it detect?

(vi) How many errors will it correct?

24.   One method of encoding a rectangular array of digits is to add a parity check digit to each of the rows and then add a parity check digit to each of the columns (including the column of row checks). For example, in the array in Figure 14.08, the check digits are shaded and the check on checks is crosshatched. This idea is sometimes used when transferring information to and from magnetic tape. The same principle is used in accounting. Show that one error can be corrected and describe how to correct that error. Will it correct two errors? What is the maximum number of errors it will detect?

   Figure 14.08

25.   Let V be a vector space over $\mathbf{Z}_p$, where p is a prime. Show that every subgroup is a subspace. Is this result true for a vector space over any Galois field?

26.   Show that the Hamming distance between vectors has the following properties:

(i) $d(\mathbf{u}, \mathbf{v}) = d(\mathbf{v}, \mathbf{u})$.

(ii) $d(\mathbf{u}, \mathbf{v}) + d(\mathbf{v}, \mathbf{w}) \geqslant d(\mathbf{u}, \mathbf{w})$.

(iii) $d(\mathbf{u}, \mathbf{v}) \geqslant 0$ with equality if and only if $\mathbf{u} = \mathbf{v}$.

(This shows that d is a metric on the vector space.)

27.   We can use elements from a finite field $\mathbf{GF}(q)$, instead of binary digits, to construct codes. If $G = \left(\dfrac{P}{I_k} \right)$ is a generator matrix, show that $H = (I_{n-k} | - P)$ is the parity check matrix.

28.   Using elements of $\mathbf{Z}_5$, find the parity check matrix of the $(7,4)$-code generated by $1 + 2x + x^3 \in \mathbf{Z}_5[x]$.

29.   Find the generators of the two- and three-error-correcting BCH codes of length 15 by starting with the primitive element β in $\mathbf{GF}(16)$ where $\beta^4 = 1 + \beta^3$.

30.   Find the generator polynomial of a single-error-correcting BCH code of length 7.

31.   Let α be a primitive element in $\mathbf{GF}(32)$ where $\alpha^5 = 1 + \alpha^2$. Find an irreducible polynomial in $\mathbf{Z}_2[x]$ with α^3 as a root.

32. Find the generator polynomial of a double-error-correcting BCH code of length 31.

33. A linear code is called *cyclic* if a cyclic shift of a code word is still a code word; in other words, if $a_1 a_2 \ldots a_n$ is a code word, then $a_n a_1 a_2 \ldots a_{n-1}$ is also a code word. Show that a binary (n, k) linear code is cyclic if and only if the code words, considered as polynomials, form an ideal in $\mathbf{Z}_2[x]/(x^n - 1)$.

Bibliography and References

The following bibliography contains a selection of general books on modern algebra as well as books and articles on particular topics.

Modern Algebra in General

[1] Birkhoff, Garrett and Saunders Maclane, *A Survey of Modern Algebra*, 3rd ed. Macmillan, New York, 1965.
[2] Dean, Richard A., *Elements of Abstract Algebra*, Wiley, New York, 1966.
[3] Fraleigh, John B., *A First Course in Abstract Algebra*, Addison-Wesley, Reading, Mass., 1967.
[4] Herstein, I. N., *Topics in Algebra*, 2nd ed. Wiley, New York, 1975.
[5] Maclane, Saunders and Garrett Birkhoff, *Algebra*, Macmillan, New York, 1967.
[6] McCoy, Neal H., *Fundamentals of Abstract Algebra*, Allyn and·Bacon, Boston, 1972.
[7] Paley, Hiram and Paul M. Weichsel, *A First Course in Abstract Algebra*, Holt, Reinhart and Winston, New York, 1966.
[8] Weiss, Edwin, *First Course in Algebra and Number Theory*, Academic Press, New York, 1971.

Applied Modern Algebra

[9] Birkhoff, Garrett and Thomas C. Bartee, *Modern Applied Algebra*, McGraw-Hill, New York, 1970.

[10]   Bobrow, Leonard S. and Michael A. Arbib, *Discrete Mathematics*, Saunders, Phila-
       delphia, 1974.
[11]   Preparata, Franco P. and Raymond T. Yeh, *Introduction to Discrete Structures*, Addi-
       son-Wesley, Reading, Mass., 1973.
[12]   Stone, Harold S., *Discrete Mathematical Structures and their Applications*, Science
       Research Associates, Chicago, 1973.

History of Modern Algebra

[13]   Kline, Morris, *Mathematical Thought from Ancient to Modern Times*, Oxford University
       Press, New York, 1972; Chapter 49.

Boolean Algebra

[14]   Harrison, Michael A., *Introduction to Switching and Automata Theory*, McGraw-Hill,
       New York, 1965.
[15]   Heath, F. G., "Large-Scale Integration in Electronics," *Scientific American*, **222**, 22–31
       (February 1970).
[16]   Hittinger, William C., "Metal-Oxide-Semiconductor Technology," *Scientific American*,
       229, 48–57 (August 1973).
[17]   Mendelson, Elliott, *Schaum's Outline of Boolean Algebra and Switching Circuits*,
       McGraw-Hill, New York, 1970.
[18]   Whitesitt, J. E., *Boolean Algebra and its Applications*, Addison-Wesley, Reading, Mass.,
       1961.

Groups

[19]   Baumslag, Benjamin and Bruce Chandler, *Schaum's Outline of Group Theory*,
       McGraw-Hill, New York, 1968.
[20]   Budden, F. J., *The Fascination of Groups*, Cambridge University Press, Cambridge,
       1972.
[21]   Hall, Marshall, Jr., *Theory of Groups*, Macmillan, New York, 1959.
[22]   Lederman, W., *Introduction to Group Theory*, Oliver and Boyd, Edinburgh, 1973.
[23]   Rotman, Joseph J., *The Theory of Groups: An Introduction*, 2nd ed. Allyn and Bacon,
       Boston, 1973.

Symmetry Groups

[24]   Benson, C. T. and L. C. Grove, *Finite Reflection Groups*, Bogden and Quigley,
       Tarrytown-on-Hudson, N. Y., 1971.
[25]   Coxeter, H. S. M., *Introduction to Geometry*, 2nd ed. Wiley, New York, 1969.

[26] Cundy, H. Martyn and A. P. Rollett, *Mathematical Models*, 2nd ed. Oxford University Press, Oxford, 1961.

[27] Dyson, Freeman J., "Mathematics in the Physical Sciences," *Mathematics in the Modern World, Readings from Scientific American*, Freeman, San Francisco, 1968, pp. 248–257.

[28] Lomont, J. S., *Applications of Finite Groups*, Academic Press, New York, 1959.

[29] Shapiro, Louis W., " Finite Groups acting on Sets with Applications," *Mathematics Magazine*, **46**, 136–147 (1973).

[30] Thompson, D'Arcy, *On Growth and Form*, abridged ed. Cambridge University Press, Cambridge, 1966.

[31] Weyl, Hermann, *Symmetry*, Princeton University Press, Princeton, 1952.

[32] Yale, Paul B., *Geometry and Symmetry*, Holden-Day, San Francisco, 1968.

Polya-Burnside Enumeration

[33] Liu, C. L., *Introduction to Combinatorial Mathematics*, McGraw-Hill, New York, 1968.

Finite-State Machines

[34] Hartmanis, J. and R. E. Stearns, *Algebraic Structure Theory of Sequential Machines*, Prentice-Hall, Englewood Cliffs, N. J., 1970.

[35] Kohavi, Zvi, *Switching and Finite Automata Theory*, McGraw-Hill, New York, 1970.

Rings

[36] Adamson, Iain T., *Elementary Rings and Modules*, Oliver and Boyd, Edinburgh, 1972.

[37] Burton, David M., *A First Course in Rings and Ideals*, Addison-Wesley, Reading, Mass., 1970.

[38] McCoy, Neal H., *The Theory of Rings*, Macmillan, New York, 1964.

Convolution Fractions

[39] Erdelyi, Arthur, *Operational Calculus and Generalized Functions*, Holt, Rinehart and Winston, New York, 1962.

[40] Marchand, Jean-Paul, *Distributions*, North-Holland, Amsterdam, 1962.

[41] Mikusinski, Jan, *Operational Calculus*, 2nd ed. International Series on Pure and Applied Mathematics, Volume 8, Pergamon Press, London, 1969.

Residue Arithmetic

[42] Szabo, Nicholas S., and Richard I. Tanaka, *Residue Arithmetic and its Applications to Computer Technology*, McGraw-Hill, New York, 1967.

Fields

[43] Adamson, Iain T., *Introduction to Field Theory*, Oliver and Boyd, Edinburgh, 1964.
[44] Gaal, Lisl, *Classical Galois Theory with Examples*, 2nd ed. Chelsea, New York, 1973.
[45] Stewart, Ian, *Galois Theory*, Chapman and Hall, London, 1973.

Latin Squares

[46] Ball, W. W. Rouse, *Mathematical Recreations and Essays*, rev. ed. Macmillan, London, 1939.
[47] Bose, Ray Chandra, "On the Application of the Properties of Galois Fields to the Problem of Construction of Hyper-Graeco-Latin Squares," *Sankhya* (*The Indian Journal of Statistics*), **3**, 323–338(1938).
[48] Denes, J. and A. D. Keedwell, *Latin Squares and their Applications*, Academic Press, New York, 1974.
[49] Horadam, A. F., *A guide to Undergraduate Projective Geometry*, Pergamon Press Australia, Rushcutters Bay, N.S.W., 1970.
[50] Mann, H. B., *Analysis and Design of Experiments*, Dover, New York, 1949.

Geometrical Constructions

[51] Courant, Richard and Herbert Robbins, *What is Mathematics?*, Oxford University Press, London; 1941, Chapter III.
[52] Kalmanson, Kenneth, "A Familiar Constructibility Criterion," *American Mathematical Monthly*, **79**, 227–278 (1972).
[53] Kazarinoff, Nicholas D., *Ruler and the Round*, Prindle, Weber and Schmidt, Boston, 1970.
[54] Klein, Felix, *Famous Problems of Elementary Geometry*, Dover, New York, 1956.

Coding Theory

[55] Berlekamp, Elwyn R., *Algebraic Coding Theory*, McGraw-Hill, New York, 1968.
[56] Blake, Ian F. and Ronald C. Mullin, *The Mathematical Theory of Coding*, Academic Press, New York, 1975.
[57] Cullman, G., *Codes Détecteurs et Correcteurs d'Erreurs*, Dunod, Paris, 1967.
[58] Lin, Shu, *An Introduction to Error-Correcting Codes*, Prentice-Hall, Englewood Cliffs, N. J., 1970.
[59] Peterson, W. Wesley, "Error-Correcting Codes," *Scientific American*, **206**, 96–108 (February 1962).
[60] Peterson, W. W. and D. T. Brown, "Cyclic Codes for Error-Detection," *Proceedings of the Institute of Radio Engineers*, **49**, 228–235 (1961).
[61] Peterson, W. Wesley and E. J. Weldon, Jr., *Error-Correcting Codes*, 2nd ed. M.I.T. Press, Cambridge, 1972.

Film on Coding Theory

[62] *Error-Correcting Codes*, b & w, 25 min., Open University, Milton Keynes (American distributors: Harper and Row, New York).

Answers to the Odd-Numbered Exercises

Chapter 2, page 44

1. Always true.

3. When $A \cap (B \, \Delta \, C) = \emptyset$.

5. When $A \cap (B \, \Delta \, C) = \emptyset$.

13. $\#(A \cup B \cup C \cup D) = \#A + \#B + \#C + \#D - \#(A \cap B)$
    $- \#(A \cap C) - \#(A \cap D) - \#(B \cap C) - \#(B \cap D) - \#(C \cap D)$
    $+ \#(A \cap B \cap C) + \#(A \cap B \cap D) + \#(A \cap C \cap D) + \#(B \cap C \cap D)$
    $- \#(A \cap B \cap C \cap D)$.

15. 4.

17. Yes; $\mathcal{P}(\emptyset)$.

25.

A	B	(i)	(ii)	(iii)	(iv)
T	T	T	T	T	T
T	F	F	F	T	T
F	T	T	T	F	F
F	F	T	T	T	T

(i) and (ii) are equivalent and (iii) and (iv) are equivalent.

27. (i) is a contradiction and (ii), (iii) and (iv) are tautologies.

29.

31. $B' \wedge C'$.

33. $A \vee (B \wedge A'); \ (A \wedge B) \vee (A \wedge B') \vee (A' \wedge B); \ A \vee B$.

35. $(A \vee B) \wedge (A' \vee B) \wedge (A' \vee B'); \ A' \wedge B; \ A' \wedge B$.

39.

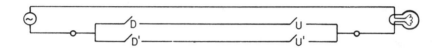

41. $(A \wedge (B \vee C)) \vee (B \wedge C)$.   43. $(A \wedge B) \vee (A' \wedge B') \vee C$.

45. $A \vee C \vee D$.

47. Orange: $(A' \wedge B' \wedge ((C' \wedge D) \vee (C \wedge D'))) \vee (((A' \wedge B) \vee (A \wedge B')) \wedge C' \wedge D')$.   Green: $A \wedge B \wedge C \wedge D$.

49. Let the result of multiplying AB by CD be EF. Then the circuit for E is $(A' \wedge B \wedge C) \vee (A \wedge ((B \vee C') \wedge D)) \vee (B' \wedge C \wedge D'))$, and the circuit for F is $((A \wedge C) \vee (B \wedge D)) \wedge (A' \vee B' \vee C' \vee D')$.

51. $(A \vee B) \wedge (A' \vee B'); \ (A \vee B) \wedge (A \vee B') \wedge (A' \vee B')$.

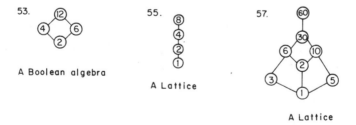

53.

A Boolean algebra

55.

A Lattice

57.

A Lattice

61. Yes.   63. $d = a \wedge b' \wedge c'$.

Chapter 3, page 78

1.

·	e	g	g^2	g^3	g^4
e	e	g	g^2	g^3	g^4
g	g	g^2	g^3	g^4	e
g^2	g^2	g^3	g^4	e	g
g^3	g^3	g^4	e	g	g^2
g^4	g^4	e	g	g^2	g^3

3. See Table 8.3.
5. Abelian group.   7. Abelian group.
9. Not a group; the operation is not closed.
11. Abelian group.   13. Abelian group.
15. Group.   25. No.
27. $\mathcal{D}_2$.   29. $\mathcal{D}_6$.
31. $\mathcal{C}_6$.
33. This is the group O(2) we meet on page 113.
35. **Z**, generated by a glide reflection.
37.

$$\{e,g^3\}\overbrace{}^{\mathcal{C}_6}\{e,g^2,g^4\}$$
$$\{e\}$$

39. No.
41. $f:\mathbf{Z}\to\mathbf{Q}$ defined by $f(n)=cn$ for any $c\in\mathbf{Q}$.
43. No; $\mathbf{Q}^*$ has an element of order 2, whereas **Z** does not.
45. The identity has order 1; $(12)\circ(34)$, $(13)\circ(24)$, $(14)\circ(23)$ have order 2, and all the other elements have order 3.
47.

·	1	-1	i	$-i$	j	$-j$	k	$-k$
1	1	-1	i	$-i$	j	$-j$	k	$-k$
-1	-1	1	$-i$	i	$-j$	j	$-k$	k
i	i	$-i$	-1	1	k	$-k$	$-j$	j
$-i$	$-i$	i	1	-1	$-k$	k	j	$-j$
j	j	$-j$	$-k$	k	-1	1	i	$-i$
$-j$	$-j$	j	k	$-k$	1	-1	$-i$	i
k	k	$-k$	j	$-j$	$-i$	i	-1	1
$-k$	$-k$	k	$-j$	j	i	$-i$	1	-1

The identity 1, has order 1; -1 has order 2; all the other elements have order 4.

53. $\{(1),(123),(132)\}$.

55. $\begin{pmatrix} 1 & 2 & 3 & 4 \\ 1 & 3 & 4 & 2 \end{pmatrix}$.

57. (12435).

59. (165432) is of order 6 and is odd.

61. $(1526)\circ(34)$ is of order 4 and is even.

63. $\begin{pmatrix} 1 & 2 & 3 & 4 & 5 \\ 2 & 3 & 4 & 5 & 1 \end{pmatrix}$.

65. (132).

67. $\{(1),(12),(34),(12)\circ(34),(13)\circ(24),(14)\circ(23),(1324),(1423)\}$.

69. $\{(1),(13),(24),(13)\circ(24),(12)\circ(34),(14)\circ(23),(1234),(1432)\}$.

71. $\phi(n)$, the number of positive integers less than n that are relatively prime to n.

73. 52; 8.

75. $\{e\}$.

81. Achievable.

83. Achievable.

85. $\mathcal{S}_3$.

87. $\mathcal{S}_2$.

89. F is not Abelian; $y^{-1}x^{-1}y^{-1}xx$.

Chapter 4, page 108

1. Equivalence relation whose equivalence classes are the integers.

3. Not an equivalence relation.

5.

Left Cosets	Right Cosets
$H=\{(1),(12),(34),(12)\circ(34)\}$	$H=\{(1),(12),(34),(12)\circ(34)\}$
$(13)H=\{(13),(123),(134),(1234)\}$	$H(13)=\{(13),(132),(143),(1432)\}$
$(14)H=\{(14),(124),(143),(1243)\}$	$H(14)=\{(14),(142),(134),(1342)\}$
$(23)H=\{(23),(132),(234),(1342)\}$	$H(23)=\{(23),(123),(243),(1243)\}$
$(24)H=\{(24),(142),(243),(1432)\}$	$H(24)=\{(24),(124),(234),(1234)\}$
$(1324)H=\{(1324),(14)\circ(23),$	$H(1324)=\{(1324),(13)\circ(24),$
$(13)\circ(24),(1423)\}$	$(14)\circ(23),(1423)\}$

7. Not a morphism.

9. A morphism; $\mathrm{Ker}f=4\mathbf{Z}$, and $\mathrm{Im}f=\{(0,0),(1,1),(0,2),(1,3)\}$.

11. Not a morphism.   19. No.

21. $f:\mathcal{C}_3\rightarrow\mathcal{C}_4$ defined by $f(g^r)=e$.

23. $f_k:\mathcal{C}_6\rightarrow\mathcal{C}_6$ defined by $f_k(g^r)=g^{kr}$ for $k=0,1,2,3,4,5$.

25. Not isomorphic; $\mathcal{C}_{60}$ contains elements of order 4, whereas $\mathcal{C}_{10}\times\mathcal{C}_6$ does not.

27. Not isomorphic; $\mathcal{C}_n\times\mathcal{C}_2$ is commutative, whereas $\mathcal{D}_n$ is not.

29. Not isomorphic; $(1+i)/\sqrt{2}$ has order 8, whereas $\mathbf{Z}_4\times\mathbf{Z}_2$ contains no element of order 8.

33. $\mathcal{C}_{10}$ and $\mathcal{D}_5$.   39. $(\mathbf{R}_{>0},\cdot)$.

49. $G_2\cong\mathcal{S}_3$.

55. 5 is a generator of $\mathbf{Z}_6^*$, and 3 is a generator of $\mathbf{Z}_{17}^*$.

Chapter 5, page 130

1. $\mathcal{C}_2$ and $\mathcal{C}_2$.
3. $\mathcal{C}_2$ and $\mathcal{C}_2 \times \mathcal{C}_2$.
5. $\mathcal{C}_3$ and $\mathcal{D}_3$.
7. $\mathcal{C}_9$ and $\mathcal{C}_9$.
9. $\mathcal{D}_4$.
11. $\mathcal{S}_4$.
13. $\mathcal{S}_4$.
15. $\mathcal{S}_4$.
17. $\mathcal{A}_5$.
19. $\mathcal{A}_5$.
21. $\mathcal{A}_5$.
25. $\mathcal{S}_4$.

27. $\begin{bmatrix} -1 & 0 & 0 \\ 0 & -1 & 0 \\ 0 & 0 & 1 \end{bmatrix}$ and $\begin{bmatrix} 0 & 0 & 1 \\ 1 & 0 & 0 \\ 0 & 1 & 0 \end{bmatrix}$.

29. $\mathcal{D}_6$.
31. $\mathcal{C}_2$.

33. $\mathcal{D}_4$ generated by $\begin{bmatrix} -1 & 0 & 0 \\ 0 & 1 & 0 \\ 0 & 0 & -1 \end{bmatrix}$ and $\begin{bmatrix} 0 & -1 & 0 \\ 1 & 0 & 0 \\ 0 & 0 & 1 \end{bmatrix}$.

Chapter 6, page 145

1. 3.
3. 38.
5. 78.
7. 35.
9. 333.
11. $(n^6 + 3n^4 + 8n^2)/12$.
13. 1.
15. 96.
17. 30.
19. 396.
21. 126.
23. 96.

Chapter 7, page 160

1. Monoid with identity 0.
3. Semigroup.
5. Neither.
7. Semigroup.
9. Semigroup.
11. Monoid with identity 1.
13. Neither.
15.

GCD	1	2	3	4
1	1	1	1	1
2	1	2	1	2
3	1	1	3	1
4	1	2	1	4

17.

$\cdot$	e	c	c^2	c^3	c^4
e	e	c	c^2	c^3	c^4
c	c	c^2	c^3	c^4	c^2
c^2	c^2	c^3	c^4	c^2	c^3
c^3	c^3	c^4	c^2	c^3	c^4
c^4	c^4	c^2	c^3	c^4	c^2

19. No; $01 \neq 1$ in the free semigroup.

29.

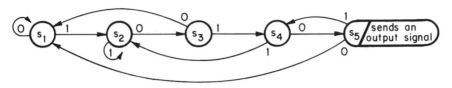

31. A congruence relation with quotient semigroup $= \{2\mathbf{N}, 2\mathbf{N}+1\}$.

33. Not a congruence relation.

35.

★	[0]	[1]	[00]	[10]	[01]	[010]
[0]	[00]	[01]	[0]	[010]	[1]	[10]
[1]	[10]	[1]	[1]	[10]	[01]	[010]
[00]	[0]	[1]	[00]	[10]	[01]	[010]
[10]	[1]	[01]	[10]	[010]	[1]	[10]
[01]	[010]	[01]	[01]	[010]	[1]	[10]
[010]	[01]	[1]	[010]	[10]	[01]	[010]

37. 24.

39.

★	[Λ]	[α]	[β]	[γ]	[αβ]	[αγ]
[Λ]	[Λ]	[α]	[β]	[γ]	[αβ]	[αγ]
[α]	[α]	[Λ]	[αγ]	[αγ]	[β]	[γ]
[β]	[β]	[β]	[β]	[γ]	[β]	[γ]
[γ]	[γ]	[γ]	[γ]	[β]	[γ]	[β]
[αβ]	[αβ]	[αβ]	[αβ]	[αγ]	[αβ]	[αγ]
[αγ]	[αγ]	[αγ]	[αγ]	[αβ]	[αγ]	[αβ]

41.

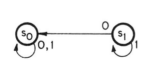

$\{[0],[1]\}$.

43.

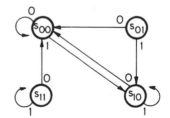

$\{[0],[1],[10]\}$.

45. The monoid contains 27 elements.

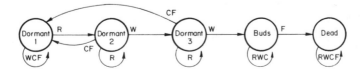

Chapter 8, page 188

1.

+	0	1	2	3
0	0	1	2	3
1	1	2	3	0
2	2	3	0	1
3	3	0	1	2

·	0	1	2	3
0	0	0	0	0
1	0	1	2	3
2	0	2	0	2
3	0	3	2	1

3. A ring.
5. Not a ring; not closed under multiplication.
7. Not a ring; not closed under addition.
9. A ring.
11. Not a ring; distributive laws do not hold.
17. A subring.
19. Not a subring; not closed under addition.
21. Neither.   23. Both.
25. Integral domain.   29. [2], [4], [5], [6], [8].
31. Any nonempty proper subset of X.
33. Nonzero matrices with zero determinant.
37. $f(x) = [x]_6$.
39. $f(x,y) = (x,y)$, (y,x), (x,x) or (y,y).
47. (ii) -1 and 0.
55. The identity is $D_n(x) = (1/2\pi) + (1/\pi)(\cos x + \cos 2x + \cdots + \cos nx)$.
     The ring is not an integral domain.

Chapter 9, page 217

1. $\frac{3}{2}x^2 + \frac{5}{4}x - \frac{15}{8}$ and $\frac{45}{8}x + \frac{7}{8}$.
3. $x^4 + x^3 + x^2 + x$ and 1.
5. $3 - i$ and $4 + 2i$, or $4 - i$ and $1 - 2i$, or $4 - 2i$ and $-3 + i$.
7. $\text{GCD}(a,b) = 3$, $s = -5$, $t = 4$.
9. $\text{GCD}(a,b) = 1$, $s = -(2x+1)/3, t = (2x+2)/3$.
11. $\text{GCD}(a,b) = 2x + 1$, $s = 1$, $t = 2x + 1$.
13. $\text{GCD}(a,b) = 1$, $s = 1$, $t = -1 + 2i$.

15. $x = -6, y = 5$.   17. $x = -14, y = 5$.
19. [23].   21. [17].
23. No solutions.   25. $(x-1)(x^4 + x^3 + x^2 + x + 1)$.
27. $(x^2 + 2)(x^2 + 3)$.   29. $x^4 - 9x + 3$.
31. $x^3 - 4x + 1$.
33. $(x - \sqrt{2})(x + \sqrt{2})(x - i\sqrt{2})(x + i\sqrt{2})(x - 1 - i)(x - 1 + i)(x + 1 - i)$
    $\cdot(x + 1 + i)$.
35. $(x^2 - 2)(x^2 + 2)(x^2 - 2x + 2)(x^2 + 2x + 2)$.
37. $x^5 + x^3 + 1$, $x^5 + x^2 + 1$, $x^5 + x^4 + x^3 + x^2 + 1$, $x^5 + x^4 + x^3 + x + 1$,
    $x^5 + x^4 + x^2 + x + 1$, $x^5 + x^3 + x^2 + x + 1$.
39. $x^3 + 2$.
41. $\text{Ker}\psi = \{q(x)\cdot(x^2 - 2x + 4)|q(x)\in\mathbf{Q}[x]\}$ and $\text{Im}\psi = \mathbf{Q}(\sqrt{3}\,i) = \{a + b\sqrt{3}\,i|a,b\in\mathbf{Q}\}$.
43. Irreducible by Eisenstein's Criterion.
45. Irreducible, since it has no linear factors.
47. Reducible; any polynomial of degree >2 in $\mathbf{R}[x]$ is reducible.
49. No.   55. No.
61. $x \equiv 40 \bmod 42$.   63. $x \equiv 22 \bmod 30$.
67. 65.

Chapter 10, page 232

1. $((0,0))$, $((0,1))$, $((1,0))$, $\mathbf{Z}_2 \times \mathbf{Z}_2$.
3. (0) and $\mathbf{Q}$.
5. $(p(x))$ where $p(x) \in \mathbf{C}[x]$.
7. The quotient ring is a field.

+	(3)	(3)+1	(3)+2
(3)	(3)	(3)+1	(3)+2
(3)+1	(3)+1	(3)+2	(3)
(3)+2	(3)+2	(3)	(3)+1

·	(3)	(3)+1	(3)+2
(3)	(3)	(3)	(3)
(3)+1	(3)	(3)+1	(3)+2
(3)+2	(3)	(3)+2	(3)+1

9. The ideal $((1,2))$ is the whole ring $\mathbf{Z}_3 \times \mathbf{Z}_3$. The quotient ring is not a field.

+	((1,2))
((1,2))	((1,2))

·	((1,2))
((1,2))	((1,2))

11. $8x+2$ and $14x+97$.   13. x^2+x and x^2.

17. (i) 6,   (ii)   36,   (iii) x^2-1, $(a)\cap(b)=(\text{LCM}(a,b))$.

33. No.   35. The whole ring.

37.

$$Z_8$$
$$|$$
$$([2]_8)$$
$$|$$
$$([4]_8)$$
$$|$$
$$([0]_8)$$

39. Irreducible; Z_{11}.   41. Reducible.

43. Irreducible; $Q(\sqrt[4]{2}\,)$.   45. Irreducible; $Q(\sqrt{2},\sqrt{3}\,)$.

47. Not a field; contains zero divisors.

49. A field by Corollary 10.18.   51. A field by Theorem 10.19.

53. Not a field; $x^2+1=(x+2)(x+3)$ in $Z_5[x]$.

55. A field isomorphic to $Q[x]/(x^4-11)$.

59. (0) and (x^n) for $n\geqslant 0$; (x) is maximal.

Chapter 11, page 252

1. $\mathbf{GF}(5)=Z_5=\{0,1,2,3,4\}$.

+	0	1	2	3	4
0	0	1	2	3	4
1	1	2	3	4	0
2	2	3	4	0	1
3	3	4	0	1	2
4	4	0	1	2	3

·	0	1	2	3	4
0	0	0	0	0	0
1	0	1	2	3	4
2	0	2	4	1	3
3	0	3	1	4	2
4	0	4	3	2	1

3. $\mathbf{GF}(9) = \mathbf{Z}_3[x]/(x^2+1) = \{a\alpha + b \mid a,b \in \mathbf{Z}_3, \alpha^2 + 1 = 0\}$.

+	0	1	2	α	$\alpha+1$	$\alpha+2$	2α	$2\alpha+1$	$2\alpha+2$
0	0	1	2	α	$\alpha+1$	$\alpha+2$	2α	$2\alpha+1$	$2\alpha+2$
1	1	2	0	$\alpha+1$	$\alpha+2$	α	$2\alpha+1$	$2\alpha+2$	2α
2	2	0	1	$\alpha+2$	α	$\alpha+1$	$2\alpha+2$	2α	$2\alpha+1$
α	α	$\alpha+1$	$\alpha+2$	2α	$2\alpha+1$	$2\alpha+2$	0	1	2
$\alpha+1$	$\alpha+1$	$\alpha+2$	α	$2\alpha+1$	$2\alpha+2$	2α	1	2	0
$\alpha+2$	$\alpha+2$	α	$\alpha+1$	$2\alpha+2$	2α	$2\alpha+1$	2	0	1
2α	2α	$2\alpha+1$	$2\alpha+2$	0	1	2	α	$\alpha+1$	$\alpha+2$
$2\alpha+1$	$2\alpha+1$	$2\alpha+2$	2α	1	2	0	$\alpha+1$	$\alpha+2$	α
$2\alpha+2$	$2\alpha+2$	2α	$2\alpha+1$	2	0	1	$\alpha+2$	α	$\alpha+1$

$\cdot$	0	1	2	α	$\alpha+1$	$\alpha+2$	2α	$2\alpha+1$	$2\alpha+2$
0	0	0	0	0	0	0	0	0	0
1	0	1	2	α	$\alpha+1$	$\alpha+2$	2α	$2\alpha+1$	$2\alpha+2$
2	0	2	1	2α	$2\alpha+2$	$2\alpha+1$	α	$\alpha+2$	$\alpha+1$
α	0	α	2α	2	$\alpha+2$	$2\alpha+2$	1	$\alpha+1$	$2\alpha+1$
$\alpha+1$	0	$\alpha+1$	$2\alpha+2$	$\alpha+2$	2α	1	$2\alpha+1$	2	α
$\alpha+2$	0	$\alpha+2$	$2\alpha+1$	$2\alpha+2$	1	α	$\alpha+1$	2α	2
2α	0	2α	α	1	$2\alpha+1$	$\alpha+1$	2	$2\alpha+2$	$\alpha+2$
$2\alpha+1$	0	$2\alpha+1$	$\alpha+2$	$\alpha+1$	2	2α	$2\alpha+2$	α	1
$2\alpha+2$	0	$2\alpha+2$	$\alpha+1$	$2\alpha+1$	α	2	$\alpha+2$	1	2α

5. $x^3 + x + 4$.

7. Impossible.

9. $x^2 + 2$.

11. $x^4 - 16x^2 + 16$.

13. $8x^6 - 9$.

17. 3.

19. 2.

21. 2.

23. ∞.

25. ∞.

27. $(1 - \sqrt[3]{2} + \sqrt[3]{4})/3$.

29. $-(1 + 6\omega)/31$.

31. $\alpha^4 + \alpha$.

33. 2; not a field.

35. 7; a field

37. 0; a field.

39. 0; not a field.

43. $m = 2, 4, p^r$ or $2p^r$, where p is an odd prime; see Weiss [8; Th. 4-6-10].

47. $\alpha = \sqrt{2} + \sqrt{-3}$.

49. All elements of $\mathbf{GF}(32)$ except 0 and 1 are primitive.

51. $x^3 + x + 1$.

57. No solutions.

59. $x = 1$ or $\alpha + 1$.

61. 5.

63. The output has cycle length 7 and repeats the sequence 1101001, starting at the right.

Chapter 12, page 269

1.

a	b	c	d	e	f	g
b	c	d	e	f	g	a
c	d	e	f	g	a	b
d	e	f	g	a	b	c
e	f	g	a	b	c	d
f	g	a	b	c	d	e
g	a	b	c	d	e	f

3. Use $\mathbf{GF}(8) = \{0, 1, \alpha, 1+\alpha, \alpha^2, 1+\alpha^2, \alpha+\alpha^2, 1+\alpha+\alpha^2\}$ where $\alpha^3 = \alpha + 1$.

7. No.

9.

		Week			
		1 ↓	2 ↓	3 ↓	4 ↓
	→	A	B	C	D
Shelf	→	B	C	D	A
height	→	C	D	A	B
	→	D	A	B	C

A, B, C and D are the four brands of cereal.

11.

		Week				
		1 ↓	2 ↓	3 ↓	4 ↓	5 ↓
M	→	A	B	C	D	0
T	→	B	C	D	0	A
W	→	C	D	0	A	B
T	→	D	0	A	B	C
F	→	0	A	B	C	D

A, B, C, and D are the four different types of music, and 0 refers to no music.

15. $y = \alpha x$.

17. $y = (\alpha + 2)x + 2\alpha$.

21. 1.

23.

1	6	11	16
12	15	2	5
14	9	8	3
7	4	13	10

1	6	11	16
15	12	5	2
8	3	14	9
10	13	4	7

25.

1	10	19	28	37	46	55	64
35	44	49	58	7	16	21	30
29	22	15	8	57	50	43	36
63	56	45	38	27	20	9	2
52	59	34	41	24	31	6	13
18	25	4	11	54	61	40	47
48	39	62	53	12	3	26	17
14	5	32	23	42	33	60	51

Chapter 13, page 284

1. Constructible.
3. Constructible.
5. Not constructible.
7. No.
11. Yes.
13. No.
15. Yes; $\frac{\pi}{21} = \frac{1}{4}\left(\frac{\pi}{3} - \frac{\pi}{7}\right)$.
23. Yes.
25. No.
27. No.
29. Yes.
31. No.
33. Yes.

Chapter 14, page 314

1. 010, 001, 111.
3. The checking is done modulo 9, using the fact that any integer is congruent to the sum of its digits modulo 9.
5. (1 2 3 4 5 6 7 8 9 10) modulo 11. It will detect one error but not correct any.
7. 000000, 110001, 111010, 001011, 101100, 011101, 010110, 100111.
9. 101, 001, 100.
11. Minimum distance = 3. It detects two errors and corrects one error.

$$H = \begin{bmatrix} 1 & 0 & 0 & 0 & 0 & 0 & 0 & 1 & 0 \\ 0 & 1 & 0 & 0 & 0 & 0 & 1 & 0 & 1 \\ 0 & 0 & 1 & 0 & 0 & 1 & 0 & 1 & 1 \\ 0 & 0 & 0 & 1 & 0 & 0 & 1 & 0 & 0 \\ 0 & 0 & 0 & 0 & 1 & 1 & 1 & 0 & 0 & 1 \end{bmatrix}.$$

13. $G^T = (1\ 1\ 1\ 1),\ H = \begin{bmatrix} 1 & 0 & 0 & 1 \\ 0 & 1 & 0 & 1 \\ 0 & 0 & 1 & 1 \end{bmatrix}.$

15. $G^T = \begin{bmatrix} 1 & 0 & 1 & 0 & 0 & 1 & 0 & 0 & 0 \\ 0 & 1 & 0 & 1 & 0 & 0 & 1 & 0 & 0 \\ 0 & 0 & 1 & 0 & 1 & 0 & 0 & 1 & 0 \\ 1 & 0 & 1 & 1 & 0 & 0 & 0 & 0 & 1 \end{bmatrix}$,

$H = \begin{bmatrix} 1 & 0 & 0 & 0 & 0 & 1 & 0 & 0 & 1 \\ 0 & 1 & 0 & 0 & 0 & 0 & 1 & 0 & 0 \\ 0 & 0 & 1 & 0 & 0 & 1 & 0 & 1 & 1 \\ 0 & 0 & 0 & 1 & 0 & 0 & 1 & 0 & 1 \\ 0 & 0 & 0 & 0 & 1 & 0 & 0 & 1 & 0 \end{bmatrix}$.

17. 110, 010, 101, 001, 011.

19.

Syndrome	Coset Leader
00	000
01	010
10	100
11	001

21.

Syndrome	Coset leader	Syndrome	Coset leader	Syndrome	Coset leader
00000	000000000	01011	000010100	10110	000000001
00001	000010000	01100	011000000	10111	000010001
00010	000100000	01101	010000010	11000	110000000
00011	000110000	01110	001000100	11001	000000111
00100	001000000	01111	000000110	11010	100000100
00101	000000010	10000	100000000	11011	000001110
00110	001100000	10001	000001010	11100	000000101
00111	000100010	10010	100100000	11101	000010101
01000	010000000	10011	000000011	11110	000001100
01001	010010000	10100	000001000	11111	000011100
01010	000000100	10101	000011000		

23. (i) 56, (ii) 7, (iii) $2^7 = 128$, (iv) 8/9, (v) It will detect single, double, triple, and any odd number of errors, (vi) 1.

25. No.

29. $x^8 + x^4 + x^2 + x + 1$ and $x^{10} + x^9 + x^8 + x^6 + x^5 + x^2 + 1$.

31. $x^5 + x^4 + x^3 + x^2 + 1$.

Glossary of Symbols

$\leqslant$	Less than or equal	25, 32
$\Rightarrow$	Implies	21
$\Leftrightarrow$	If and only if	21
$\cong$	Isomorphic	65, 174
$\equiv \bmod n$	Congruent modulo n	84
$\equiv \bmod H$	Congruent modulo H	85
$\#X$	Number of elements in X	13, 60
$\#(G:H)$	Index of H in G	87
R^*	Invertible elements in the ring R	202
a'	Complement of a in a Boolean algebra	15, 30
a^{-1}	Inverse of a	4, 52
$\overline{A}$	Complement of the set A	9
V^T	Transpose of the matrix V	113
$\square$	End of a proof or example	10
(a)	Ideal generated by a	222
$(a_1 a_2 \ldots a_n)$	n-cycle	69
$\begin{pmatrix} 1 & 2 \ldots n \\ a_1 a_2 \ldots a_n \end{pmatrix}$	Permutation	68
$\begin{pmatrix} n \\ r \end{pmatrix}$	Binomial coefficient $n!/r!(n-r)!$	139
$\mathbf{F}(a)$	Smallest field containing $\mathbf{F}$ and a	238
$\mathbf{F}(a_1, \ldots, a_n)$	Smallest field containing $\mathbf{F}$ and $a_1, \ldots, a_n$	238
(n,k)-code	Code of length n with messages of length k	287
$(X, \star)$	Group or monoid	5, 52, 147
$(R, +, \cdot)$	Ring	166
$(K, \wedge, \vee, ')$	Boolean Algebra	15
$[x]$	Equivalence class containing x	83
$[x]_n$	Congruence class modulo n containing x	108
$R[x]$	Polynomials in x with coefficients from R	178
$R[[x]]$	Formal power series in x with coefficients from R	182
$R[x_1, \ldots, x_n]$	Polynomials in $x_1, \ldots, x_n$ with coefficients from R	180
$[\mathbf{K}:\mathbf{F}]$	Degree of $\mathbf{K}$ over $\mathbf{F}$	236
X^Y	Set of functions from Y to X	148
$R^{\mathbf{N}}$	Sequences of elements from R	180
$\langle a_i \rangle$	Sequence whose ith term is a_i	180
$G \times H$	Direct product of G and H	98, 176
S/E	Quotient set	83
G/H	Quotient group or set of right cosets	89
R/I	Quotient ring	223
$a \mid b$	a divides b	26, 198
$l /\!/ m$	l is parallel to m	263
Ha	Right coset of H containing a	85
aH	Left coset of H containing a	88
$I + r$	Coset of I containing r	223

Index